Principles and Techniques in

HISTOLOGY,
MICROSCOPY
and
PHOTOMICROGRAPHY

SECOND EDITION

Principles and Techniques in

HISTOLOGY, MICROSCOPY and PHOTOMICROGRAPHY

SECOND EDITION

DR Singh MBBS MS PhD

Emeritus Professor
Department of Anatomy
King George's Medical University
Lucknow, UP

Former

Professor and Head, Department of Anatomy
King George's Medical College, Lucknow University, Lucknow

Professor and Head, Department of Anatomy
BP Koirala Institute of Health Sciences, Dharan, Nepal

Seema Dental College, Rishikesh, Uttarakhand
Nepalgunj Medical College, Kathmandu University, Nepal

CBS

CBS Publishers & Distributors Pvt Ltd

New Delhi • Bengaluru • Chennai • Kochi • Kolkata • Mumbai
Hyderabad • Jharkhand • Nagpur • Patna • Pune • Uttarakhand

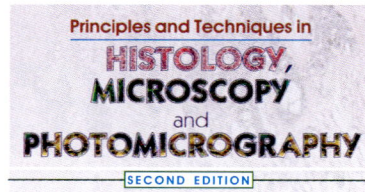

Principles and Techniques in
HISTOLOGY,
MICROSCOPY
and
PHOTOMICROGRAPHY
SECOND EDITION

ISBN: 978-93-87085-80-0

Second Edition: 2018
First Edition: 2003
Reprint: 2006, 2010, 2012, 2013, 2015, 2016

Published by Satish Kumar Jain and produced by Varun Jain for
CBS Publishers & Distributors Pvt Ltd
4819/XI Prahlad Street, 24 Ansari Road, Daryaganj, New Delhi 110 002, India.
Ph: 23289259, 23266861, 23266867   Fax: 011-23243014   Website: www.cbspd.com
e-mail: delhi@cbspd.com; cbspubs@airtelmail.in.
Corporate Office: 204 FIE, Industrial Area, Patparganj, Delhi 110 092
Ph: 4934 4934   Fax: 4934 4935   e-mail: publishing@cbspd.com; publicity@cbspd.com

Branches

- **Bengaluru:** Seema House 2975, 17th Cross, K.R. Road,
  Banasankari 2nd Stage, Bengaluru 560 070, Karnataka
  Ph: +91-80-26771678/79   Fax: +91-80-26771680   e-mail: bangalore@cbspd.com
- **Chennai:** 7, Subbaraya Street, Shenoy Nagar, Chennai 600 030, Tamil Nadu
  Ph: +91-44-26680620/26681266   Fax: +91-44-42032115   e-mail: chennai@cbspd.com
- **Kochi:** Ashana House, No. 39/1904, AM Thomas Road, Valanjambalam, Ernakulam 682 016,
  Kochi, Kerala
  Ph: +91-484-4059061-65   Fax: +91-484-4059065   e-mail: kochi@cbspd.com
- **Kolkata:** 6/B, Ground Floor, Rameswar Shaw Road, Kolkata-700 014, West Bengal
  Ph: +91-33-22891126, 22891127, 22891128   e-mail: kolkata@cbspd.com
- **Mumbai:** 83-C, Dr E Moses Road, Worli, Mumbai-400018, Maharashtra
  Ph: +91-22-24902340/41   Fax: +91-22-24902342   e-mail: mumbai@cbspd.com

Representatives

• **Hyderabad**	0-9885175004	• **Jharkhand**	0-9811541605	• **Nagpur**	0-9021734563
• **Patna**	0-9334159340	• **Pune**	0-9623451994	• **Uttarakhand**	0-9716462459

Printed at: Nutech Print Services, Faridabad, India

to
_____

my wife, Rekha, for her constant encouragement;
my children, for their heartedly inspiration;
my revered teachers, for their helpful guidance
my professional colleagues, for their critical suggestions
and
my past and present students, for their love and affection

— this book is for you

Foreword

The dawn of the twenty-first century heralds the coming age of molecular medicine and the era of predictive medicine. Physical examination (resolution 0.2 mm) alone no longer suffices for clinching a diagnosis and the light microscopy (resolution 0.2 μm) needs to be strengthened by ultrastructural analysis (up to 0.2 nm) for the evaluation of disorders at the molecular level. The greater use of the manifold techniques of light microscopy as well as electron microscopy entails a thorough in-depth knowledge and understanding of the equipment and methods basic to the aforementioned tools. This book

(1936–2013)

is primarily addressed to the fulfilment of the objectives listed above. It is common knowledge that a thorough grasp of the tools of the trade equips one with capabilities to maximize their skillful utilization with finesse. Kudos to Prof Dhanraj Singh for his timely realization of this dire need and for painstaking following it up with singular drive and devotion towards the creation of this reservoir of technical details which has grown to almost encyclopedic proportions. The author has superbly integrated his artistic acumen and skills of photography with the recently acquired mastery of computer graphics for creating this *vade mecum* of microscopy not only to address undergraduates and postgraduates of anatomy and pathology but also to histology and histopathology technicians, oral path dentist, veterinarians, paramedics in general.

Padmashree Prof (Dr) Mahdi Hasan

MBBS MS (Hons) PhD DSc FAMS FICS FRMS FNASc FNA

Recipient, Dr BC Roy National Award
Founder Director of Interdisciplinary Brain Research Centre
Ex-Professor and Chairman, Department of Anatomy, and
Principal and Dean, JN Medical College
Aligarh Muslim University, Aligarh
Founder Fellow, Indian Academy of Neurosciences
Adviser, WHO

Preface to the Second Edition

The publication of the second edition of *Principles and Techniques in Histology, Microscopy, and Photomicrography* was a great feeling since I received many comments and suggestions from the readers. While revising the book I have considered all these and it has greatly helped me in further developing the book.

In the second edition I have attempted to provide students with the most contemporary and useful text possible and in a more clear and concise manner without compromising on the information required by the postgraduate students. To facilitate understanding, I have restructured the contents to build a logical flow of information at the end of all main topics. Also a summary has been provided at the end of each chapter.

New tables have been added in the 'Techniques in Histology' section, which also includes now colour pictures of histologic slides on important components. The text in chapters on the general guidelines for photographic processes and technical details of photomicrography (Part 3 in the first edition) though has become old-fashioned, it has been retained for dual reasons: *First* to maintain the originality of the book; and *second* because of the fact that not all the laboratories concerning photomicrography are provided with the modern sophisticated equipment including digital SLR cameras needed for photomicrography. A new chapter on 'Digital Photomicrography' has been added to update the relevant information required for better equipped postgraduate students. Common questions asked during the *viva voce* of practical examination of histological techniques (in almost all university examinations) with the answers have been provided for the benefit of the students and also to make the book examination-friendly. The book is likely to help the students succeed in postgraduate examination.

A clear understanding of the subject, its retention and reproducibility, and usefulness for examination, the second edition has evolved into a truly integrated, and student-friendly textbook for histology students. It can also serve as a useful companion for medical undergraduates pursuing MBBS.

I would appreciate any comments or suggestions concerning any of the sections in the book. You may contact me at *dr.drsingh40@gmail.com.*

DR Singh

Preface to the First Edition

Histology provides a basis for understanding the internal structure of body organs. With increasing importance of histology in the anatomy curriculum, it becomes necessary for histologists to facilitate understanding of the basic principles in examination of a stained section with an optical light microscope. There is some diversity of opinion as to what features of histology should be taught, and to what extent. The purpose of this book is to give the student a concise account of the relevant details which have been brought up to date. The present book departs from other books in that it has combined animal tissue preparation for obtaining stained paraffin sections, proper use of the light microscope, and finally to get the images photomicrographed. Techniques in histology, microscopy, and photomicrography have been so overwhelming that the medical student often has difficulty in grasping the basic steps about the actual procedures. Since photomicrography has become of great practical value, both in teaching and research, Part 3 of this book is provided with various formulas for the final processing of photographic emulsions for the desired reproduction of results to large audiences; and also for scientific publications. The book is freely illustrated to enable the reader visualize summarized concepts and processes. While writing this book, an attempt was to keep in mind the often-quoted Chinese proverb, "A little picture is worth a million words". It is hoped, however, that detail has not been neglected too much and that the text is sufficiently readable although precise. In summary, the book is bound to satisfy the students' need in learning and, despite its brevity, create sufficient confidence to the self-dependent in the way of learning histological techniques.

DR Singh

Acknowledgements

The assistance, cooperation, and professionalism of the editorial staff of the CBS Publishers & Distributors, New Delhi, is greatly acknowledged. A special appreciation is extended to Mr Satish Jain, CMD, who encouraged me to get my book revised for the second edition; the first edition of which was published in 2003. The challenging task was initiated with my first meeting with Mr YN Arjuna, Senior Vice President—Publishing, Editorial, and Publicity, who deserves my special thanks and sincere regards for his many helpful suggestions and much consideration and cooperation throughout the completion of this edition. I also appreciate Mr Prasenjit Paul, Mrs Sunita Rautela and Mr Rohan Ram who worked extremely hard to organize and finalize the illustrations and the proofs.

I am greatly indebted to my faculty colleagues and postgraduate students of my alma mater—Department of Anatomy, King George's Medical College, Lucknow, and postgraduate students at Nepalgunj Medical College, Chisapani, Nepal, as well, for their suggestions to bring out the long-awaited second edition of this book.

I put on record my full appreciation and sincere regards to my revered teachers in anatomy—(late) Prof Dharam Narayan, (late) Prof AC Das, and (late) Prof Mahdi Hasan at King George's Medical College, Lucknow (now upgraded as KG's Medical University), who taught me anatomy during 1959–1961. For the long association with my alma mater for more than five decades, I owe all appreciation for the Department of Anatomy, where actually I learned the subject of anatomy. Prof A Halim, *ex*-Professor and Head, Department of Anatomy, has been one of the best teachers during my undergraduate medical career and later as my role model in teaching when I joined as faculty in the Department of Anatomy.

Finally, I owe much to my wife Rekha and children who always encouraged me in all my academic pursuits. It is impossible to put into words the sum total of my gratitude to them.

DR Singh

Contents

Part 1 — Techniques in Histology

Part 2 — Microscopy and Micrometry

Part 3	Photomicrography

Techniques in Histology

1

Overview of Histology and Approaches to its Study

INTRODUCTION TO HISTOLOGY

Histology (derived from two Greek words: *histos* meaning "tissue", and *logia*, a suffix generally used to denote the study of a subject or the branch of knowledge of a discipline "science") is the study of the microscopic anatomy of cells and tissues of plants and animals. Tissue was first used to describe the different textures of body parts being dissected by an anatomist. *Microanatomy* is the study of the structure (anatomy) of small (micro) things. Small things—cells and their arrangement—constitute tissues and the association among these form organs. Histology is a discipline that performs a scientific study dealing with the fine detail structure of biological cells and their formation into tissues and organs using microscopes to look at specimens of tissues that have been carefully prepared using special processes called **histological techniques**.

Histology is the study of the tissues of the body. But since the tissues are composed of cells and their products, histology also deals with the study of the structure and activities of the cell. Histology is a structural science and serves to complete the anatomical knowledge gained from dissection. Thus, it is intimately related to physiology and pathology. However, a thorough knowledge of normal histology is essential for the understanding of the altered structure seen in various conditions of disease. The term histology is used in a broad sense for the study of tissues as revealed by various techniques utilizing the methods of cytology, histochemistry, electron microscopy, and tissue culture. In a restricted sense, histology is a structural science and serves to complete the anatomical knowledge gained from dissection. Since tissues are composed of cells and their products, the structure of cells must necessarily form the basis of histology. All the technologies employed for the study of histology bring an insight into the details inside a cell with the use of microscopes. The simplest and most common method in the study of histology by a student is to use his microscope for the examination of preparations termed *sections*. A section is a very thin slice of tissue, laid flat on a sheet of glass called a *slide*, stained and mounted in a medium of suitable refractive index and covered with another circular or square very thin piece of glass often called a *cover slip*. Because there are both living and non-living components in tissues of our body, it is logical to employ methods which would reveal the salient features of both these components. The methods utilized for the study of cell micro-anatomy fall basically in two groups: (i) Those employed for the living cells, and (ii) Those employed for the dead cells that have been fixed. In the modern histology lab immuno-logical and molecular (DNA) techniques are frequently utilized to provide accurate tumor identification which will aid the clinician in selecting a mode of therapy that offers that greatest probability of cure.

Histological Classification of Animal Tissues

Tissues are divided into four groups (epithelial tissue, connective tissue, muscle tissue, and nervous tissue). These groups are further subdivided into many subgroups of these four basic tissue types (for example, blood cells are classified as connective tissue, since they generally originate inside bone marrow). Also, for another example, the epithelial tissue is subdivided into **covering and/or lining epithelia** (outer layer of skin, inner surface of heart and blood vessels, inner surface of respiratory cavities, etc.) and **glandular epithelia** (most of the glands in the body).

- **Epithelium**: The lining of glands, bowel, skin, and some organs like the liver, lung, and kidney.
- **Endothelium**: The lining of blood and lymphatic vessels.
- **Mesothelium**: The lining of pleural and pericardial spaces.

- **Mesenchyme**: The cells filling the spaces between the organs, including fat, muscle, bone, cartilage, and tendon cells.
- **Blood cells**: The red and white blood cells, including those found in lymph nodes and spleen.
- **Neurons**: Any of the conducting cells of the nervous system.
- **Germ cells**: Reproductive cells (spermatozoa in men, oöcytes in women).
- **Placenta**: An organ characteristic of true mammals during pregnancy, joining mother and offspring, providing endocrine secretion and selective exchange of soluble, but not particulate, blood-borne substances through an apposition of uterine and trophoblastic vascularized parts.
- **Stem cells**: Cells with the ability to develop into different cell types.

Note

Tissues from plants, fungi, and microorganisms can also be examined histologically. Their structure is very different from animal tissues.

Short Definitions of Histology and Related Sciences

Some short definitions of the term histology are as follows:

- **Histology** is the scientific and microscopic study of cells and tissues of plants and animals using special staining techniques combined with light and electron micro-scopy. It is often carried out by examining a thin slice (called a "section") of tissue under a light microscope or an electron microscope. In order to distinguish different biological structures more easily and accurately histo-logical stains are often used to add colors to or enhance the colors of certain types of biological structures differently from other types of structures that may be located next to and/or in contact with each other.

Histology is an essential tool of biology, medicine and veterinary medicine.

- **Cell biology** is the study of living cells, their DNA and RNA and the proteins they express.
- **Anatomy** is the study of organs visible by the naked eye.
- **Morphology** studies entire organisms by whole mounts (Fig. 1.1).
- **Cytology** is the microscopic study of loose cells or clusters obtained from bodily secretions, aspirations, scrapes, swipes, or washings (Fig. 1.2). Cytology is a specialized department whose primary role is to detect cancerous and precancerous lesions on Pap smears and cancerous lesions in non-gynecological specimens including fine needle aspiration (FNA) biopsies, urine samples, sputum samples, CSF and other body fluids. All scientific and technical staff

are highly trained with strict quality control procedures rigorously implemented to ensure high standards of reporting. This may involve a slide being examined under the microscope by several cytologists and pathologists before a final diagnosis is made.

Brief Historical Account

Marie François Xavier Bichat (1771–1802), born at Thoirette in Jura, France, was a French anatomist and physiologist who is best remembered as the **Father of modern histology and** descriptive anatomy. Despite working without a microscope (because he distrusted it), he was the first to introduce the notion of tissues as distinct entities, and maintained that diseases attacked tissues rather than whole organs. Bichat (Fig. 1.3) made significant contributions to medicine and physiology and opined that the diverse body

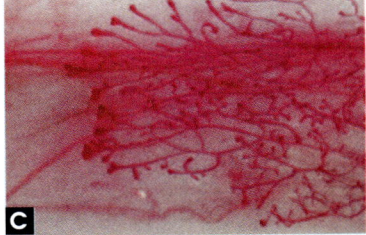

Fig. 1.1: Whole mounts: (A) of Fasciola hepatica, (B) of an insect, and (C) of mouse mammary gland as viewed with a microscope.

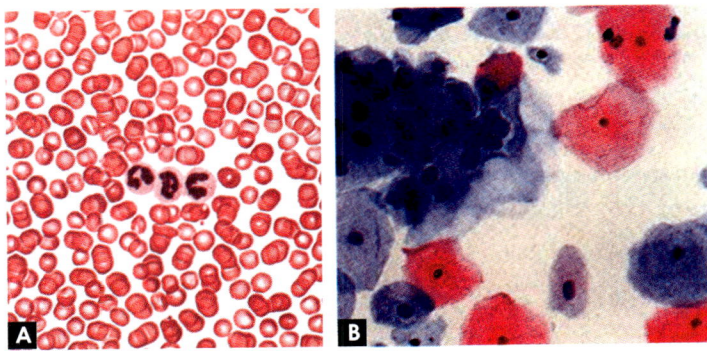

Fig. 1.2: Microscope views of smear preparations of (A) human blood from a normal individual, and (B) cervical smear from a case of uterine cervix malignancy

Fig. 1.3: Marie François Xavier Bichat (1771–1802) is regarded as Father of modern histology descriptive anatomy and pathology.

organs contain particular tissues or *membranes*. He described 21 such membranes, including connective tissue, muscle tissue, and nerve tissue. His analyses did not include any acknowledgement of cellular structure. Nonetheless, he formed an important bridge between the *organ pathology* of Giovanni Battista Morgagni and the *cell pathology* of Rudolf Ludwig Carl Virchow.

In the 19th century, histology was an academic discipline in its own right. The 1906 Nobel Prize in Physiology or Medicine was awarded to histologists Camillo Golgi and Santiago Ramon y Cajal. They had dueling interpretations of the neural structure of the brain based in differing interpretations of the same images. Cajal won the prize for his correct theory and Golgi for the staining technique he invented to make it possible.

Uses of Histology

The knowledge about histology may be utilized for applying it to the following causes:

• *Education:* Histology slides are often used in teaching laboratories to help students learn about the microstructures of human (and animal) *biological tissues.*

• *Diagnosis for treatment:* Biological tissue samples taken from a patient (that is, a specific person or animal's body) may be studied in detail to enable medical or veterinary experts to learn more about the patient's condition and hence perhaps understand its causes and make recommendations for treatment or management of the condition. Although the study of the microstructure of *diseased* cells and tissues (e.g. to help inform decisions about treatment in clinical medicine) is an aspect or use of histology because it uses histological techniques, study of diseased tissues is more accurately called *histopathology.*

• *Forensic investigations:* Forensic histology, immunohistochemistry and cytology involving microscopic study of biological tissues using various stains can help clarify the cause of sudden unexpected deaths and other issues in forensic science.

• *Autopsy:* Biological tissues from a deceased person or animal can be studied using histological techniques enabling experts (e.g. pathologists to diagnose unexplained death of a person) to learn about the circumstances and possibly cause of death. *Histologists* are people who have and use the special skills necessary to process samples of biological tissues in histology laboratories. Tissue obtained from the patient (or sometimes from a suspect in the case of forensic science labs) is processed using a series of techniques to prepare appropriate very tiny samples of tissue then mount them on slides and stain the tiny sample of tissue on each slide using chemicals called histology stains that have been carefully selected in order to help the people who will look at the slides to distinguish between the different types of biological material within the tiny sample on the slide.

HISTOCHEMISTRY, HISTOPATHOLOGY, AND CYTOCHEMISTRY

Histochemistry is the aspect of histology study of the identification and distribution of *chemical compounds* within and between biological cells using histological techniques such as histology stains, indicators and *light* (optical) and electron microscopy. In order to fully understand these definitions of histochemistry, it is helpful to recall the definitions of *histology* and *chemistry*. Histology is the microscopic study of the structure of biological *cells* and *tissues* using special techniques including histology stains, combined with light and electron microscopy. Chemistry is the science of *matter* and the changes that occur between different kinds of matter, especially chemical reactions, during which types of matter are rearranged into other types of matter, e.g. liquid water splitting into the gases *hydrogen* and *oxygen*. So, whereas histology in general is the study of biological cells and tissues in fine microscopic detail using special histological techniques, histochemistry is concerned specifically with the chemicals within, between, and forming the biological cells and tissue themselves.

Histopathology (a sub-discipline within *pathology*) is the microscopic study of diseased biological tissue. Medical specialists who study and interpret diseased tissues in microscopic detail are called histopathologists. In addition, *histotechnicians* and *histotechnologists* are members of a laboratory team who employ technology to diagnose diseases, and to conduct research.

Histotechnologists play a fundamental role, in the allied health profession, by preparing suitable thin slices of human, animal or plant tissue for microscopic examination. This is an important part of the intricate process of scientific investigation used in establishing and confirming patient diagnosis (Fig. 1.4). Because of the histotechnologist's skillful application of sophisticated laboratory techniques, the seemingly invisible world of tissue structure becomes visible under a microscope. The tasks performed by the histotechnologist require patience, mechanical ability, knowledge of biology, immunology, molecular biology, anatomy and chemistry. It requires five basic

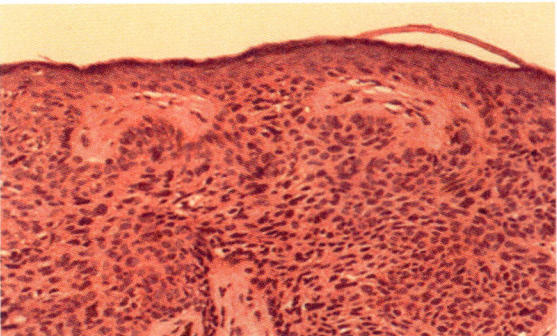

Fig. 1.4: A histopathology slide from tiny biopsies to entire organs that have been surgically removed. The thin sections are routinely stained with hematoxylin and eosin, and reviewed under a microscope by specialist pathologists.

steps, each an integral part of the histotechnologist's job. Histopathologists also provide frozen section analysis in operating theatres in hospitals allowing a prompt examination of small selected pieces of tissue while the patient is still in the operating room. The diagnosis of this tissue will predetermine the extent or continuance of the operation.

Cytochemistry is the method for detecting different substances in tissue sections (Fig. 1.5). Most based on a specific chemical reaction or high-affinity interactions between molecules.

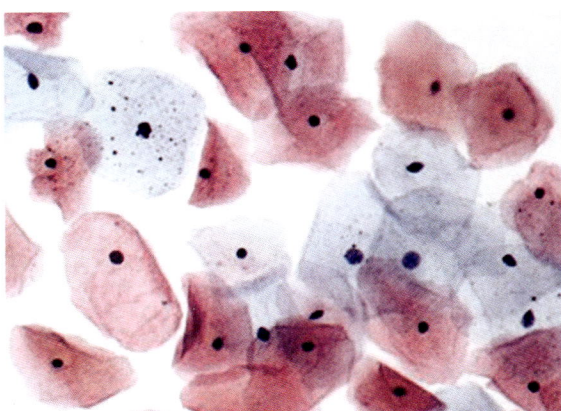

Fig. 1.5: Cells on a Pap smear are carefully examined by several specialists to detect cancerous and pre-cancerous lesions in non-gynecological specimens (of body fluids) before a final diagnosis is made.

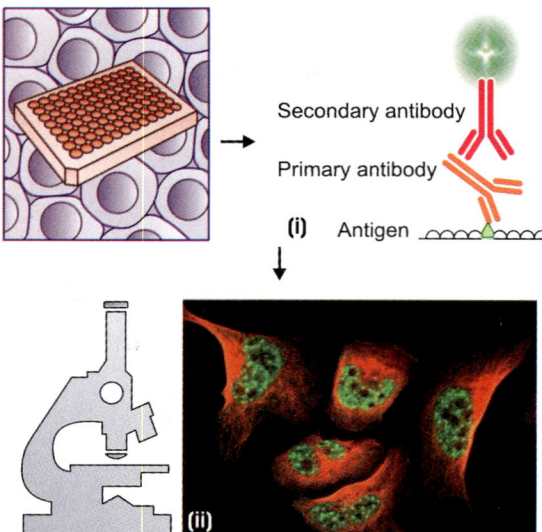

Fig. 1.6: Steps in immunocytochemistry: (i) Cell-seedling, fixation and immunostaining; (ii) imaging, annotation and image analysis. RNA binding motif protein 25 (RBM25) is localized in the nuclear speckles (green). Microtubules are stained in red.

Usually produce a colored or electron-dense compound.

Examples are:

• Phosphatases, peroxidases and dehydro-genases used to react with specific chemical bonds.

• Lipid soluble dyes like Oil red O and Sudan black used to detect lipids

• Periodic-Schiff (PAS) reaction transforms glycol groups into aldehydes for detection of polysaccharides.

• **Immunocytochemistry** depends on a specific interaction between an antigen and its antibody. Labeled antibodies can identify and localize specific proteins (Fig. 1.6).

METHODS OF STUDYING HISTOLOGY

The methods utilized for the study of cell microanatomy fall basically in *two* groups:

  i. those employed for the living cells and tissues, and

  ii. those employed for the dead 'fixed' specimens

Study of Living Cells and Tissues

Living cells are usually more difficult to prepare and the preparations can be retained temporarily, that too for a relatively short period only. The first human cells studied in the living condition were blood cells. Due to the fluid nature of the blood, a thin film of blood can be made between a glass slide and a cover slip, edges are sealed to prevent evaporation and the microscope stage is warmed to average body temperature. By this procedure, the blood cells are still surrounded by their natural environment—the plasma. Thus, the conditions inside the body are nearly achieved. However, the cells of organized tissues cannot be studied by this procedure as they are too thick for examination by transmitted light. Hence the cells from specimens, like spleen and lymph nodes are obtained from the cut surfaces of organs and tissues, mixed with an isotonic saline and examined in the same manner as the thin film of blood. Very small pieces from the subcutaneous tissues are generally examined after being teased to form thin 'spreads' on a glass slide. This depicts the normal arrange-ment of fibers, but only some cells because a few cells are destroyed. When the structure of a tissue is to be quickly studied, as for the histopathological diagnosis of a patient being operated, some quicker methods are followed. By these methods, the preparations cannot be preserved for recording purposes and the tissue deteriorates after sometime. Such pre-parations are termed temporary preparations, and these depend on the point whether the material is a liquid or a solid.

Study of Dead (or 'Fixed') Specimens

A *temporary* or a *permanent* preparation may be needed to demonstrate the structural details.

1. Temporary Preparations

Temporary preparation from a solid material may be made for histological examination by selection of any of the following procedures.

a. *Scraping*: The material is scraped with a glass slide or cover slip, and the scraped out tissue is spread on a slide as smear, fixed, and examined under microscope, either as such or after staining.

b. *Teasing*: The tissue is placed over a glass slide with a little normal saline solution. By means of mounted needles, the tissue is spread, fixed, and examined.

c. *Maceration*: In this method, the tissue is dissociated by chemical agents such as 33 per cent alcohol, and nitric acid 1 in 4 dilutions with glycerine–water mixture. After keeping the tissue in any of the above solutions, for about 24 hours, 35 per cent potassium hydroxide solution is used to macerate a tissue.

d. *Sectioning* (or *slicing*): Thin slices of the tissue are cut either by hand or with a knife, or by a freezing microtome (or cryostat if available) after the material has been treated in gum solution.

Besides the above four conventional methods for making the temporary preparations of a solid material, there are a few other methods as indicated below:

• *Tissue culture is* a method of biological research in which fragments of tissue from an animal or plant are transferred to an artificial environment in which they can continue to survive and function. The cultured tissue may consist of a single cell, a population of cells, or a whole or part of an organ. Cells in culture may multiply; change size, form, or function; exhibit specialized activity (muscle cells, for example, may contract); or interact with other cells (Fig. 1.7). In practice, the term "cell

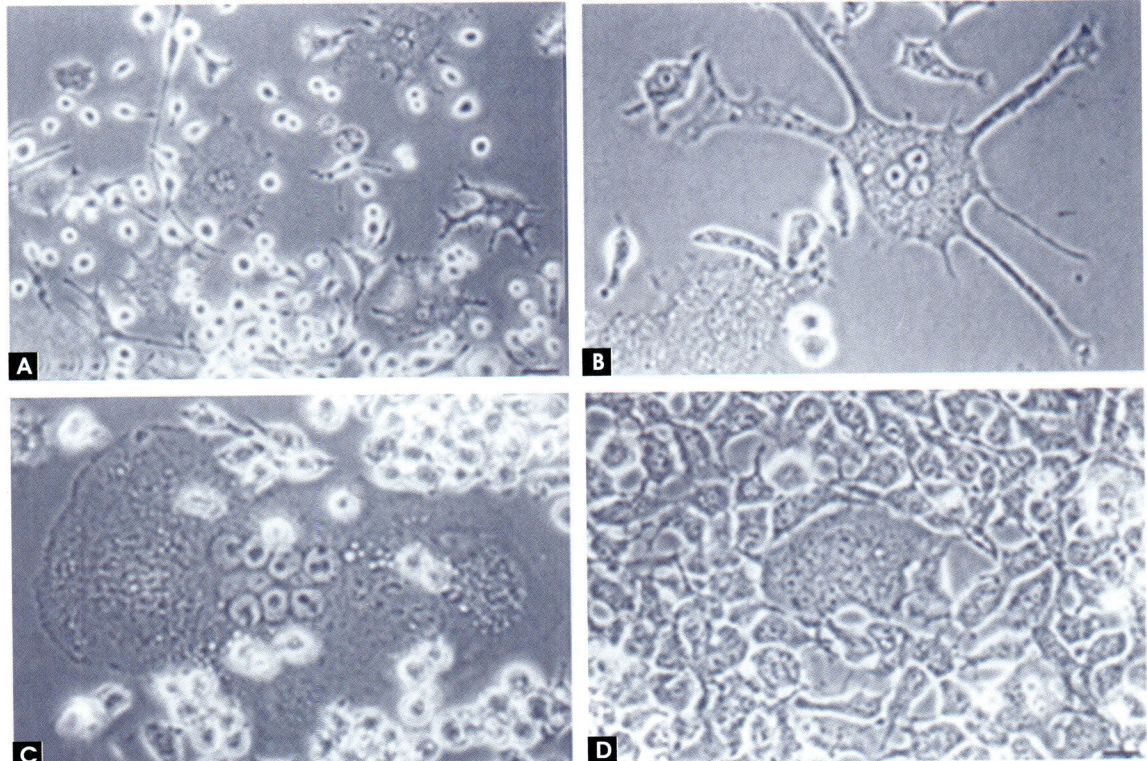

Fig. 1.7: Photomicrographs depict formation of multinucleated giant cells in a tissue culture. Photographs were taken with a 20X objective in A and a 40X objective in B–D. Bars: (A) 50 µm; (B–D) 25 µm.

culture" now refers to the culturing of cells derived from multicellular eukaryotes, especially animal cells, in contrast with other types of **culture** that also grow cells, such as plant **tissue culture**, fungal **culture**, and microbiological **culture** (of microbes).

- *Micro incineration* (in which microscopic sections are incinerated in a crucible and the 'ash' left behind is examined for qualitative chemical analysis of inorganic constituents)
- *Freezing–drying* method of tissue fixation in which a specimen may be frozen rapidly using carbon dioxide snow and then dehydrated in a vacuum. Sections may then be cut for microscopic examination.

2. Permanent Preparations

When a permanent histological preparation is required, the tissue needs to be what is called *fixed* or *dead*. A fixed tissue after subsequent processing can be sectioned into thin slices and many morphological and histochemical staining methods can also be employed.

It may be noted that only a fixed material can be examined with an electron microscope. Also, tissues fixed, unlike the fresh ones, many years earlier can be examined. However, a small delicate tissue for better handling and to avoid its distortion during section cutting, is placed in a supporting medium which varies in its consistency at varying temperature, liquid at a higher temperature, and solid at a lower temperature, for smooth cutting. When the liquid embedding medium is allowed to solidify in a special container, a small *block* of the given specimen is formed. A block is sectioned with an instrument called the *microtome* (Chapter 8), the tissue together with its supporting medium cuts as one mass. This procedure prevents distortion making thin sections that can be examined with a *light microscope* (Chapter 16, Part 2). In routine work, *paraffin-wax method of embedding* is followed and only this method is described in detail in the present book. Specimens are usually received in fixative (preservative) but

sometimes arrive fresh (Fig. 1.8) and must be immediately fixed.

Before specimens are accepted by a laboratory the identification (labeling) and accompanying documentation will be carefully checked, all details recorded and "specimen tracking "commenced. It is vital that patient or research specimens are properly identified and the risk of inaccuracies minimized. The preparation of sections is the most technically complicated of these methods as it requires specialized equipment and considerable expertise. Although the methodology for preparing sections from both animal and plant material is similar, the following description relates to animal (human) tissues.

a. The tissue can be *rapidly frozen* and kept frozen while sections are cut using a cryo-stat microtome (a microtome in a freezing chamber). These are called "**frozen sections**". Frozen sections can be prepared very quickly and are therefore used when an intra-operative diagnosis is required to guide a surgical procedure or where any type of interference with the chemical makeup of the cells is to be avoided (as in some histochemical investigations).

b. Alternatively specimens can be infiltrated with a liquid agent that can subsequently be converted into a solid that has

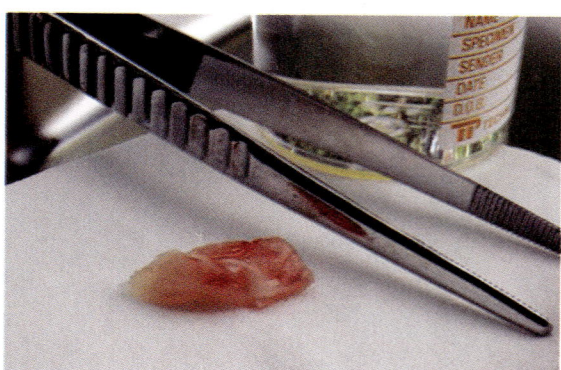

Fig. 1.8: A fresh, unfixed specimen after surgical removal. To prevent degeneration or drying-out the specimen should be fixed as soon as possible.

appropriate physical properties that will allow thin sections to be cut from it. Various agents can be used for infiltrating and supporting specimens including epoxy and methacrylate resins but paraffin wax based histological waxes are the most popular for routine light microscopy. This produces so-called "paraffin sections". These sections are usually prepared with a "rotary" micro-tome. "Rotary" describes then cutting action of the instrument. In all histopathology laboratories paraffin sections are routinely prepared from almost every specimen and used in diagnosis.

3. Cinematography

Cinematography is a technique to maintain permanent records of cell activity by taking the cinematograph taken through the objectives of a microscope. The technique of television microscopy (Fig. 1.9) is the only available technique for the direct quantitative measurement of skin capillary blood flow. Television microscopy has applications in both research and clinical practice, and has attracted considerable interest from clinical groups attempting to study capillary blood flow. The critical requirements for a functioning system are difficult to ascertain from the published literature.

The exposures are taken at intervals of several seconds and thrown on the screen at the usual speed. This method is extremely useful for noticing certain cellular stages like the process of cell division by mitosis, the shifting of the nucleus during the prophase, and separation of chromosomes. These steps cannot be appreciated by any other means of cell examination.

TISSUE PREPARATION STEPS FOR LIGHT MICROSCOPY

Routine Standard Techniques

To process any given specimen for microscopy, following *four* types of preparations can be made:

1. *Whole-mounts:* The whole-mounts are prepared in those cases where an entire organism or structure is small enough or

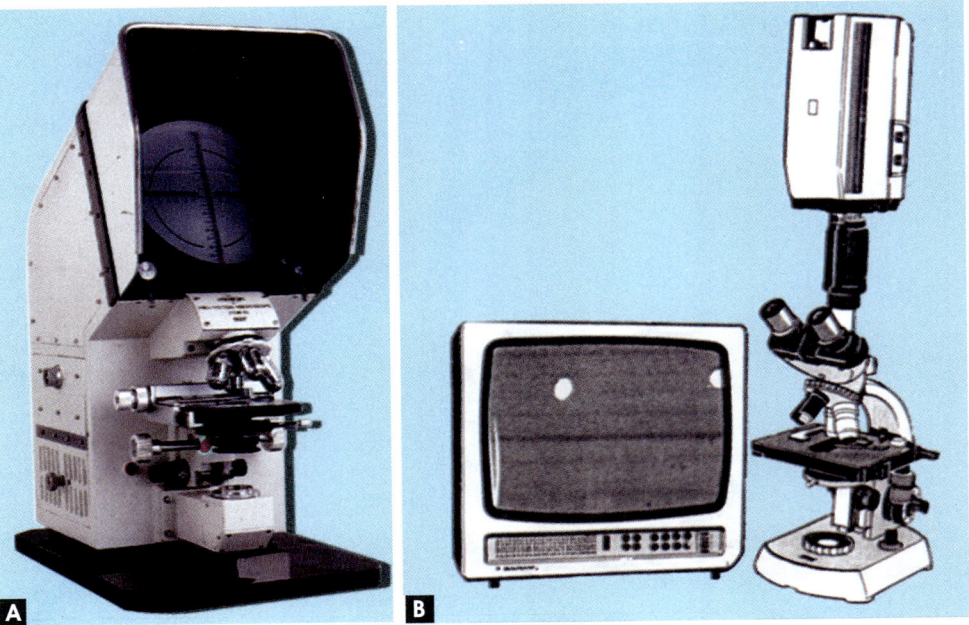

Fig. 1.9: (A) Projection microscope and (B) equipment for television microscopy as tools for histology teaching

thin enough to be placed directly onto a microscope slide. For example, a small unicellular or multicellular organism or a membrane that can be stretched thinly on to a slide (*see* Fig. 1.1). The whole-mounts preserve the structural relationships between individual cells and extracellular components.

2. *Squash preparations:* In a 'squash' preparation whole-mounts cells are intentionally squashed or crushed onto a slide to reveal their contents, e.g. botanical specimens such as root tips where cells are disrupted to reveal chromosomes (Fig. 1.10).

3. *Smears:* Smears are made in those specimens that consist of:
   a. Cells suspended in a fluid (e.g. blood, semen, cerebrospinal fluid, or a culture of micro-organisms)
   b. A situation where individual cells have been scraped brushed or aspirated from a surface
   c. From within an organ (exfoliative cytology) Smears are the basis of the well-known "Pap test" that is used to screen for cervical cancer in women. However, smears and squash preparations provide detail about individual cells and relative

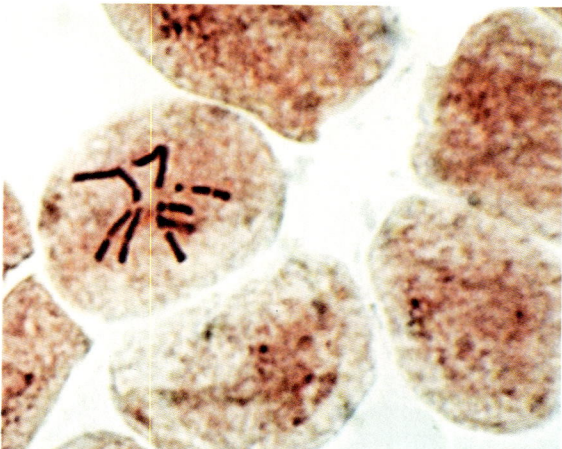

Fig. 1.10: Different phases in a squashed root tip of a plant.

cell numbers, but structural relationships are lost.

4. *Section preparation:* Most fresh tissue is very delicate, easily distorted and damaged and it is thus impossible to prepare thin sections (slices) from it unless it is supported in some way whilst it is being cut. Usually the specimen also needs to be preserved or "fixed" before sections are prepared.

Broadly there are two strategies that can be employed to provide this support. The section cutting is done in such situations where the specimen cannot be examined by the above three methods. Sections very thin slices can be cut from them, mounted on slides, and stained are prepared using an instrument called a "microtome" (Chapter 8).

EQUIPMENT IN A HISTOLOGY LABORATORY

Sterilization and Cleaning of Glassware

Sterilization

The sterilization can be done by the following methods:

1. **Dry Heat** sterilization has a limited value. Prolonged exposure may cause damage.
2. **Hot air oven** where heat is transferred by convection, conduction or radiation.
   a. The temperature of 100° C for one hour can destroy the non-sporing organism.
   b. Fungal spores need 115° C for one hour.
   c. While for other all bacteria 160°C temperature is needed for one hour.
3. **Incineration** where the flame is an effective way of sterilization. Flame heat is needed for the loop for culture.
4. **Moist heat** is the most reliable method of sterilization. This is the most lethal agent to kill microorganism. Microbial death is due to coagulation and denaturation of the protein and enzyme.
5. **Boiling** is not effective to kill spore bearing bacteria and for surgical instruments.

6. **Steam sterilization** or tyndallization is exposure to steam at 100° C for 90 minutes. This good means to sterile the media which contain sugar.

7. **The autoclave** is heating water under pressure which boils at progressively higher temperature. This method is good for rubber material and surgical instruments.

8. **Membrane filters** are Millipore filters. Filters with the pore of 0.22 micrometer are sufficient for the bacteria.

9. **Seitz filter** are disposable asbestos pad filter.

10. **Flaming** when the material is wetted by alcohol and then flamed. This method is rapid.

11. **Ultraviolet light** causes damage to bacteria.

12. **Radiation** in form of beta and gamma X-rays used for the surgical pads.

13. **Supersonic and ultrasonic** waves, 9000 cycles per second or above are used to rupture and disintegration of the cells.

Cleaning

Different glassware in a histology laboratory, brushes commonly used, and bottle brush, bent for cleaning laboratory glassware are shown in Fig. 1.11.

Cleaning of Micro slides and Cover slips

Micro Slides

The slides available in gross (1 gross = 12 dozens) are usually rectangular in shape. Their thickness varies (Fig. 1.9). Thin slides are less than 1 mm in thickness; the medium ones

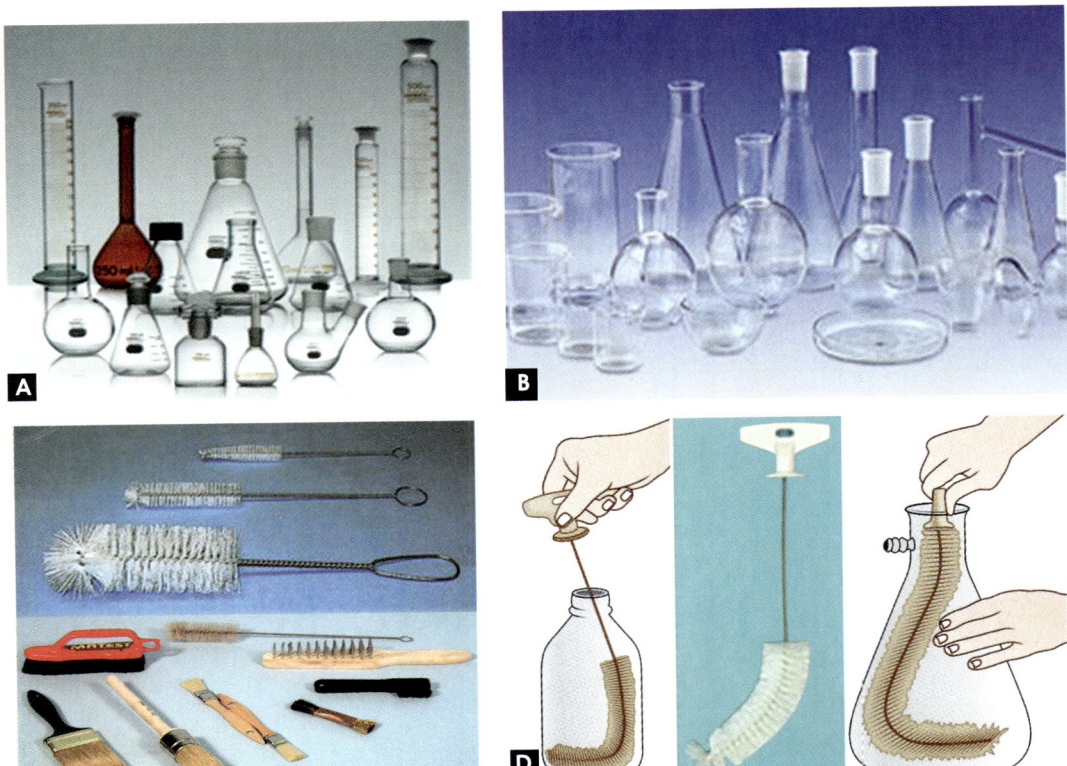

Fig. 1.11: (A and B) Different glassware in a histology laboratory. (C) Laboratory brushes commonly used, (D) Bottle brush, bent for cleaning laboratory glassware.

are between 1 and 1.25 mm; and the thick ones are more than 1.25 mm thick. These slides are available in different sizes.

For very large sections 82 × 82 mm slides or even photographic emulsion plates are used. The most common size of slides in laboratory is 76 mm × 25 mm (3" × 1"). Other sizes are 76 × 50 mm; 76 × 32 mm and 76 × 40 mm, i.e. 3" × 2", 3" × 1.25" and 3" × 1.5" respectively.

Cover Slips

Unlike glass slides, the cover slips are purchased in ounces or grams. These are circular, square, or rectangular in shape; and their sizes are also variable. According to their thickness, the cover slips are numbered from 0 to 3. Number '0' and '1', suitable for oil-immersion and photomicrographic work are 0.07 to 0.13 mm and 0.13 to 0.17 mm thick respectively. For ordinary microscopy, cover slips number '2' (0.17 to 0.25 mm thick), and number '3' (0.25 mm and above) are used. The average size of cover slips is 32 mm at its diameter, range being 0.155 to 0.185 mm. The more viscous the mounting medium used, the thinner should be the cover slip.

SUMMARY

What is Histology, Histochemistry, and Histopathology?

Histology is a structural scientific study of the microscopic anatomy of cells and tissues of animals and plants gained from dissection. *Marie Francois Xavier Bichat* is remembered as **Father of Modern Histology** and Descriptive Anatomy. Histologically the tissues are in *four* groups: Epithelial tissue (covering and lining the body tubes/tubular structures), connective tissue (mainly filling the spaces between the organs), muscular tissue (with inherent contractility), and nervous tissue (with characteristic property of irritability for conduction of impulses).

Histochemistry is the aspect of histology study of the identification and distribution of *chemical compounds* within and between biological cells using histological techniques such as histology stains, indicators and *light* (optical) and electron microscopy.

Histopathology (a sub-discipline within *pathology*) is the microscopic study of diseased biological tissue. Histopathologists also provide frozen section analysis in operating theatres in hospitals allowing a prompt examination of small selected pieces of tissue.

Histological Techniques

The term is generally used for performing a scientific study to enable a specimen prepared in such a way so as to study the detailed structure of its constituent cells and/or tissues by cutting it into sufficient thin slices—called *sections* with the help of an instrument called a *microtome*.

1. **Study of Living Cells and Tissues**: Blood cells still surrounded by their natural environment—the plasma. However, cells of organized tissues cannot be studied by this procedure because of their excessive thickness to allow light transmitted through them.

2. **Study of Dead or 'Fixed' Cells and Tissues**
   a. Specimen may be *rapidly frozen* and kept frozen while sections are cut
   b. Specimen can be *infiltrated with a liquid agent* that can subsequently be converted into a solid *block*. Thin *sections* can be cut later at will.

Preparations from a solid material may be made for histological examination by any of the following procedures:

a. *Scraping*: The material is scraped with a glass slide or cover slip, and the scraped-out tissue is spread on a slide as a smear, fixed, and examined under a microscope, either as such or after staining.

b. *Teasing*: The tissue is placed over a glass slide with a little normal saline solution. By means of mounted needles, the tissue is spread, fixed, and examined.

c. *Maceration*: In this method, the tissue is dissociated by chemical agents such as 33 per cent alcohol, and nitric acid 1 in 4 dilutions with glycerine-water mixture. After keeping the tissue in any of the above solutions, for about 24 hours, 35 per cent potassium hydroxide solution is used to macerate a tissue.

d. *Sectioning* (or *slicing*): Thin slices of the tissue are cut either by hand or with a knife, or by a freezing microtome (or cryostat if available) after the material has been treated in gum solution.

Steps in Section Preparation for Light Microscopy

- **Fixation**—carried out preferably by perfusion method in a living anaesthetized animal to make preparation almost as in life
- **Dehydration**—through increasing graded strengths of ethanol
- **Clearing** or **dealcoholization**—for removal of alcohol and transfer the specimen in a medium acting as a paraffin wax solvent
- **Infiltration**—with molten paraffin wax
- **Embedding**
- **Paraffin block making**
- **Section cutting** (or microtomy) *see* Chapter 9 for procedure and precautions
- **Ribbon of sections**/or single section flattened on water bath
- **Affix** on clean (pg. 13) egg-albumin (adhesive) coated glass sides
- **Slides**
   - For normal routine work 76 × 25 mm slides are universally used.
   - 1.0–1.2 mm thick slides are preferred because these do not break easily.
- Slides are air-dried and stored till routine/or other selective staining is done
- For details of staining steps—*refer* to Chapter 10; *see* Chapter 11 for special staining procedures
- Mount, cover, and label stained sections (*refer* to Chapter 12).
- Examine the stained section with proper use of light microscope (Part-II, Chapter 17).

Sterilization and Cleaning of Glassware

Sterilization

The sterilization can be done by following methods:

1. **Dry heat** sterilization has limited value. Prolonged exposure may cause damage.
2. **Hot air oven** where heat is transferred by convection, conduction or radiation.
   a. The temperature of 100° C for one hour can destroy the non-sporing organism.
   b. Fungal spores need 115° C for one hour.
   c. While for all other bacteria 160° C temperature is needed for one hour.
3. **Incineration** where the flame is an effective way of sterilization. Flame heat is needed for the loop for culture.
4. **Moist heat** is the most reliable method of sterilization. This is the most lethal agent to kill microorganism.
   - Microbial death is due to coagulation and denaturation of the protein and enzyme.
5. **Boiling** is not effective to kill spore bearing bacteria and for surgical instruments.
6. **Steam sterilization** or tyndallization is exposure to steam at 100° C for 90 minutes. This good means to sterile the media which contain sugar.
7. **The autoclave** is heating water under pressure which boils at progressively higher temperature. This method is good for rubber material and surgical instruments.
8. **Membrane filters** are Millipore filters. Filters with the pore of 0.22 micrometer are sufficient for the bacteria.
9. **Seitz filter** are disposable asbestos pad filter.
10. **Flaming** when the material is wetted by alcohol and then flamed. This method is rapid.
11. **Ultraviolet light** causes damage to bacteria.
12. **Radiation** in the form of beta and gamma X-rays used for the surgical pads.
13. **Supersonic and ultrasonic** waves, 9000 cycles per second or above are used to rupture and disintegration of the cells.

Cleaning

1. Cleaning of large glassware

For cleaning use sterilized disposable plasticware or submerge the glassware in a cleaning solution:

- Potassium dichromate .... 2 g
- Distilled water .... 200 ml

Dissolve dichromate in distilled water; when cool, *add very slowly*

- Sulfuric acid (concentrated) .... 9 parts
- Potassium dichromate solution (2%) .... 1 part

For removal of grease from the glassware used in a histology laboratory, following procedure is advised:

- Soak for several hours in a mixture of sulfuric acid and excess amount of potassium dichromate
- Wash in running tap-water, preferably overnight
- Clean with suitable brush (*see* Fig. 1.10)
- Rinse in distilled water.

2. Cleaning of glass slides and cover slips

Slides are to be properly cleaned; otherwise the attached sections on them tend to fall during subsequent steps of routine histological procedures.

a. **Soak in any one of the following solutions**
i.	Old xylene	1 part
	Methylated spirit	1 part
ii.	Acid alcohol	0.5% or 1.0%
	Alcohol (70%)	1000 cc
	HCl (conc.)	5 cc (for 0.5%) or 10 cc (for 1.0%)

   iii. Concentrated aqueous solution of sodium hydroxide

   iv. Concentrated nitric acid (HNO_3)

v.	Sulfuric acid or nitric acid (conc.)	1 part
	Potassium dichromate	9 parts

b. Wash in running tap-water
c. Rinse in distilled water
d. Rinse in 0.5 per cent acid alcohol
e. Dip in 90 per cent alcohol
f. Wipe dry each slide with a soft, lintless cloth.

2

Some Basic Things to Know

A student working in a standard histology laboratory has to deal with a number of *chemicals* used generally as some kind of *solution*. Often solutions are used on the basis of percentages, but sometimes their strengths are required in a given *normalcy* or *molarity*. Also, buffers of a particular *pH* are to be obtained. Solutions are to be kept generally in the *glassware* used in any scientific work. Sections of a given biological samples are attached on *glass slides* and mounted under *cover slips* before being examined with an optical light microscope. It is needless to emphasize that the glassware and microscope lenses should only be used absolutely clean. Hence, it is always beneficial to familiarize with some terms and units of measurements that are not much frequently used in daily practice.

UNITS AND SYMBOLS OF MEASUREMENT

The units of measurements to describe different magnitudes of length, volume, and weight are commonly encountered. These are recommended for learning in accordance to their prefixes and abbreviations.

Atomic Mass

The standard unit for measuring the mass of atoms and their subatomic particles is a **dalton,** also known as an *atomic mass unit (amu)*. A neutron has a mass of 1.008 daltons, and a proton has a mass of 1.007 daltons. The mass of an electron, at 0.0005 dalton, is almost 2000 times smaller than the mass of a neutron or proton. The **atomic mass** (also called the *atomic weight*) of an element is the average mass of all its naturally occurring isotopes. Typically, the atomic mass of an element is close to the mass number of its most abundant isotope.

Ions, Molecules and Compounds

As we discussed, atoms of the same element have the same number of protons. The atoms of each element have a characteristic way of losing, gaining, or sharing their electrons when interacting with other atoms to achieve stability. The way that electrons behave enables atoms in the body to exist in electrically charged forms called ions, or to join with each other into complex combinations called molecules. If an atom either *gives up* or *gains* electrons, it becomes an ion. An **ion** is an atom that has a positive or negative charge because it has unequal numbers of protons and electrons. *Ionization* is the process of giving up or gaining electrons.

An ion of an atom is symbolized by writing its chemical symbol followed by the number of its positive (+) or negative (–) charges. Thus, Ca_2^+ stands for a calcium ion that has two positive charges because it has lost two electrons. When two or more atoms *share* electrons, the resulting combination is called a **molecule**. A *molecular formula* indicates the elements and the number of atoms of each element that make up a molecule. A molecule may consist of two atoms of the same kind, such as an oxygen molecule (Fig. 2.1A). The molecular formula for a molecule of oxygen is O_2. The subscript 2 indicates that the molecule contains two atoms of oxygen. Two or more different kinds of atoms may also form a molecule, as in a water molecule (H_2O). In H_2O, one atom of oxygen shares electrons with two atoms of hydrogen.

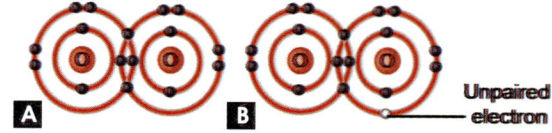

Fig. 2.1: Atomic structures of: (A) An oxygen molecule (O_2) and (B) a superoxide free radical (O_2^-). Note that a free radical has an unpaired electron in its outermost electron shell.

A **compound** is a substance that contains atoms of two or more different elements. Most of the atoms in the body are joined into compounds. Water (H_2O) and sodium chloride (NaCl), common table salt, are compounds. However, a molecule of oxygen (O_2) is not a compound because it consists of atoms of only one element.

A **free radical** is an atom or group of atoms with an unpaired electron in the outermost shell. A common example is superoxide, which is formed by the addition of an electron to an oxygen molecule (Fig. 2.1B). Having an unpaired electron makes a free radical unstable, highly reactive, and destructive to nearby molecules. Free radicals become stable by either giving up their unpaired electron to, or taking on an electron from, another molecule. In so doing, free radicals may break apart important body molecules.

Base units			Prefixes			
Unit	Quantity	Symbol	Prefix	Multiplier		Symbol
Meter	Length	m	tera-	$10^{12} = 1,000,000,000,000$		T
Kilogram	Mass	kg	giga-	$10^{9} = 1,000,000,000$		G
Second	Time	s	mega-	$10^{6} = 1,000,000$		M
Liter	Volume	l	kilo-	$10^{3} = 1,000$		k
Mole	Amount of matter	mol	hecto-	$10^{2} = 100$		h
			deca-	$10^{1} = 10$		da
			deci-	$10^{-1} = 0.1$		d
			centi-	$10^{-2} = 0.01$		c
			milli-	$10^{-3} = 0.001$		m
			micro-	$10^{-4} = 0.000,001$		μ
			nano-	$10^{-9} = 0.000,000,001$		n
			Pico-	$10^{-12} = 0.000,000,000,001$		p

US to SI (metric) conversion—measurements

When you know	Multiply by	To find
Inches	2.54	Centimeters
Feet	30.48	Centimeters
Yards	0.91	Meters
Miles	1.61	Kilometers
Ounces	28.35	Grams
Pounds	0.45	Kilograms
Tons	0.91	Metric tons
Fluid ounces	29.57	Milliliters
Pints	0.47	Liters
Quarts	0.95	Liters
Gallons	3.79	Liters

SI (metric) to US conversion—measurements

When you know	Multiply by	To find
Millimeters	0.04	Inches
Centimeters	0.39	Inches
Meters	3.28	Feet
Kilometers	0.62	Miles
Liters	1.06	Quarts
Cubic meters	35.32	Cubic feet
Grams	0.035	Ounces
Kilograms	2.21	Pounds

Note

In light microscopy work the most common prefix is *micro-* (conventionally referred to as micron). On the other hand, in electron microscopy the linear measurements are scaled in angstrom unit (Å). One angstrom equals to one-tenth of a nanometer. Nowadays, according to SI (System Internationale) units, micron is called *micrometer*, and the term *nanometer* is preferred in place of angstrom units. As distances become shorter, the number of zeros after the decimal point becomes larger. One nanometer is a millionth of a millimeter (10^{-9} meter).

For ready reference the following equivalents are useful.

1 Meter (m)	=	100 Centimeters (cm)
1 Centimeter (cm)	=	10 Millimeters (mm)
1 Millimeter (mm)	=	1000 Micrometers (µm)
1 Micrometer (µm)	=	100 Nanometers (nm)
1 Nanometer (nm)	=	10 Angstrom (Å)
1 Angstrom (Å)	=	10^{-1} nm = 10^{-4} µm = 10^{-7} mm = 10^{-8} cm = 10^{-10} m

Some equivalents of metric, United States, and English measures with values rounded off to two decimal places are given below.

Length	Volume	Weight	Temperature
1 Meter = 39.37 inches	1 Liter = 1.06 US liquid quart	1 Kilogram = 2.20 pounds	100° Centigrade = 180° Fahrenheit
1 Inch = 2.54 cm = 1/12 foot	1 US liquid quart = 32 fluid ounces = 1/4 US gallon = 0.95 liter	1 Pound = 16 ounces = 453.6 grams	
	1 Milliliter = 0.03 fluid ounce; 1 Fluid ounce = 29.57 milliliters	1 Grain = 65 milligrams	

Majority of the conversions listed above are simple being direct, but conversions from temperatures of one scale to the other are to be done as follows.

To convert from Centigrade to Fahrenheit: The given temperature (°C) is multiplied by 9/5 and 32 is added to get temperature in (°F). F= C × (9/5) + 32

To convert from Fahrenheit to Centigrade: From the given temperature (°F) 32 is subtracted and then the obtained figure is multiplied by 5/9 to get temperature in (°C). C= [F−32] × 5/9

SOLUTION STRENGTHS OF CHEMICALS

Although dyes or stains are chemicals, but the term *chemical* is used generally to designate acids, bases, salts, solvents, and organic compounds other than dyes. In a true sense water belongs to this category, but familiarity with it tends to prevent it being considered as a chemical. Unless the cost of chemically pure (CP) or reagent grade chemicals is prohibitive, one should always use pure forms. The chemicals, used in a histology laboratory, are in form of *solutions*. It is well known that a solution is a homogeneous mixture of two or more substances on molecular level. The constituent of a mixture present in smaller amount is called the *solute* and the one present in larger amount is called the *solvent*. Generally, the concentration of a solution is defined as the amount of the solute present in a given amount of solution.

$$\text{Concentration} = \frac{\text{Quantity of solute}}{\text{Volume of solution}}$$

Depending upon the low and high concentrations of solute, solutions are termed dilute and concentrated respectively. The solutions may be prepared according to the following methods.

Solution Strengths Based on Percentage

The more commonly used reagents are in different concentrations made on the basis of percentage, which may be calculated on the basis of *weights* or *volumes*. A 10 per cent solution of sodium chloride would ordinarily be made by weighing 10 grams of dry salt and adding it to 100 ml of distilled water. A 5 per cent solution of hydrochloric acid might be made either by adding 5 ml of acid to 95 ml of water, or the final volume of the solution made up to 100 ml by placing the 5 ml acid in a volumetric flask and adding water to the mark. Although both the methods are satisfactory for general purposes, the latter method is regarded as more accurate for a volume to volume relationship (abbreviated v/v).

Percents by Weight

All men weigh 200 pounds. All women weigh 125 pounds. What is the percent by weight of woman in married couples? A married couple is one man and one woman (No political implications intended). The total weight is 325 pounds. The formula for percent is:

$$\frac{\text{Target}}{\text{Total}} \times 100 = \text{Percent}$$

In this case the woman is the target, so the weight of the woman goes on top, and the total weight goes on the bottom of the fraction.

The weights of atoms are the atomic weights. What is the percentage of chloride in potassium chloride? The atomic weight of potassium is 39.10 g/mol. The atomic weight of chlorine is 35.45 g/mol. So the formula weight of potassium chloride is 74.55 g/mol. The chloride is the target and the potassium chloride is the total. 35.45 g/mol × 100% = 47.55198% or 47.6% to three significant figures.

$$\frac{35.45 \text{ g/mol}}{74.55 \text{ g/mol}} \times 100 = 47.552\% = 47.6\%$$

Solution Strengths Based on Normality

A student or technician working in a histology laboratory has to handle a situation when dilutions of certain acids and bases are required to be prepared on the basis of normality. Hence, a basic knowledge about the term *normal* is essential. Also, some other terms in relation to the strengths of solutions, commonly of salts, used will be described here. Commonly they are utilized in relation to the buffer solutions.

Molarity, Molality and Normality

The quantitative relationship between chemical substances in a reaction is known as **stoichiometry**. Avogadro was a pioneer in this field of chemistry (Fig. 2.2).

Avogadro hypothesized that there was a specific number that would represent the number of atoms or molecules in a **mole** of

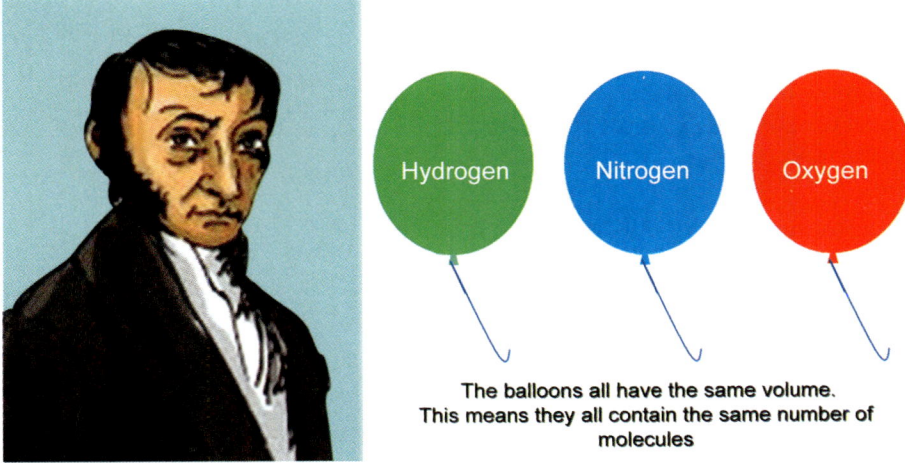

Fig. 2.2: Amedeo Avogadro: Avogadro's law states that "Two equal volumes of gas, at the same temperature and pressure, contain the same number of molecules".

that atom or molecule. The weight of that unit known as a **mole** would be equivalent to the atomic or molecular weight of the atom or molecule in grams. According to this theory, one mole of carbon-12 would have a mass of 12 grams because carbon-12 has an atomic weight of 12 (6 neutrons and 6 protons). One mole of hydrogen would weigh one gram and would contain the same number of atoms as one mole of carbon. The magical number was, in fact, discovered to be **6.024E** (Avogadro's number, number of units in one mole of any substance (defined as its molecular weight in grams), equal to **6.022140857 × 10²³**). It was named the **Avogadro number** in honor of the Father of stoichiometry even though he did not actually determine the exact number. The modern definition of a mole is as follows:

One atom or even ten atoms are too small for an individual to measure out in a lab. A mole of a substance equals the **gram-formula mass** or the **gram-molar mass**. This equals the sum of all of the masses of all the elements in the formula of the substance. Basically, if one were to count all of the carbon atoms in one mole of carbon-12, there would be $6.02E^{23}$ atoms and it would weigh 12 grams (remember the atomic weight of carbon is 12).

Using the information above, it is possible to calculate concentrations of solutions and make up solutions of desired concentration. It is also possible to use this information to determine how much of a given base would be needed in order to neutralize a specific acid and reach a pH of 7.

There are five units of concentration that are particularly useful to chemists. The first three: **Molality**, **molarity** and **normality** are dependent upon the mole unit. The last two: Percent by volume and percent by weight have nothing to do with mole, only weight or volume of the **solute** or substance to be diluted, versus the weight or volume of the **solvent** or substance in which the solute is diluted. Percentages can also be determined for solids within solids.

Besides such solutions, which are prepared percentagewise, there are certain other ways of expressing the strengths of some chemicals. These terms are given below:

1. **Molarity:** It is defined as the number of moles of solute per liter of the solution. The symbol for molarity is M, read as *molar*. The molar unit is probably the most commonly used chemical unit of measurement. The unit of molarity is *mol per liter*.

$$\text{Molarity } (M) = \frac{\text{Moles of solute (n)}}{\text{Volume (V) in liters of solvent}}$$

If one gram mole of a substance is dissolved in a solvent and the solution is made up of one liter mark of the flask, such solution is called a *molar solution*.

When the amount of solute (x) is given in grams (and not in moles) and its molecular weight is MW, the following expression is helpful in deriving the moles of solute (*n*):

$$n = \frac{\text{Amount of solute (x) in grams}}{\text{Molecular weight (MW) of solute in grams}} \text{ moles}$$

The 'mol' is defined as the *gram-molecular weight* of a substance. A milli mol (m mol) equals 10^{-3} mol, and a micromole (μ mol) as 10^{-6} mol.

Molarity is the number of moles of a solute dissolved in a liter of solution. A molar solution of sodium chloride is made by placing 1 mole of a solute into a 1 liter volumetric flask. (Taking data from the example above, we will use 58 grams of sodium chloride). Water is then added to the volumetric flask up to the one liter line. The result is a one molar solution of sodium chloride.

Why Do We Need Moles?

The view from the atom is very different from the view of trillions and trillions of atoms. The mass action of the atoms that we see on our "macro" view of the world is the result of the action of an incredibly large number of atoms averaged in their actions. The most usual way we count the atoms is by weighing them. The mass of material as weighed on a balance and the atomic weight of the material being weighed is the way we have of knowing how many atoms or molecules we are working with. Instead of counting rice grains, we buy kilograms or pounds of rice and have an idea of how many rice grains are in the container. There are less than one hundred naturally occurring elements. Each element has a characteristic atomic weight.

2. Molality: It is defined as the *number of moles of solute per kilogram of solvent*. The symbol for molality of a substance is '*m*', which may be expressed as follows:

$$\text{Molality } (m) = \frac{\text{Moles of solute (n)}}{\text{Mass of solvent (S) in kilograms}}$$

A solution obtained by dissolving one gram mole of the solute in 1000 grams of solvent is called *one molal* or *1 m solution*. Therefore, it is noteworthy that molarity is defined in terms of volume of a solution, while the molality is defined in terms of the mass of the solvent.

The molal unit is not used nearly as frequently as the molar unit. Be careful not to confuse molality and molarity. Molality is represented by a small "m," whereas molarity is represented by an upper case "M." Note that the solvent must be weighed unless it is water. One liter of water has a specific gravity of 1.0 and weighs one kilogram; so one can measure out one liter of water and add the solute to it. Most other solvents have a specific gravity greater than or less than one. Therefore, one liter of anything other than water is not likely to occupy a liter of space. To make a one molal aqueous (water) solution of sodium chloride (NaCl), measure out one kilogram of water and add one mole of the solute, NaCl to it. The atomic weight of sodium is 23 and the atomic weight of chlorine is 35. Therefore the formula weight for NaCl is 58, and 58 grams of NaCl dissolved in 1kg water would result in a 1 molal solution of NaCl.

3. Normality: This is another term frequently used to express concentrations of solutions of acids and bases. The normality of a solution (symbol, N) is defined as the *number of gram equivalents of a solute per liter of the solution*. If one gram equivalent of an acid, base or salt is dissolved in water and the solution is made up to one liter, such solution is called *normal solution*. The normality may be expressed by the following equation:

$$Normality\ (N) = \frac{\dfrac{Equivalents\ (Eq)}{of\ solute}}{Liters\ of\ solution} = \frac{Grams\ per\ liter}{\dfrac{Equivalent}{weight}}$$

There is a relationship between normality and molarity. Normality can only be calculated when we deal with reactions, because normality is a function of equivalents.

Equivalent weight = Molar mass/(H^+ per mole)

Equivalent = Mass of compound/Equivalent weight

And Normality = (Equivalents of X)/Liter

One should remember that Normality = molarity x n (where n = the number of protons exchanged in a reaction).

4. **Osmolarity**: It is defined as the *number of osmols per liter of the solution*. The term osmolarity is used when expressing the concentrations of the amounts of the osmotically active particles. The unit of osmolarity is osmol (symbol, *Osm*) which equals the molecular weight of any given substance in grams divided by the number of freely moving particles each molecule of that substance liberates in a given solution. Since the osmotically active substances in the body are dissolved in water, and the density of water is 1, the osmolal concentrations can be expressed *Osm per liter* of water.

$$Osmolarity\ (Osm) = \frac{Number\ of\ osmols}{Volume\ of\ solution\ (in\ liters)}$$

5. **Osmolality**: It is defined as the *number of osmols per kilogram of the solvent*. The *osmolal concentration* of a substance in a fluid is measured by the degree to which it depresses the freezing point, 1 mol per liter of an ideal solute depresses the freezing point by 1.86°C.

$$Osmolality = \frac{Number\ of\ osmols}{Volume\ of\ solution\ (in\ liters)}$$

As in the case of molarity and molality, the osmolarity is affected by the volume of the various solutes in the solution and the temperature, while the osmolality is not.

pH AND ITS MEANING

The word 'pH' is frequently used while preparation for solutions of various fluids is done. It is necessary to understand its meaning. The term means *power of hydrogen ions* and it can be derived mathematically from an exponent (or power) of the hydrogen ions present. For example, in the equation

$$\frac{(C_{H^+}) \times (C_{OH^-})}{C_{H_2O}} = K$$

for the ionic equilibrium of water, C designates the concentration of hydrogen ions, hydroxyl ions, and undissociated water respectively, and K is a constant. It has been established that K for water, at room temperature, is very close to 10^{-14}. Since both hydrogen and hydroxyl ions are derived in equal numbers from the dissociation of water, the concentration of these would be 10^{-7} each. These negative exponents are somewhat cumbersome to handle mathematically. Hence, for convenience, *the term pH is defined as the logarithm of the reciprocal of the hydrogen ion concentration*. Thus, $CH^+ \times 10^{-7}$ becomes pH 7.0, which is a symbol of neutrality. The pH of a given solution can be measured by an instrument known as a pH meter (Fig. 2.3).

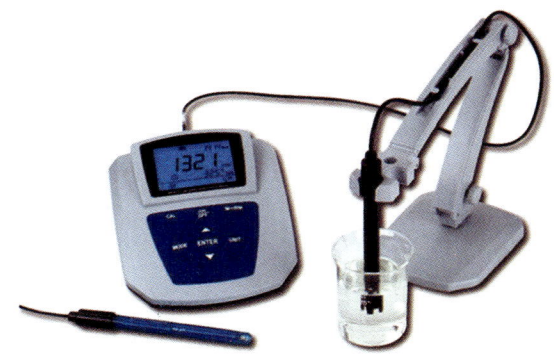

Fig. 2.3: Diagram showing a pH meter.

While pH zero represents a hydrogen ion activity of 1 normal (1 N) and pH 14 an hydroxyl ion activity of 1 normal (1 N). A 0.1 N solution of hydrochloric acid (HCl) has a pH of approximately 1.0, and a 0.1 N solution of sodium hydroxide (NaOH) or potassium hydroxide (KOH) has a value that approaches pH 13.0. A *strong acid* or *strong base* is one that ionizes freely; examples are mineral acids like hydrochloric acid and sulphuric acid. On the other hand, *weaker* acids (for example, acetic acid and lactic acid) ionize only to a limited extent. Each increase of one point of the pH scale represents a tenfolds increase in ionization (log 10 = 1). A calculation will show that the acetic acid is only about 1/50th as strong as hydrochloric acid even their normality is same. Most of the biological fluids have pH ranging between 5 and 9 except gastric juice which has a pH near 2. However, there are higher degrees of acidity than pH 0 and that of alkalinity beyond pH 14, but it is unlikely that such extremes will be encountered in histological studies of biological tissue samples (Fig. 2.4).

BUFFERS SOLUTIONS IN COMMON USAGE

A *buffer* solution is defined as one that will tolerate the addition of 'H' ions or of 'OH' ions in considerable quantity without affecting a marked change in its pH value. Preparation of buffers is required in enzyme studies of biological specimens. Two *stock solutions*, 'A' and 'B' are prepared separately and kept in clean glass container till the buffer is required. These should be preferably stored in a refrigerator. A *working solution* of desirable pH is prepared by mixing together recommended volumes of 'A' and 'B' stock solutions. Formulae of the two commonly used buffers are given below. For other buffers, reader is advised to consult any standard book of biochemistry.

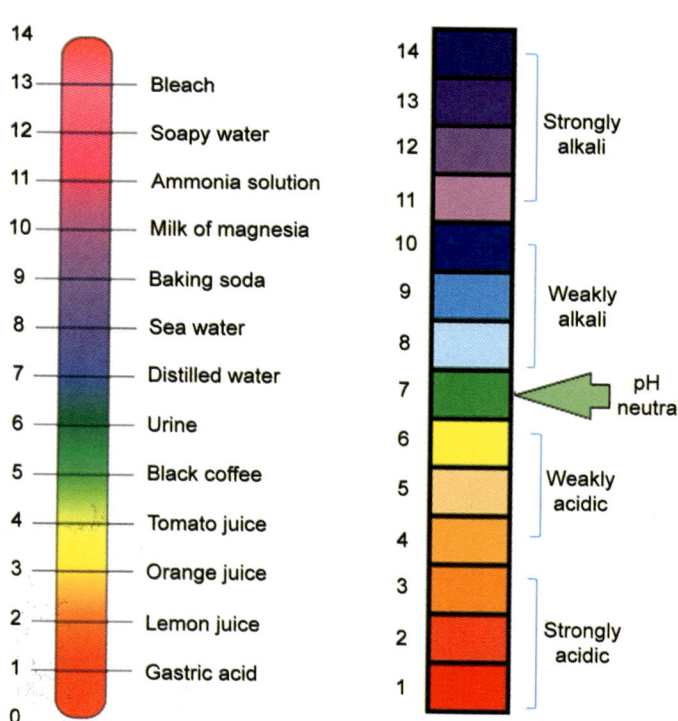

Fig. 2.4: pH scales—acids and bases, the other indicators that can be used are the pH meter (Fig. 2.3), which provides the most accurate readings, phenolphthalein and methyl orange.

Phosphate Buffer

Stock Solutions

A. 0.2 M solution of monobasic sodium phosphate (dissolve 27.8 g salt in 1000 ml water)

B. 0.2 M solution of dibasic sodium phosphate (dissolve 53.65 g of $Na_2HPO_4.7H_2O$ or 71.7 g of $Na_2HPO_4.12H_2O$ in 1000 ml water).

Working Solution

A working solution of the desired pH can be made by mixing of x ml of solution (A) with y ml of solution (B) and the mixture diluted with distilled water to a total of 200 ml as shown in Table 2.1.

Table 2.1	Quantities of solutions A (x ml) and B (y ml) to obtain desired pH	
x	y	pH
93.5	6.5	5.7
92.0	8.0	5.8
90.0	10.0	5.9
87.7	12.3	6.0
85.0	15.0	6.1
81.5	18.5	6.2
77.5	22.5	6.3
73.5	26.5	6.4
68.5	31.5	6.5
62.5	37.5	6.6
56.5	43.5	6.7
51.0	49.0	6.8
45.0	55.0	6.9
39.0	61.0	7.0
33.0	67.0	7.1
28.0	72.0	7.2
23.0	77.0	7.3
19.0	81.0	7.4
16.0	84.0	7.5
13.0	87.0	7.6
10.5	90.5	7.7
8.5	91.5	7.8
7.5	93.0	7.9
5.3	94.7	8.0

Cacodylate Buffer

Stock Solutions

A. 0.2 M solution of sodium cacodylate $Na(CH_3)_2AsO_2.3H_2O$. Dissolve 42.8 g of salt in 1000 ml distilled water.

B. 0.2 N HCl

Working Solution

50 ml of solution (A) is mixed with x ml of solution (B) and mixture diluted to 200 ml for preparing buffer of the desired pH.

x	pH
93.5	5.7
2.7	7.4
4.2	7.2
6.3	7.0
9.3	6.8
13.3	6.6
18.3	6.4
23.8	6.2
29.6	6.0
34.8	5.8
39.2	5.6
43.0	5.4
45.0	5.2
47.0	5.0

PROPERTIES OF LIGHT

The light is a form of energy that has the following properties:

1. Finite velocity 3×10^{10} cm/sec
2. Produces heating effect, and like heat, light is a form of energy.
3. It exerts pressure upon the surface on which it falls.

Since light is a form of energy, it must be transmitted from one place to the other. There are two theories about the light:

1. *Newton's Emission or Corpuscular Theory*
a. Light is carried by streams of very small material particles (travelling with finite velocity, having zero mass) called **corpuscles**.

b. It does not require any medium for its propagation

c. Corpuscles travel in straight line and are so minute that they readily pass through the particles of matter

Newton suggested **theory of fits**—according to which when corpuscles reach the surface, sometimes they are in a state of **fit for easy reflection**, and sometimes in a **fit for easy transmission**. He had objection to the wave theory lay in the fact that light travels in straight lines and does not bend round corners like sound waves.

2. *Wave Theory of Light*: Put forward by Huygeus in 1678.

a. Light energy moves by wave motions without actually travelling the matter.

b. This process requires an intervening medium as a vehicle called ether.

Reflection of light: When light travelling in a homogeneous medium is incident on a second medium, some of it is sent back into the first medium according to definite laws and is said to be reflected.

Refraction of light: When a ray of light passes from one transparent medium to another, in which it has a different velocity, there occurs a change in the direction of propagation of light except when it strikes the surface of separation of two media normally. This bending of light is called **refraction** (Fig. 2.5).

An **incident ray** (PE) travels through air medium and strikes a glass prism (with an **angle of the prism** A of 60°) at a point E making an **angle of incidence** 'i' and slows down. It gets refracted by bending towards the *normal* (N'). The **refracted ray** EF, making an **angle of refraction** 'r' with the normal travels within the glass and exits at a point F. Getting exit and entering again into the air medium the **emergent ray** FS bends again and speeds up again (but this time bending away from another normal M') making an **angle of emergence** 'e'. If the incident and emergent rays are projected inside the prism, an **angle of deviation** 'D' is formed.

In reality, when beams of white light pass through a prism the light resolves into different colors and creates a **spectrum** of colors (Fig. 2.6). This happens because each wavelength of light has a slightly different refractive index for a given material. The higher energy (shorter wavelength) or blue light is refracted more than the lower energy (longer wavelength) or red light. The various wavelengths that make up white light (e.g. sunlight) all get refracted by a different amount and so the light is **dispersed** into its constituent colors. This produces the familiar **rainbow of colors** or spectrum.

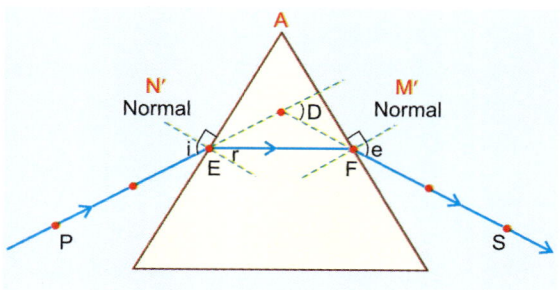

Fig. 2.5: Diagram showing the phenomenon of refraction of light.

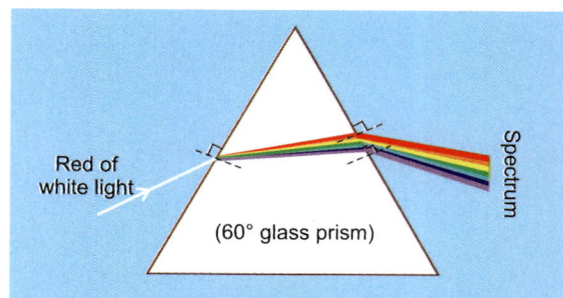

Fig. 2.6: Diagram shows dispersion of light through a glass prism. Note that a white light is a mixture of seven different colors. Each color is refracted to a different degree by the glass—that is why the colors disperse into a spectrum of light (shown on the right). Remember that violet really disperses most, but **BBB** (**B**lue **B**ends **B**est) is more memorable. The red light bends less because it is refracted less.

Relationship between Angle of Incidence and Angle of Refraction

The relationship between these angles (Fig. 2.6) is explained by **Snell's Law** which states that the ratio of the sine of angle of incidence to the angle of refraction for any two media is constant for a light of definite color. This constant is called **refractive index** of the second medium in respect to the first.

$$\sin i/\sin r = 1\,\mu\,2$$

Where, 1 and 2 indicate that light passes from one to two.

When the path is changed

$$2\,\mu\,1 = \sin r/\sin i \quad \text{Hence, } 1\,\mu\,2 = 1/2\,\mu\,1$$

or $\quad 1\,\mu\,2/2\,\mu\,1 = 1$ ... (2.1)

Dispersion of light: When a beam of light (white) is refracted through a prism, it is split up into its constituent colors (R, O, Y, G, B, I, V), each having a different *refrangibility*. This separation of colors by a prism is known as **dispersion** and the band of colors is called the **spectrum**. The angle between the direction of any two colors is called the angular dispersion between them

$$d = (\mu - 1)\,A$$

The ratio between the angular dispersion and the mean deviation when light is deviated

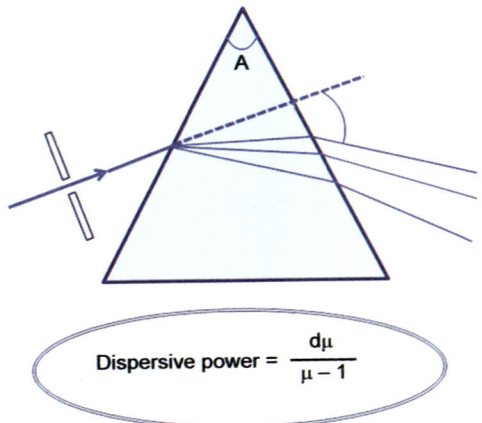

$$\text{Dispersive power} = \frac{d\mu}{\mu - 1}$$

Fig. 2.7: Diagram showing equation for dispersive power of a glass prism.

at minimum deviation by a prism of small angle is called the dispersive power of the medium of which the prism is made (Fig. 2.7).

While this dispersion is nice in a prism it causes problems in lenses. **Lenses** are designed to focus light perfectly but because of the different degrees of refraction of the various wavelengths of light, the different colors get focused at different points. This blurs the image and creates colored outlines. The defect is called **chromatic aberration**.

Interference of light: It is modification in the distribution of light energy obtained by superimposition of two or more waves. The theory of interference was put forward by Thomas Young in 1801. The theory states that when there are two adjacent sources giving out continuous waves of the same wavelength and amplitude and having the same phase or a constant difference in phase—then the distribution of energy is no longer uniform. At the points where the crest of one wave upon the trough of the other and vice versa, the resultant amplitude is reduced to zero and the energy is minimum. Also, at other points where the crest of one wave falls upon the crest of the other (or trough of one falls upon the trough of other)—the resultant amplitude is increased and the energy becomes maximum (Fig. 2.8).

The wavelength of light is 6×10^{-5} cm that of sound is approximately 10^7 larger. It is for this property that while a small obstacle will cut off a beam of light we need comparatively a very large obstacle (nearly 10^7 larger) to achieve the same effect in sound.

Grimaldy noted that with a small source of light the shadow of an object was larger than that given by geometrical construction. He also noted that shadow was not sharp and well-defined but on close examination it was found to be surrounded by a system of fringes. These fringes or bands outside the geometrical shadow gradually get closer together until they fade away into uniform illumination. This phenomenon of bending of light around

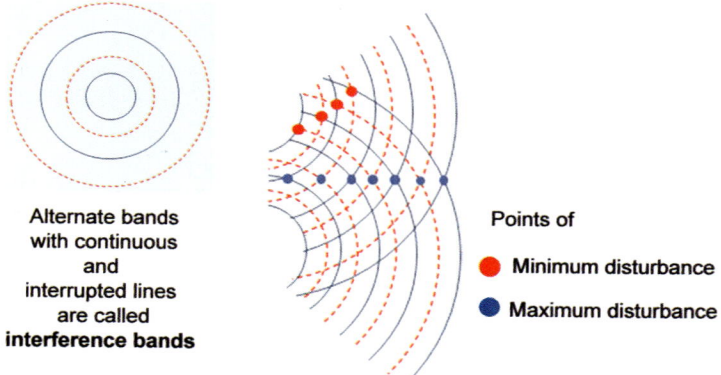

Alternate bands
with continuous
and
interrupted lines
are called
interference bands

Points of

● **Minimum disturbance**

● **Maximum disturbance**

Fig. 2.8: Diagrammatic representation of inteference bands in understanding the principles in specific type of microscopy

corners and the spreading of light waves into geometrical shadow of an object is called 'diffraction'.

ELECTROMAGNETIC SPECTRUM

What is a Spectrum?

A *spectrum* is an array of waves spread out according to the increase or decrease of some of their properties. The waves are produced by a vibrating object, and can be represented by a wavy curve with tops called *crests* and the bottoms *troughs* (Figs 2.9 and 2.10). The *wavelength* is defined as the distance between *two successive crests or troughs of a wave*. It is generally denoted by the Greek letter λ (lambda), and expressed in centimeters or meters or in Angstrom (Å) units.

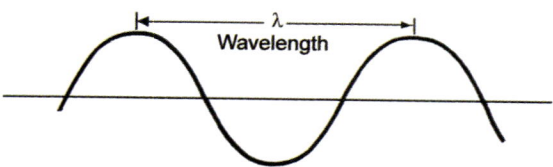

Fig. 2.9: Diagram shows a typical wave in motion. The distance between two crests (or troughs) on the same side of the baseline is termed the wavelength which is depicted by the Greek letter lambda (λ).

History of Electromagnetic Spectrum Discovery

For most of history, visible light was the only known part of the electromagnetic spectrum. The ancient Greeks recognized that light travelled in straight lines and studied some of its properties, including reflection and refraction. Over the years the study of light continued and during the 16th and 17th centuries there were conflicting theories which regarded light as either a wave or a particle.

The first discovery of electromagnetic radiation other than visible light came in 1800, when William Herschel discovered infrared radiation. He was studying the temperature of different colors by moving a thermometer through light split by a prism (Fig. 2.10). He noticed that the highest temperature was beyond red. He theorized that this temperature change was due to "calorific rays" which would be in fact a type of light ray that could not be seen. The next year, Johann Ritter worked at the other end of the spectrum and noticed what he called "chemical rays" (invisible light rays that induced certain chemical reactions) that behaved similar to visible violet light rays, but were beyond them in the spectrum. They were later renamed ultraviolet radiation.

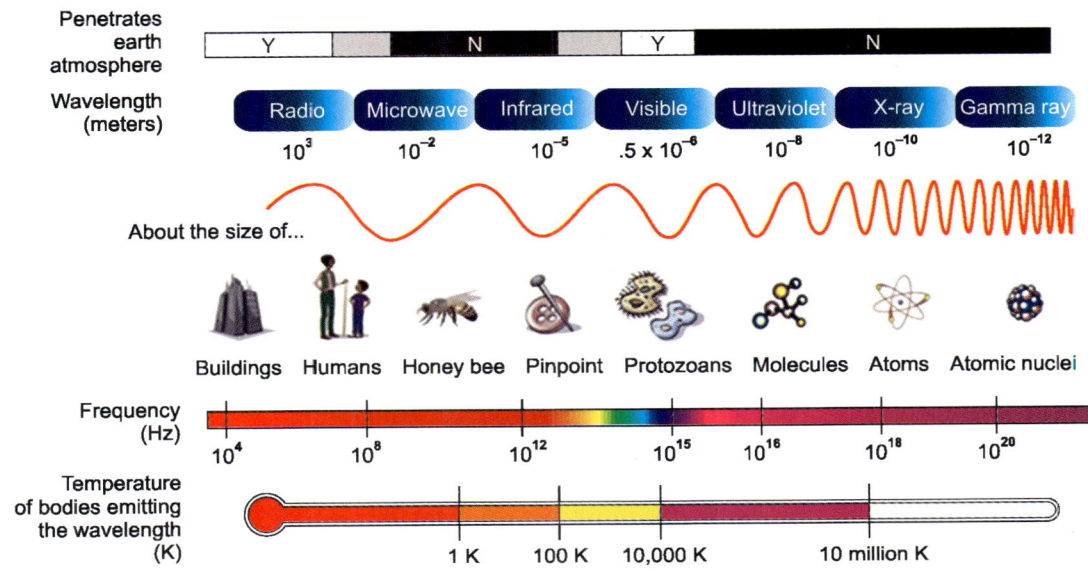

Fig. 2.10: Diagram shows relationship between the sizes of various living and/or non-living objects with the wavelength and frequency of different type of waves in the nature.

Electromagnetic radiation (had been first linked to electromagnetism in 1845, when Michael Faraday noticed that the polarization of light travelling through a transparent material responded to a magnetic field (*see* Faraday effect). During the 1860s James Maxwell developed four partial differential equations for the electromagnetic field. Two of these equations predicted the possibility of and behaviour of waves in the field. Analyzing the speed of these theoretical waves, Maxwell realized that they must travel at a speed that was about the known speed of light. This startling coincidence in value led Maxwell to make the inference that light itself is a type of electromagnetic wave.

The *frequency* is the number of waves that pass a given point in one second; and is denoted by another Greek letter ν (nu). The unit of frequency is *hertz* (Hz). A wave of high frequency has a shorter wavelength in contrast to a wave of low frequency which has a longer wavelength (Fig. 2.10). The *speed* (or *velocity*) of a wave is the distance through which a particular wave travels in one second. The relationship between the speed, frequency, and wavelength of a wave is given by the following expression:

$$\text{Speed (c)} = \text{Frequency (ν)} \times \text{Wavelength (λ)}$$

The electromagnetic radiations include a range of wavelengths and this arrangement of wavelengths is referred to as the *electromagnetic radiation spectrum* or simply *electromagnetic spectrum* (Fig. 2.10). It is interesting to note that the visible range of spectrum comprises only a small range of the electromagnetic spectrum. It extends from 4000 Å (violet) to 8000 Å (red). The violet part of the typical VIBGYOR has wavelengths between 4000 and 4250 Å with higher frequency. The red strip of VIBGYOR, has lower frequencies and the wavelength ranges between 6500 and 7500 Å.

The **electromagnetic spectrum** is the range of all possible frequencies of electromagnetic radiation. The "electromagnetic spectrum" *of an object* has a different meaning, and is instead the characteristic distribution of electromagnetic radiation emitted or absorbed by that particular object (Fig. 2.11).

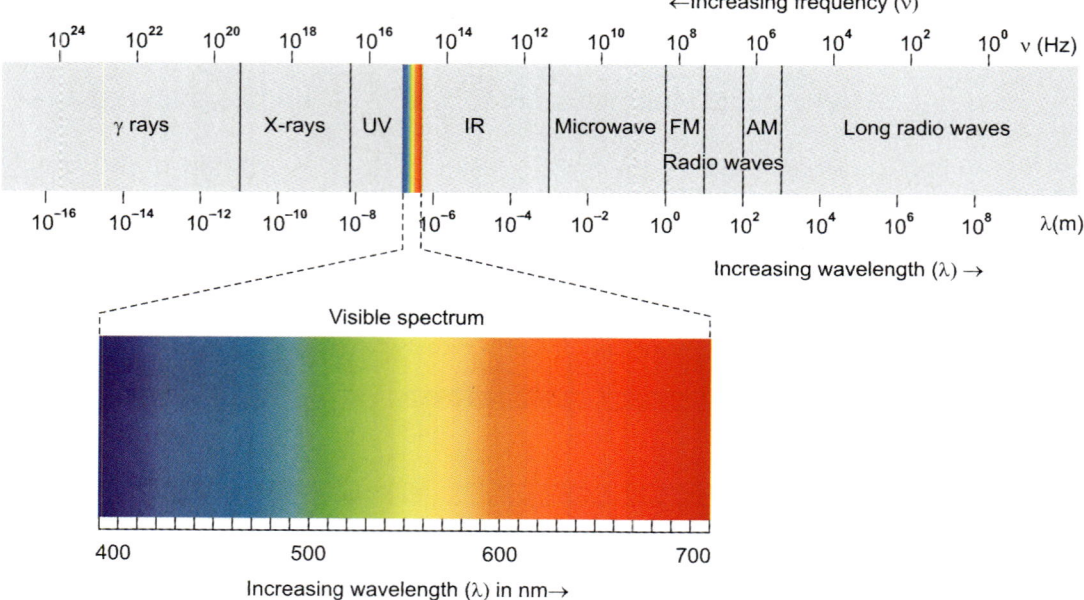

Fig. 2.11: Diagram depicting that the visible spectrum occupies a very small range of increasing wavelengths (λ) of electromagnetic spectrum (from 400 nm or 4000 Å to 700 nm or 7000 Å). The visible spectrum is sandwiched between ultraviolet and infrared parts.

The electromagnetic spectrum extends from below the low frequencies used for modern radio communication to gamma radiation at the short-wavelength (high-frequency) end, thereby covering wavelengths from thousands of kilometers down to a fraction of the size of an atom. The limit for long wavelengths is the size of the universe itself, while it is thought that the short wavelength limit is in the vicinity of the Planck length. Until the middle of last century it was believed by most physicists that this spectrum was infinite and continuous.

Regions of Electromagnetic Spectrum

A discussion of the regions (or bands or types) of the electromagnetic spectrum is given in Fig. 2.12. Note that there are no precisely defined boundaries between the bands of the electromagnetic spectrum; rather they fade into each other like the bands in a rainbow (which is the sub-spectrum of visible light). Radiation of each frequency and wavelength (or in each band) will have a mixture of properties of two regions of the spectrum that bound it. For example, red light resembles infrared radiation in that it can excite and add energy to some chemical bonds and indeed must do so to power the chemical mechanisms responsible for photosynthesis and the working of the visual system. The types of electromagnetic radiation are broadly classified into the following classes:

1. Gamma radiation
2. X-ray radiation
3. Ultraviolet radiation
4. Visible radiation
5. Infrared radiation
6. Terahertz radiation
7. Microwave radiation
8. Radio waves

This classification goes in the increasing order of wavelength (Fig. 2.11), which is characteristic of the type of radiation. While, in general, the classification scheme is accurate, in reality there is often some overlap between neighboring types of electromagnetic energy.

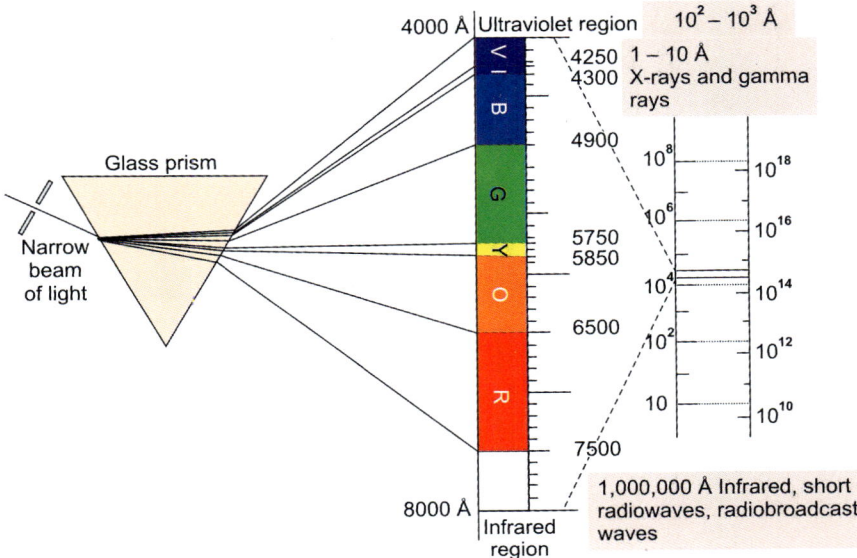

Fig. 2.12: Diagram shows regions of electromagnetic spectrum with boundaries of different colors in VIBGYOR.

For example, SLF radio waves at 60 Hz may be received and studied by astronomers, or may be ducted along wires as electric power, although the latter is, in the strict sense, not electromagnetic radiation at all (*see* near and far field).

Eight most abundant elements in human body		Sixteen trace elements in human body			
Element	% of total no. of atoms	Element	Symbol	Element	Symbol
H : Hydrogen	63.0	Iron	Fe	Chromium	Cr
O : Oxygen	25.5	Copper	Cu	Fluorine	F
C : Carbon	9.5	Manganese	Mn	Selenium	Se
N : Nitrogen	1.4	Cobalt	Co	Silicon	Si
Ca : Calcium	0.31	Iodine	I	Tin	Sn
P : Phosphorus	0.22	Molybednum	MO	Boron	B
Cl : Chlorine	0.08	Vanadium	V	Arsenic	As
K : Potassium	0.06	Nickel	Ni	Zinc	Zn

BIOMOLECULES

These are building-block compounds which perform specific cell functions. They are identical in all organisms. Their main functions are as follows:

i. Energy transmission
ii. Maintenance of orderliness
iii. Chemical reactions (complimentarity)
iv. Enzyme catalysed reactions (Proteins)
v. Self-regulation of cell reactions
   • Only 27 of 92 natural chemical elements are essential for different forms of life
   • Majority of them have low atomic numbers; only three have atomic numbers above 34
   • Most biomolecules are compounds of carbon

Bulk elements of organic matter (6)

- C
- H (63.0)
- O
- N
- P
- S

Elements occurring as ions (5)

- Na^+
- K^+
- Mg^{2+}
- Ca^{2+}
- Cl

HOW TO PREPARE HISTOLOGY SLIDES

There are many reasons to examine human cells and tissues under the microscope. Medical and biological research is underpinned by knowledge of the normal structure and function of cells and tissues and the organs and structures that they make up. In the normal healthy state the cells and other tissue elements are arranged in regular recognizable patterns. Changes induced by a wide range of chemical and physical influences are reflected by alterations in structure at a microscopic level and many diseases are characterized by typical structural and chemical abnormalities that differ from the normal state. Identifying these changes and linking them to particular diseases is the basis of histopathology and cytopathology, important specializations of modern medicine. Microscopy plays an important part in hematology (the study of blood), microbiology (the study of micro-organisms including parasites and viruses), and more broadly in the areas of biology, zoology and botany. In all these disciplines specimens are examined under a microscope.

The knowledge of how to prepare histology slides is not usually required for first-level courses in Anatomy and Physiology and Human Biology. It is, however, useful to have a general awareness of the steps involved in preparing histology slides.

It may be useful for school and college students to know a bit about how to prepare histology slides as part of their general appreciation of laboratory biology and techniques used in medical research. Familiarity with terminology used in the discipline of histology is also useful when reading around the subject or communicating with other professionals working in various fields within health sciences, e.g. during work experience.

The five main stages in the preparation of histology slides are:

1. Fixing
2. Processing
3. Embedding
4. Sectioning
5. Staining

Each of these stages is described briefly in appropriate chapters of this book.

SUMMARY

Dalton is a term used for the mass of atoms and their subatomic particles

Ion of an atom is written chemical symbol of its positive (+) or negative (−) charges.

Molecule, results when two or more atoms *share* electrons; or a molecule may result by combination of two or more different kinds of atoms.

Compound is a substance with atoms of two or more different elements.

Free radical is an atom or group of atoms with an unpaired electron in the outermost shell (for example, superoxide)

Conversion of temperatures:
1. (°F). $F = C \times (9/5) + 32$
2. (°C). $C = [F -32] \times 5/9$

Solution Strengths of Chemicals:
1. *Concentration* = Quantity of solute/Volume of solution
2. *Percent* = Target/Total weight × 100
3. *Molarity (M)* – number of moles of solute per liter of solution unit **(mol per liter)**
   M = moles of solute (n)/volume (V) in liters of solvent
4. *Molality (m)* = number of moles of solute per kilogram of solvent
   m = moles of solute (n)/mass of solvent (S) in kilograms
   A solution obtained by dissolving one gram mole of the solute in 1000 grams *of solvent is called* **one molal** or **1 m solution.**
5. *Normality (N)* is defined as *number of* equivalents of a solute per liter of the solution
   N = Equivalent (Eq) of solute/liters of solution
   = Grams per liter/Equivalent weight
6. *Osmolarity* **(Osm)** is defined as the *number of osmols per liter of the solution.*
   Osm = number of osmols/volume of solution (in liters)
7. *Osmolality* is defined as the *number of osmols per kilogram of the solvent.*
8. **pH** is a term used for *power of hydrogen ions* (measured with a pH meter). Most of the biological fluids have pH ranging between 5 and 9.
9. **Buffer solution** is one that tolerates the addition of H ions or OH ions in considerable quantity *without affecting* a marked change in its pH.

10. **Properties of Light**
    i. form of energy with a finite velocity 3×10^{10} cm/second
    ii. produces heating effect
    iii. exerts pressure upon the surface on which it falls
       a. *Reflection of light* is a phenomenon of light being **sent back to the same** homogenous medium from which it passes into a second medium
       b. *Refraction of light* is the bending of light, from one transparent medium to another, with a different velocity, after it has **changed its direction** of propagation except when it strikes the surface of separation of two media normally.
          Refractive index = sin i/sin r, where i is the *angle of incidence* and r is the *angle of refraction* (Fig. 2.5)
       c. *Dispersion of light* is split of a beam of white light after refraction through a glass prism, into its constituent colors (VIBGYOR), each having different *refrangibility.*
       d. *Interference of light* is the modification in the distribution of light energy obtained by superimposition of two or more waves. When there are two adjacent sources giving out continuous waves of the same wavelength and amplitude and having the same phase or a constant difference in phase, then the distribution of energy is no longer uniform.

Electromagnetic Spectrum

It is the range of all possible *frequencies* of electromagnetic radiation emitted or absorbed by a particular object.

Frequency (Hz, hertz) is the number of waves that pass a given point in one second.

Speed (or *velocity*) is the distance traveled in one second

Speed = Frequency × Wavelength

The spectrum extends from below the low frequencies to gamma radiations at short wavelength (high frequency) end, covering wavelengths from thousands of kilometers down to a fraction of the size of an atom.

Steps in preparation of tissue slides:
 i. Fixation
 ii. Dehydration
 iii. Clearing
 iv. Infiltration-embedding
 v. Section cutting (microtomy)
 vi. Staining
 vii. Mounting and labeling.

These are depicted diagrammatically below:

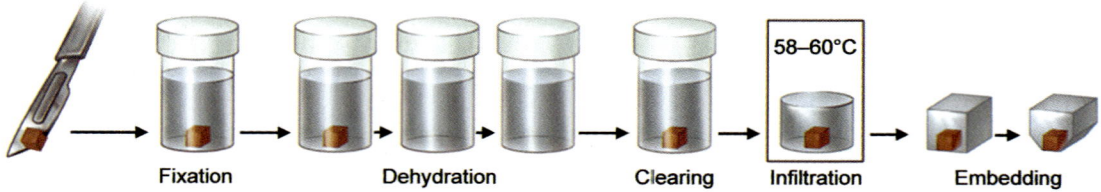

| Fixation | Dehydration | Clearing | Infiltration | Embedding |

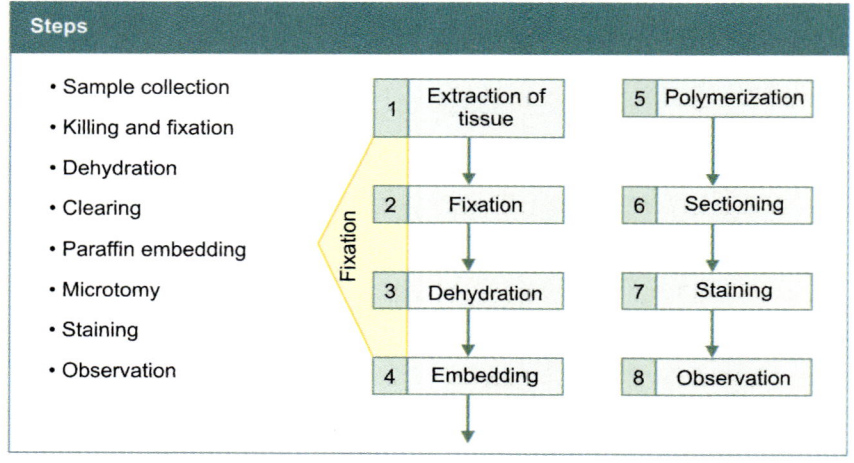

3

Grossing of Specimens and Fixation of Biological Samples

PROCUREMENT (OR SPECIMEN RECEPTION) FOR HISTOLOGICAL STUDY

Tissue specimens taken from routine surgical cases, autopsies, or other scientific investigations are examined, described and trimmed to proper size. This process is referred to as **"grossing the specimen"**. Grossing, often referred to as "cut-up", involves a careful examination and description of the specimen that will include the appearance, the number of pieces and their dimensions. Larger specimens may require further dissection to produce representative pieces from appropriate areas. For example, multiple samples may be taken from the excision margins of a tumor to ensure that the tumor has been completely removed. In the case of small specimens the entire specimen may be processed. The tissues selected for processing will be placed in cassettes (small perforated baskets) and batches will be loaded onto a tissue processor for processing through wax.

Specimens received for histological examination may come from a number of different sources (Fig. 3.1). They range from very large

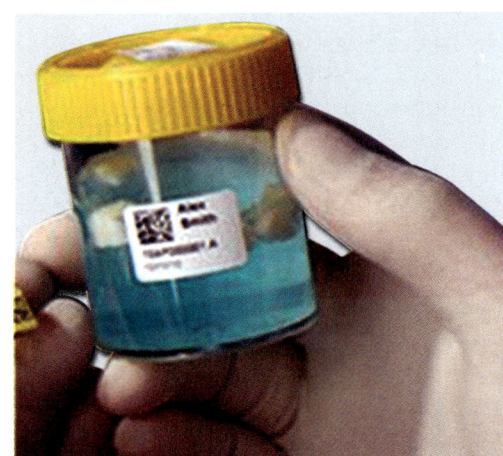

Fig. 3.1: A surgical specimen fixing in formalin and ready for grossing. Notice that there is a generous volume (×10 or even more) of fixative fluid compared to the size of the specimens removed for processing.

specimens or whole organs to tiny fragments of tissue. For example, the following are some of the specimen-types commonly received in a histopathology lab. Excision of specimens at operation (or *surgical biopsies*), is a procedure where whole organ(s) or affected areas are removed at operation. A few other common types of biopsies are:

- *Incisional biopsy* specimens, where tissue is removed for diagnosis from within an affected area
- *Punch biopsies*, where punches are used to remove a small piece of suspicious tissue for examination (often from the skin)
- *Shave biopsies*, where small fragments of tissue are "shaved" from a surface (usually skin)
- *Curettings*, where tissue is removed in small pieces from the lining of the uterus or cervix
- *Core biopsies*, where a small tissue sample is removed using a special needle sometimes through the skin (percutaneously).

Specimens are usually received in fixative (preservative) but sometimes arrive fresh and must be immediately fixed. Before specimens are accepted by a laboratory the identification (labeling) and accompanying documentation will be carefully checked, all details recorded and "specimen tracking" commenced (Fig. 3.2). It is vital that patient or research specimens are properly identified and the risk of inaccuracies minimized.

Precautions to Avoid Tissue Damage during Procurement

The following precautions must be followed to avoid damage prior to fixation (next step to be described).

Pressure

Tissue should be cut with a very sharp knife or razor, available in a good quality dissection set (Fig. 3.3). Dull cutting edge of a knife, careless use of forceps and use of scissors cause more crushing of tissues should be avoided.

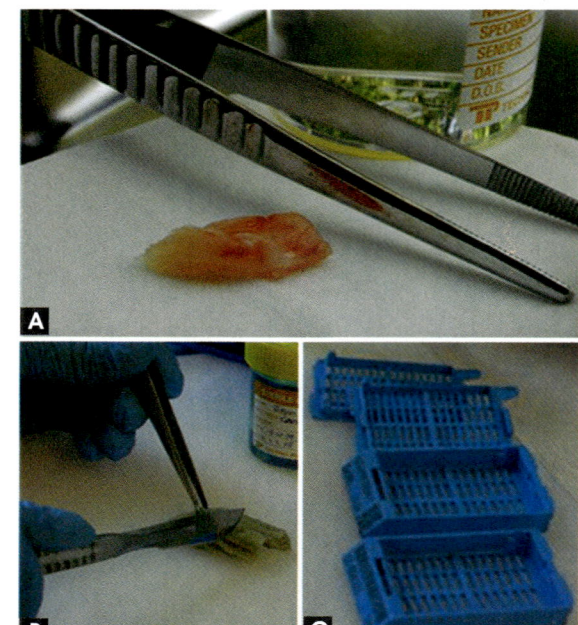

Fig. 3.2: (A) A fresh, unfixed specimen after surgical removal. To prevent degeneration or drying-out the specimen should be fixed as soon as possible. This surgical specimen of stomach has been fixed in formalin, (B) Slices about 4 mm thick are taken from appropriate areas and placed in the (C) labeled cassettes for further processing.

For minimizing the tendency of tissues to stick to the knife, it is advisable to moisten the knife's edge with some physiological saline. Tissue slices can be transferred directly from

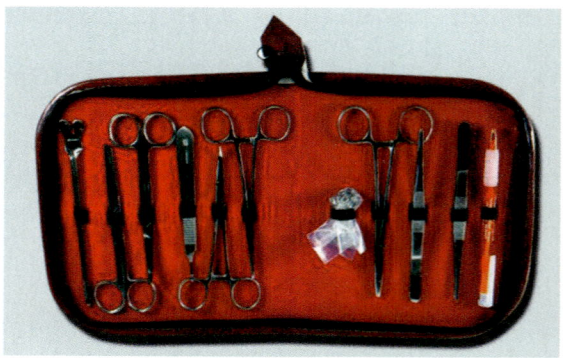

Fig. 3.3: A good quality dissection-instrument set for use of perfusion and dissection of small laboratory animals.

the knife to the fixing fluid by immersion or gently dislodging the slice by shaking. However, care should be taken not to damage the edge of the knife by contact with certain fixatives capable of corroding metal.

Osmotic Injury

The common practice of washing tissues with water (tap or distilled) is, of course, convenient but causes swelling of tissues and cells, as some of the water is imbibed by the salt-rich tissues. Therefore, the tissues excised for fixation should be rinsed, not in water, but in physiological saline (0.85–0.90% NaCl) solution.

Drying

Exposure of the gross specimen or slices or tissue to air for any considerable period causes shrinkage. They should be kept moistened preferably with physiological saline.

Excessive Thickness of Tissue Slice

If tissue slices are too thick, autolysis may become advanced inside the tissue before adequate penetration of the fixative has occurred. The thinner the slice of tissue the better and more rapid is the fixation. For general purposes tissue should not be more than 3 to 4 mm thick. The length and width of the slice make only a little difference in the rate of penetration of the fixative, but slices too large provide inconvenience in handling during later stages of processing. If the specimen is tubular, it is wise to slit it from one side for the easy access to the fixative into the lumen.

Excessive Blood and Mucous in Tissues

Both blood and mucous tend to coagulate rapidly on the surface of the tissues as soon as they come in contact with fixatives or are exposed to air. The tough coating, thus produced, on the surface of tissue slice is then penetrated very slowly by fixative. Hence, it is always advisable to wash off excess blood or mucous with physiological saline before placing slices in the fixative.

FIXATION OF SAMPLES

Fixation is usually the first step towards the preparation of a histological section from a dead biological specimen. The substances used for fixation are called fixatives. Fixation is a selective preservation of cells or tissue components in such a way that the sample can withstand subsequent steps for histological studies. It is considered good if it does not show us pictures that are only 'most life-like', but when it tells us 'most about life'. Primarily fixation is stabilization of proteins.

Although the purpose of fixatives is, as a general rule, to leave the structure of tissue as far as possible unchanged, yet this cannot apply to its chemical composition. Indeed it may be said that whole purpose of fixation is to alter the chemical composition of certain tissue constituents. Fixation is thus a progressive forward-looking process. The fixative brings about the death of a cell in such a way that structure of living cell is preserved with a minimum addition of artefacts; and the chemical composition is kept as intact as possible. In the process of fixation, there is separation of solid phase (*dispersed phase*) of the colloid from liquid phase (*dispersing phase*). Hence, an ideal fixative must precipitate proteins in the finest form and if possible, in ultramicroscopic aggregates.

A Few Notes on Fixation

Fixation is a crucial step in preparing specimens for microscopic examination. Its objective is to prevent decay and preserve cells and tissues in a "life-like" state. It does this by stopping enzyme activity, killing microorganisms and hardening the specimen while maintaining sufficient molecular structure to enable appropriate staining methods to be applied (including those involving antigen–antibody reactions and those depending on preserving DNA and RNA). The sooner fixation is initiated following separation of a specimen from its blood supply, the better the result will be. The most popular fixing agent is formaldehyde, usually in the form of a phosphate-buffered solution (often referred to as "formalin"). Ideally specimens should be fixed by immersion in formalin for six to twelve hours before they are processed.

1. Where the best possible morphology is required, animals should be anesthetized and subjected to cardiac perfusion with saline, followed by 10% formalin flush. If biochemical studies need to be performed on the tissue, a 10% formalin flush should not be used as it may interfere with subsequent analysis.

2. For routine stains where perfusion is not required, tissue is sectioned and drop—fixed in a 10% formalin solution. Fixative volume should be 20 times that of tissue on a weight per volume; use 2 ml of formalin per 100 mg of tissue.

3. Due to the slow rate of diffusion of formalin (0.5 mm hr), tissue should be sectioned into 3 mm slices on cooled brain before transfer into formalin. This will ensure the best possible preservation of tissue and offers rapid uniform penetration and fixation of tissue within 3 hours.

4. Tissue should be fixed for a minimum 48 hours at room temperature.

5. After 48 hours of fixation, move tissue into 70% ethanol for a long-term storage.

6. Keep fixation conditions standard for a particular study in order to minimize variability. (Although set times are best, tissue may be fixed for substantially longer periods without apparent harm.)

The usual fixative for paraffin embedded tissues is neutral buffered formalin (NBF). This is equivalent to 4% paraformaldehyde in a buffered solution plus a preservative (methanol) which prevents the conversion of formaldehyde to formic acid. Because of the preservative, NBF has a shelf life of months, whereas 4% PF must be made fresh. Optimal histology requires adequate fixation, about 48 hr at room temperature for thinly sliced tissues. Inadequately fixed tissues will become dehydrated during tissue processing, resulting in hard and brittle specimens. Alcohol-based fixatives generally do not give good morphology but may be useful in special cases. A particular challenge for the histopathology is immunostaining fixed specimens. In many cases formaldehyde fixation will prevent recognition of epitopes by the primary antibody. Occasionally, "antigen retrieval" procedures will improve results but usually frozen sections are a better bet. An alternative approach, suitable for thin or porous tissues, is to perform immunohistochemistry on fresh tissues and then post-fix and embed the tissues in paraffin.

Why Fix Tissue?

Preserve structure, to make the structural components of the tissue more durable so that the tissue can be manipulated in various ways. Fixed material is dead. Fixative used will depend on type of tissue to be fixed.

Aims of Fixation

1. To prevent autolysis and bacterial attack.

2. To fix the tissues so they will not change their volume and shape during processing.

3. To prepare tissue and leave it in a condition which subsequently allow clear staining of sections.

4. To leave tissue as close as their living state as possible and no small molecules should be lost. Fixation is coming by reaction between the fixative and protein which form a gel, so keeping everything as their *in vivo* relation to each other.

5. The fixation solutions used for microscopy penetrate rapidly to prevent postmortem changes in the cells. These coagulate the cell contents into insoluble substances.

6. Protect tissues against shrinkage and distortion during subsequent processing.

7. Allow cell parts to become selectively and clearly visible when stained.

Storage in 70% Ethanol

After adequate fixation tissues are transferred to 70% ethanol and may be stored at 4°C.

Methods of Fixation

Broadly speaking the methods of fixing a tissue can be divided in two broad categories.

1. *Fixation by immersion* (*in vitro fixation*): This method of fixation is usually followed when small pieces of tissues are to be sectioned for routine histological work. Generally, 5 × 2.5 cm glass specimen tubes are suitable to keep the sample in the fixative used.

2. *Fixation by perfusion* (*in vivo fixation*): In this method the fixation is usually done by intravenous cannulation either directly through heart (intracardiac, transcardiac, or ventricular) or through a blood vessel (vascular) of a living but anaesthetised animal. *In vivo* fixation method is advocated for high quality research work where very good fixation is desirable. The procedure of perfusion-fixation results in excellent fixation of practically all cellular organelles and is the method of choice for electron microscopic studies.

Transcardial Perfusion

The subject animal (in this case, an albino rat) is deeply anesthetized before surgery is performed to expose the heart (Fig. 3.4). A perfusion needle is placed in the left ventricle and a saline solution is then pumped at constant pressure through the circulatory system before exiting through an incision made to the right atrium. This clears the subject of blood. A paraformaldehyde (PFA) solution is then accordingly pumped through the cleared circulatory system, quickly and consistently preserving the tissue in a life-like state. The fixed tissue can then be harvested and stored for later experiment. For experimental needs, whole animal perfusion fixation yields the best possible preservation of the brain for immunohistochemistry.

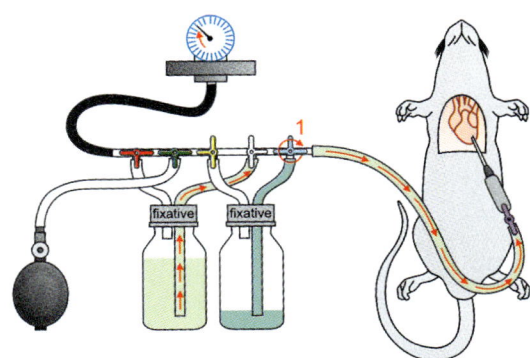

Fig. 3.4: In intracardiac perfusion fixation a needle tip is introduced into the left ventricle for circulation of fixative/buffer through a two-way device under controlled pressure.

For fixation of specimens which contain dense connective tissue in preponderance, a recommended step in fixation is called *hyaluronidase* pre-treatment. In those histology laboratories where sophisticated equipment for perfusion set-up as depicted in Fig. 3.4 is not available, a self-designed apparatus (Fig. 3.5) may be made with a little efforts and at minimal cost.

Chemical Fixation

Chemical fixatives such as formaldehyde or other chemicals are used to preserve tissue from degradation, and to maintain the structure of the cell and of sub-cellular components such as cell organelles (e.g. nucleus, endoplasmic reticulum, mitochondria). In this case biological structures are preserved (both chemically and structurally) in a state as close to that of the living tissue as possible.

The most common fixative for light microscopy is 10% neutral buffered formalin (4% formaldehyde in phosphate buffered saline).

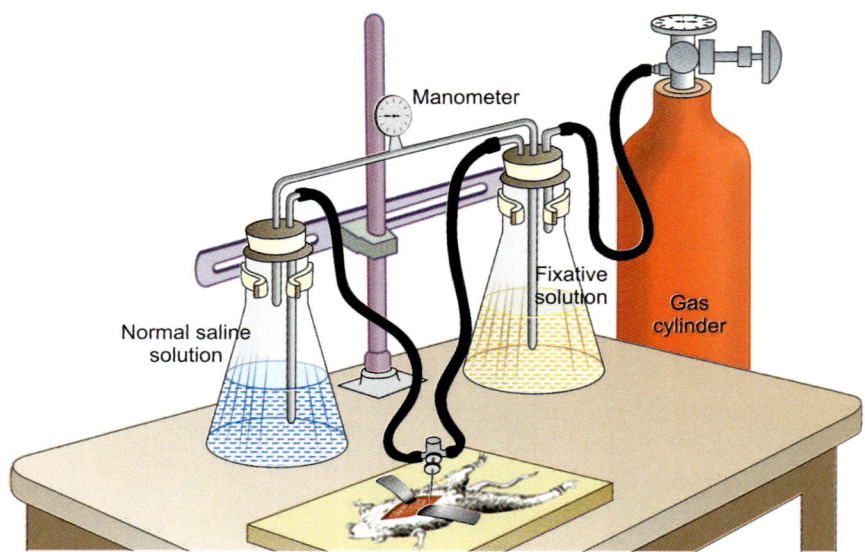

Fig. 3.5: A self-designed perfusion fixation 'set-up' for use on small laboratory mammals.

These fixatives preserve tissues or cells mainly by irreversibly cross-linking proteins. The main action of these aldehyde fixatives is to cross-link amino groups in proteins through the formation of methylene bridges ($-CH_2-$), in the case of formaldehyde. This process preserves the structural integrity of the cells and tissue but damages the biological functionality of proteins, particularly enzymes, and also denatures them to a certain extent. This can be detrimental to certain histological techniques.

Formalin fixation leads to degradation of mRNA (messenger RNA), miRNA (mitochondrial RNA), and DNA in tissues. However, extraction, amplification and analysis of these nucleic acids from formalin-fixed, paraffin-embedded tissues are possible using appropriate protocols.

Due to the toxic characteristics of paraformaldehyde, all work must be done under the fume hood, and gloves and a lab coat are donned at all times. And in any lab that works with both live tissue and fixed tissue, a special set of tools should preferably be put aside for each, to ensure that no cross-contamination occurs.

Frozen Section Fixation

Frozen section is a rapid way to fix and mount histology sections. It is used for surgically excised tumors, and allows rapid determination of margin (that the tumor has been completely removed). It is done using a refrigeration device called a cryostat. The frozen tissue is sliced using a microtome, and the frozen slices are mounted on a glass slide and stained the same way as other methods. Unfixed frozen sections can also be used for studies requiring enzyme localization in tissues and cells. It is necessary to fix tissue for certain stain such as antibody-linked immunofluorescence staining. It can also be used to determine if a tumor is malignant when it is found incidentally during surgery on a patient.

Small pieces of tissue (typically 5 mm × 5 mm × 3 mm) are placed in a cryoprotective embedding medium, then snap frozen in isopentane (an alkane) cooled by liquid nitrogen. Tissue is then sectioned in a freezing microtome or cryostat. Sections are then fixed by immersion in a specific fixative or series of fixatives for carefully controlled period of time.

Fixatives and Fixing Fluids

Ordinarily, the word 'fixative' is used in two different senses. *Firstly*, it refers to a pure chemical substance; *secondly*, it means a dilution of the pure chemical substance or its mixture with some other chemicals that can be used to fix the tissue.

Strictly speaking, the word 'fixative' should be used only in the first sense that is to denote a chemical substance. On the contrary, dilutions of chemical substances either alone or in combination with mixtures of other chemicals should be referred to as 'fixing fluids'. However, the commonly used connotation has been followed in this book.

Selection of an Ideal Fixative (Fixing Fluid)

There is no ideal fixative available which fulfills all of the under-mentioned qualities. However, a good fixative is so designed as to fulfill most of the following criteria. The so-called 'ideal' fixative is the one which:

1. Produces immediate death of cells in a tissue in such a way that they retain closest possible resemblances of their life-like appearances.
2. Prevents *autolysis* (self-digestion) and *putrefaction* of the sample. Whereas autolysis is caused by the autolytic enzymes, such as cathepsins immediately following death, putrefaction is subject to attack by bacteria.
3. Reacts rapidly and completely with the tissue to stabilize it sufficiently for subsequent histological procedures to withstand chemicals during various steps.
4. Fixes all the constituents of the tissue without removing anyone of these from it.
5. Neither shrinks nor swells any tissue components.
6. Makes the specimen hard enough to be handled without over-hardening it.
7. Raises the refractive indices of some of the cell contents for their better visualization.
8. Provides full range of staining methods available after its use, often with great selectivity.
9. Has no rigid upper limit of fixing time beyond which damage to the tissue can be caused.
10. Shows no tendency to deteriorate before or after a tissue is placed in it.
11. Could be easily prepared and should also be cheap, non-toxic, non-inflammable, and non-irritant.

The different types of cytoplasmic organelles and inclusions are acted upon differently by the various fixing fluids. A selection of the most suitable fluid, therefore, is essential in order to fix properly a particular type of structure in a cell. Such a fluid or fixative is then said 'specific' for that particular structure, while for the rest of the component of that cell it acts as a 'general' fixative. The best fixation is obtained if the two fixing fluids are used one after the other. The first fluid is then said to provide *primary fixation* and the second fluid, *secondary fixation*.

Classification of Fixatives

Because the chemical substances, in solid forms, cannot diffuse into tissues and cells, they must necessarily be used as solutions. Such solutions designated as fixing fluids or fixatives may be classified in a number of ways. Some of these are oxidizing substances others reducing agents. Some are acidic and some basic in their chemical behavior. Some are used in solution form singly, others in combination. Generally, when a fixative contains a mixture of two or more chemical substances, the object is that certain ingredients of the fixative shall compensate for 'fixation deficiencies' of the others. Conventionally the fixatives are grouped as *routine* fixatives for general use, and *special* fixatives for a specific use.

On the basis of their effect on protein contents of tissues and cells, the fixatives are classified as *coagulants* (*precipitants*) and non-coagulants (non-precipitants). The coagulation of proteins may occur as a result of either:

i. Formation of an insoluble compound when a fixative (for example, mercuric chloride) combines with proteins, or

ii. When a fixative (for example, alcohol) without combining with proteins results in their denaturation that is a change (usually a decrease) in the solubility of proteins. The non-coagulant fixatives invariably form some additional compounds when they combine with proteins. The two categories of fixatives are listed in Table 3.1.

Whatever fixing fluid or combination of fluids is preferred for the fixation of a given tissue, it is important to ensure that an adequate volume of fixative solution is taken. The volume of a fixative should not be less than ten to twenty times the approximate volume of the piece of tissue. Economy at this stage is one of the four common causes of poor fixation. The other three being:

i. Too short a time for full fixation,

ii. The use of poorly penetrating rapid fixatives, and

iii. Allowing a specimen to settle to the bottom of container so closely that the fixative cannot circulate between the sample and the container.

Removal of Excess Fixative (Washing After Fixation)

The object of washing after a tissue has been fixed by a particular fixative is to remove excess fixative from the tissue before proceeding further. Thorough washing in the appropriate fluid after fixation is essential for good staining. Washing becomes still more important when mercuric chloride and acid fixatives have been used. Even after apparently thorough washing, acid has been known to come out of the tissue during subsequent storage. Therefore, as a precautionary measure rewashing of the stored material is always recommended before subjecting the tissue to the next step in the process of paraffin block-making.

Is Fixation Always Necessary?

As mentioned earlier, fixation is the first step toward the preparation of a histological section. But certain structures, e.g. chitin, cellulose, scleroproteins (nails, etc.) inorganic

Table 3.1	Names and chemical formulae of the coagulant and non-coagulant fixatives			
Coagulant fixatives			**Non-coagulant fixatives**	
Names	**Chemical formulae (percentages used)**	**Names**	**Chemical formulae (percentages used)**	
Alcohol	C_2H_5OH (70–100%)	Formalin	HCHO (10% sol.)	
Acetone	CH_3COCH_3 (100%)	Osmium tetroxide	OsO_4 (1%)	
Acetic acid	CH_3COOH (0.3–5%)	Potassium dichromate	$K_2Cr_2O_7$ (2–3%)	
Chromic acid	H_2CrO_4 (1–3%)			
Mercuric chloride (corrosive sublimate)	$HgCl_2$ (5–7%)			
Picric acid	$C_6H_2(NO_2)_3OH$ (see Bouin's fixative)			
Trichloracetic acid	CCl_3COOH			

Note: The osmium tetroxide is **not an acid**, hence the term osmic acid, though commonly used, is a **misnomer**.

Table 3.2	Fixation time, merits and demerits of important coagulant and non-coagulant fixatives		
Fixatives	Fixation time	Merits	Demerits
		Coagulant fixatives	
Acetic acid CH_3COOH	Up to 1 hour	• Best fixative for nuclei • Counteracts the shrinking effects of other fixatives	• Pronounced swelling effect mainly on white connective tissue • Distorts mitochondria and Golgi apparatus
Acetone CH_3COCH_3	Approximately 1 hour at room temperature 24 hours at 0 to $-4°$ C	• Best fixative for certain enzymes (acid phosphatase and lipase)	• None
Ethyl alcohol C_2H_5OH	15 to 30 minutes	• Best fixative for acid phosphatase and lipase	• Pronounced hardening effect on tissues • Too much shrinkage • Improper fixation of chromatin • Distorts mitochondria and Golgi apparatus
Chromic acid H_2CrO_4	Few hours	• Good fixative for carbohydrates	• Powerful and rapid hardening causes brittleness of tissue • Washing with tap-water for several hours required after dehydration
Picric acid $C_6H_2(NO_2)_3OH$	3 to 4 hours	• Alcoholic solutions are good fixatives for proteins and carbohydrates • Staining becomes better	• Much shrinkage, but less hardening • Lipids decreased and seriously altered • **Can explode when dry**
Mercuric chloride $HgCl_2$		• Good fixative for proteins	• Fairly great shrinkage effect on the cytoplasm • Staining is hindered • Corrosive for metals
		Non-coagulant fixatives	
Formalin HCHO	1 to 2 days	• Good fixative for proteins • Little shrinkage of tissues • Helps staining with H&E	• Pronounced hardening effect on tissues • Although does not harden albumen but hampers its hardening by alcohol during dehydration • Renders staining with acid dyes difficult • Not suitable for fixation of mammalian testis since causes much shrinkage (reason not known) • Pronounced irritation of conjunctiva and mucous membrane

Contd...

Table 3.2	Fixation time, merits and demerits of important coagulant... (Contd...)		
Fixatives	**Fixation time**	**Merits**	**Demerits**
Potassium dichromate $K_2Cr_2O_7$	3 to 7 days	• Excellent fixation of myelin, phospholipids, mitochondria • Improves staining of mito-chondria • Iron-containing pigments are better fixed at higher pH	• Chromatin gets dissolved • Mitochondria fixed well but tend to thicken
Osmium tetroxide OsO_4		• Valuable for fixation of cyto-plasmic organelles	• Tissues become very friable • Nuclei not fixed well • Storage is difficult because it reduces with slightest conta-mination with organic matter • Severe irritant of conjunctiva and respiratory passages • Tissue penetration extremely poor • Cot of fixative is very high • Traces cause over-blackening of tissues

crystals and silica, do not require fixation. Not only that they do not need fixation, they are not acted upon by fixatives. In case frozen sections of any tissue are to be cut, prior fixation is not essential.

Storage of Fixing Fluids

For keeping the fixatives safely in histology laboratory following precautions should be observed:

1. The fixing fluids (fixatives) should be kept in bottles of clear glass with wide mouth and adequate capacity.
2. The bottles should be provided preferably with glass-stoppers.
3. Bottles containing fixatives should be properly labeled to avoid any confusion.
4. Fixatives should be stored at a low temperature, preferably in a refrigerator (4–5° C). After use, they should not be left outside.

Artefacts Observed in Histology Studies

Although fixatives, as a general rule, leave the structure of tissue samples as far as possible unchanged, the process of fixation produces some artefacts. Artefacts are structures or features in tissue that interfere with normal histological examination. These are not always present in normal tissue and can come from outside sources. Artefacts interfere with histology by changing the tissues appearance and hiding structures. These can be divided into *two* categories which are as follows.

These artefacts may be broadly classified as *extrinsic* when there is deposition of some extraneous material, or *intrinsic* when distortion of original components in the sample occurs. The intrinsic artefacts are of two subtypes: (i) *primary*, which are produced by the fixative used itself and hence are visible; and (ii) *secondary*, which result not due to the fixative, but are produced by a subsequent treatment of the tissue.

The features and structures that have been introduced prior to the collection of the tissues are called pre-histology artefacts. A common example of these include: Ink from tattoos and freckles (melanin) in skin samples. On the contrary, post-histology artefacts can result from tissue processing. Processing commonly leads to changes like shrinkage, washing out of particular cellular components, color changes in different tissues types and alterations of the structures in the tissue. Because these are caused in a laboratory, the majority of posthistology artefacts can be avoided or removed after being discovered. A common example is mercury pigment left behind after using Zenker's fixative to fix a section.

Normal patterning of tissues and artefacts resulting from the tissue preparation process ensures that each histological section is unique. Like a piece of biological art these images provide a deep insight into the organization and function of our bodies. Histological patterns that look like everyday objects or features are emerging on social and scientific communities and even in histopathology journal articles. Histology is an area of science where art and science collide. It demonstrates that histology can be appreciated by not only the detail-oriented pathologist but also the art loving layperson and is making histology and pathology more accessible and less daunting as a complex science.

Unless a fixative produces extrinsic artefacts its prolonged use is seldom harmful. In some fixing fluids a tissue may be left undamaged for a period as long as about three years. The variation in time for keeping a tissue in a particular fixative is due to different rate of penetration of the fixative into the tissue, and is generally expressed as:

$$d = k \sqrt{t} \quad \text{or} \quad k = \frac{d}{\sqrt{t}}$$

where, d = distance penetrated (in mm); k = constant; and t = time (in hours) for which tissue was kept in a fixative.

The values of k, based at 5^2 hours for some of the fixatives are 4.65 (hydrochloric acid), 4.3 (nitric acid), 3.6 (formol), 2.75 (acetic acid), 2.2 (mercuric chloride), 1.45 (methanol), 0.85 (osmium tetroxide), and 0.8 (picric acid).

Troubleshooting in Fixation of Specimens

Problem	Prevention
Incomplete fixation	
• The cells characterized by smudge nuclei (indistinct nuclear pattern)	• Prolong fixation time
	• Thin gross section
	• Fresh formalin solution
• Nuclear bubbling	• Cassette should NOT be tightly backed
	• Agitation of cassettes in the fixative

Commonly Used Special Fixatives

Certain fixatives in the form of much used fixing fluids are often called *routine* fixatives. Most of such fixatives may be used for tissues of both animals and plants. Some are, however, considered better for animal tissues; and the noteworthy ones are mentioned below with their formula:

Formalin solution

Formalin (concentrated)	10 ml
Distilled water	90 ml

A concentration of less than 10 per cent is generally not recommended, except when the formalin solution is used in combination with other reagents. Instead of 10 per cent, strengths of 12, 15, 20 or 25 per cent formalin solution have been used.

Since formalin fixation (introduced by Blum, 1893) is the most common method employed for most of the animal tissue samples, some details about formalin are of general interest for a histologist. Formalin (also marketed as formol) is available commercially as an aqueous solution (by weight) of formaldehyde gas. Its designated strengths have different specific gravities and possess varying contents (by weight) of formaldehyde per unit volumes of commercial formalin solutions. These are depicted in Table 3.3.

Generally, a 10 per cent formalin solution is referred to as 4 per cent solution of formaldehyde, but it is erroneous, in the strict terms,

Table 3.3	Percentage strengths of formalin with their specific gravity and formaldehyde contents	
Percentage strengths of formalin (by weight)	Specific gravity	Formaldehyde contents (gm/ 100 ml)
40	1.124	44.96
39	1.120	43.67
38	1.116	42.39
37	1.111	41.12
36	1.107	39.86
35	1.103	38.61

because this strength of formalin does not consist of 4 per cent formaldehyde, but actually 4.496 (approximately 5) per cent formaldehyde concentration.

On keeping formalin solution becomes acidic due to formation of formic acid. This needs neutralization of the fixative, which can be achieved either (i) by adding 40 per cent potassium hydroxide (KOH) solution drop by drop, or (ii) by allowing the solution stand for sometimes over magnesium carbonate ($MgCO_3$) or chips of white marble. Furthermore, the fixative solution may also turn turbid especially if it is left for quite sometimes in a very cold place. This happens due to formation of a white precipitate of paraformaldehyde. In such situation the solution needs filtering.

Buffered neutral formalin (pH 7.0)

Formalin (concentrated)	100.0 ml
Disodium phosphate (anhydrous) Na_2HPO_4	6.5 g
Acid sodium phosphate, monohydrate $NaH_2PO_4.H_2O$	4.0 g
Distilled water	900.0 ml

This fixative is most widely used, because it prevents the formation of *formalin pigment* (an artefact) after formalin fixation. The fixed tissues that are not immediately required for sectioning are often allowed to keep in buffered neutral formalin for several years without much deterioration. The recom-

mended fixation time is overnight at 4°C, but the tissues may be left in the fixative for 72 hours. Marble chips or calcium carbonate (in the form of chalk dust) or magnesium carbonate in excess may be added to the stock solution of the fixative to achieve neutralisation.

Formalin-ammonium bromide (FAB)

Formalin (concentrated)	10 ml
Distilled water	85 ml
Ammonium bromide (NH_4Br)	2 g

This is recommended when sample of tissue from nervous system is to be fixed for staining of neurofibrils and oligodendrocytes. Optimum fixation time is 2–25 days.

Alcoholic formalin

Formalin (concentrated)	10 ml
Absolute alcohol	70 ml
Distilled water	20 ml

The fixative is recommended when glycogen in the animal tissues needs to be demonstrated. Fixation time is usually 8 to 24 hours.

Acetic formalin

Formalin (concentrated)	10 ml
Distilled water	88 ml
Glacial acetic acid	2 ml

This fixative is widely used for fixation of tissue samples from routine surgical biopsies. Fixation time is 18 to 24 hours.

Formalin-picric-acetic (Bouin's fluid)

The *Bouin's fluid* is subject to many modifications in proportions of the ingredients used. The most frequent combination is as follows.

Formalin (concentrated)	25 ml
Glacial acetic acid	5 ml
Saturated aqueous solution of picric acid	75 ml

The fixative is used when nuclear fixation and staining is of greater importance. But the connective tissue tends to swell during fixation. The modified Bouin's fixative makes use of 2 per cent picric acid solution in 70 per cent alcohol (instead of aqueous solution). This facilitates penetration of fatty tissues, specially the brain and spinal cord.

Fixation in the Bouin's fluid is rapid with minimum tissue distortion. The fluid does not produce undue hardness of tissue even after prolonged fixation periods. It improves staining of tissues; and the bright yellow color makes tissues easy to see both during embedding as well as section cutting. However, the fixative is not recommended for fixing kidney tissue samples, and also when demonstration of fat in the frozen sections is desired. This is so because protein-picrate complexes formed during fixation get dissolved with water.

After fixation with Bouin's fluid, tissues need not be washed with water; rather they should be treated directly with 70 per cent alcohol onwards.

Formalin-picric-trichloroacetic

Formalin (concentrated)	15.0 ml
Trichloroacetic acid	0.5 g
Picric acid solution (2% in 50% alcohol)	85.0 ml

The trichloroacetic acid reduces the tendency of connective tissue to swell up.

Zenker's fluid

Potassium dichromate ($K_2Cr_2O_7$)	2.5 g
Mercuric chloride ($HgCl_2$)	5.0 g
Sodium sulfate (optional)	1.0 g
Glacial acetic acid	5.0 ml
Water distilled	100.0 ml

The salts are dissolved with the aid of heat. Tissues are fixed for 12 to 24 hours. After fixation they are washed in running water for 12 to 24 hours and preserved in 80 per cent alcohol.

Helly's fixative or Zenker formalin

Potassium dichromate ($K_2Cr_2O_7$)	2.5 g
Mercuric chloride ($HgCl_2$)	5.0 g
Distilled water	100.0 ml

The ingredients are heated gently to dissolve. Stock solution keeps indefinitely. For use if 100 ml of the stock solution is mixed with 5 ml of formalin, Helly's fluid is made. Adding 10 ml concentrated formalin to the above mixture makes *Maximow's* and *Spuler's fluid*.

The fixation time is usually 24 to 36 hours; but may be extended up to 48 hours (not more). The fixative is an excellent nuclear and cytoplasmic fixative achieving desirable hardening of tissues with minimal distortion. It also facilitates action of decalcifying fluids. $K_2Cr_2O_7$ acts as a *lipid stabilizer*; $HgCl_2$ causes *protein coagulation*; and formalin, as *cytoplasmic fixative*.

Note

With fixation in the Helly's fluid, the processing of tissues is delayed as the tissues after fixation are to be thoroughly washed. If the washing is not proper, a yellowish-brown precipitate forms inside the tissues, the removal of which is difficult but essential before proceeding further. After thorough washing, tissues are transferred to 80 per cent alcohol. *Removal of mercurial precipitate is essential* and can be achieved by soaking tissues in 70 per cent alcohol to which 0.1 to 0.5 per cent iodine has been added.

Chamberlain's fixative

Chromic acid	1.0 g
Glacial acetic acid	2.0 ml
Water distilled	90.0 ml
Add before use:	
Osmic acid (1% aq. solu.)	8 ml

Fixation time: 2–4 hours for samples less than 1 mm size.

Remarks: Good fixative for cytological studies, chiefly for chromosomes, fats, and other lipids.

SUMMARY

Grossing the Specimen

The term *grossing* (referred to as *cut-up*) is often used for *procuring* and *trimming* of the specimens (obtained from *curettings* or core *biopsies*) to the proper size (of 3×5 mm approximately).

Factors Responsible for Damage of Tissue during Procurement

i. Pressure
ii. Osmotic injury
iii. Drying
iv. Excessive thickness of sample
v. Excessive blood and/or mucous in the sample

Fixation of sample—means selective preservation of tissue components in such a way that the sample can withstand subsequent steps of tissue-processing. Fixation is a progressive forward-looking process.

Primary fixation is **stabilization of protein components**:
a. Critical step in almost 'life-like' preparation of specimen for histological procedures.
b. Enables tissue for appropriate staining.
c. Most popular fixing fluid is the phosphate-buffered aqueous solution of formaldehyde often referred as '**formalin**'.
d. Ideal *period for immersion* in 10% formalin (or 4% formaldehyde) for 4 to 6 hrs.
e. *Volume of fixative*—preferably 20 times that of sample to be processed.
f. *Minimum fixation*–48 hrs at room temperature (27 degrees).

For *long-term storage*—keep in 70% ethanol.

Methods of Fixation

i. Fixation by *immersion* (*in vitro*)
ii. Fixation by *perfusion* (*in vivo*)
   • Transcardial perfusion
   • Vascular perfusion

On the *basis of substances* used in perfusion it may be of 2 types:
1. **Chemical fixation** (10% neutral buffered formalin)
2. **Frozen section fixation**—a refrigeration device used to fix the tissue

Classification of Fixatives

1. *Coagulants* (or precipitants): For example, alcohol, acetone, picric acid, mercuric chloride, etc.
2. *Non-coagulants* (or Nonprecipitants): For example, formalin, osmium tetroxide, potassium dichro-mate, etc.

An ideal fixative (*fixing fluid*)—is the one which mainly:
   i. produces *immediate death of cells* (thus retains cells' closest possible life-like resemblance)
   ii. prevents *autolysis* (self-digestion) and *putrefaction* (caused by bacteria, etc) of tissue
   iii. *stabilizes tissue* (by its rapid action) for subsequent procedures to withstand actions of chemicals used
   iv. neither swells nor shrinks any tissue component
   v. may be easily prepared
   vi. is *cheap, non-toxic*, and *non-irritant*, and
   vii. provides *full range of staining methods* (often with great selectivity).

Artefacts observed in a histology preparation

Artefacts are structures that interfere with normal microscopic examination. These are of two main types:
   i. **extrinsic**—deposition of some extraneous material
   ii. **intrinsic**—distortion of original tissue compo-nents
      • *primary*—produced by action of fixative used
      • *secondary*—produced by subsequent treat-ment of tissue (but not caused by fixative)

Commonly used fixatives in histology preparations

1. 10% neutral buffered formalin
2. 4% paraformaldehyde (to be made fresh, use under fume hood)
3. Zinc formalin
4. Bouin's fluid—quick acting fixative but hardens tissues after prolonged use. Specimens transferred to 70% alcohol after 6 hours of fixation

In each case use of 10 times volume of the fixative against the volume of sample is preferable.

4

Decalcification of Bone Samples

INTRODUCTION

Bone decalcification is the removal of calcium ions from the bone through histological process, thereby making the bone flexible and easy for histological processing. This is necessary in order to obtain soft sections of the bone using the microtome. Decalcification is a lengthy procedure, as bone pieces have to be left in the decalcifying agent for several days or even weeks, depending on the size of the bone. Traditional methods of handling hard tissues, i.e. bone and teeth, usually present a problem to histotechnologist. Many of the grossing and cutting-in techniques in current usage for these tissues dictate the use of gross sectioning procedures with a high-speed saw and/or long periods in a decalcifying solution, prior to reducing the specimen to a size that can be easily processed, embedded and sectioned. Frequently the poor quality thin sections obtained when these methods are employed contribute to the already difficult task. It is now possible to routinely and rapidly reduce undecalcified surgical specimens of hard tissue, to a thickness of 2–3 mm, without compromising the integrity of the tissue.

DECALCIFICATION AGENTS

After fixation, the next step for preparation of routine histological sections is dehydration. Certain structures such as bones and teeth, due to their high calcium contents, require an additional step called *decalcification* before they can be subjected for dehydration and are suitable for microtomy subsequently. The calcium contents present in such tissues not only hamper section-cutting but may cause even damage to the edge of the microtome knife. There are two categories of decalcifying agents, namely **acids** and **chelating agents**. The acids are further divided into: *weak acids* (picric, acetic and formic acid) and *strong acids* (nitric and hydrochloric acid). The acids make up a solution of calcium ions, and have some effects on the stainability of the tissue. The chelating agents, on the other hand, take up the calcium ions. Most frequently used chelating agent is ethylenediaminetetra-acetic (EDTA) acid.

Although none of these is ideal for decalcification, the selection of a 'good' quality decalcification agent may be made (Fig. 4.1).

DECALCIFYING AGENTS

Weak Acids

Weak acids such as formic acid are popular and are widely used for decalcification (Table 4.1). Formic acid can be used as a simple 10% aqueous solution or combined with formalin or with a buffer. Although it is slower than the strong acid agents it is much gentler in action and less likely to interfere with

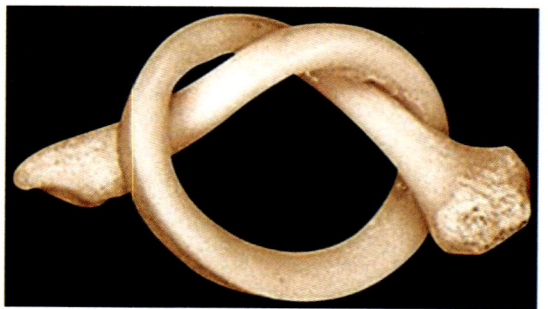

Fig. 4.1: Diagram of a decalcified bone.

nuclear staining. An example of a proprietary decalcifier based on formic acid is Surgipath's Decalcifier. It also contains formalin and is claimed to fix as well as decalcify and be gentle in action. Other acids such as trichloroacetic (TCA) acid have also been used. Picric acid, as a component of some fixatives, has weak decalcifying properties.

Chelating Agents

Chelating agents such as ethylenediamine-tetra-acetic (EDTA) acid, work by capturing the calcium ions from the surface of the apatite crystal, slowly reducing its size (Table 4.2). Because the process is very slow but very gentle (weeks may be required depending on the size of the specimen), this reagent is not suitable for urgent specimens but more appropriate for research applications (for example, decalcification of cartilages (Fig. 4.2) where very high quality morphology is required or particular molecular elements must be preserved for techniques such as IHC, FISH or PCR). It is used at a concentration of approximately 14% as a neutralized solution. The rate at which EDTA will decalcify is pH dependent. It is generally used at pH 7.0. It works more rapidly at pH 10 but some tissue elements can be damaged at alkaline pH.

Table 4.1	Weak acid decalcifiers: Their formulae with comments	
Decalcifier	**Formula**	**Comment**
Formic acid	10% in distilled water	A simple effective decalcifier.
Evans and Krajian	Formic acid 25 ml Sodium citrate 10 g Distilled water 75 ml	An effective formic acid decalcifier buffered with citrate.
Kristensen	Formic acid 18 ml Sodium formate 3.5 g Distilled water 82 ml	An effective formic acid decalcifier buffered with formate
Gooding and Stewart	Formic acid 5–25 ml 40% formaldehyde 5 ml Distilled water 75 ml	A formic acid decalcifier with added formalin, claimed to fix and decalcify.

Table 4.2	Chelating agents used as decalcifiers with formula and comments	
Decalcifier	**Formula**	**Comment**
Neutral EDTA	EDTA disodium salt 250 g Distilled water 1750 ml Bring to pH 7.0 by adding sodium hydroxide (about 25 g will be needed)	Acts slowly but causes a little tissue damage Conventional stains are largely unaffected.

Criteria of a Suitable Decalcification Agent

The following criteria should be fulfilled for selecting a 'good' decalcifying agent:

1. It should be capable of complete removal of calcium from a tissue.

2. It should not cause any damage to tissue cells and/or fibres.

3. It should not perform any kind of impairment to subsequent staining procedures.

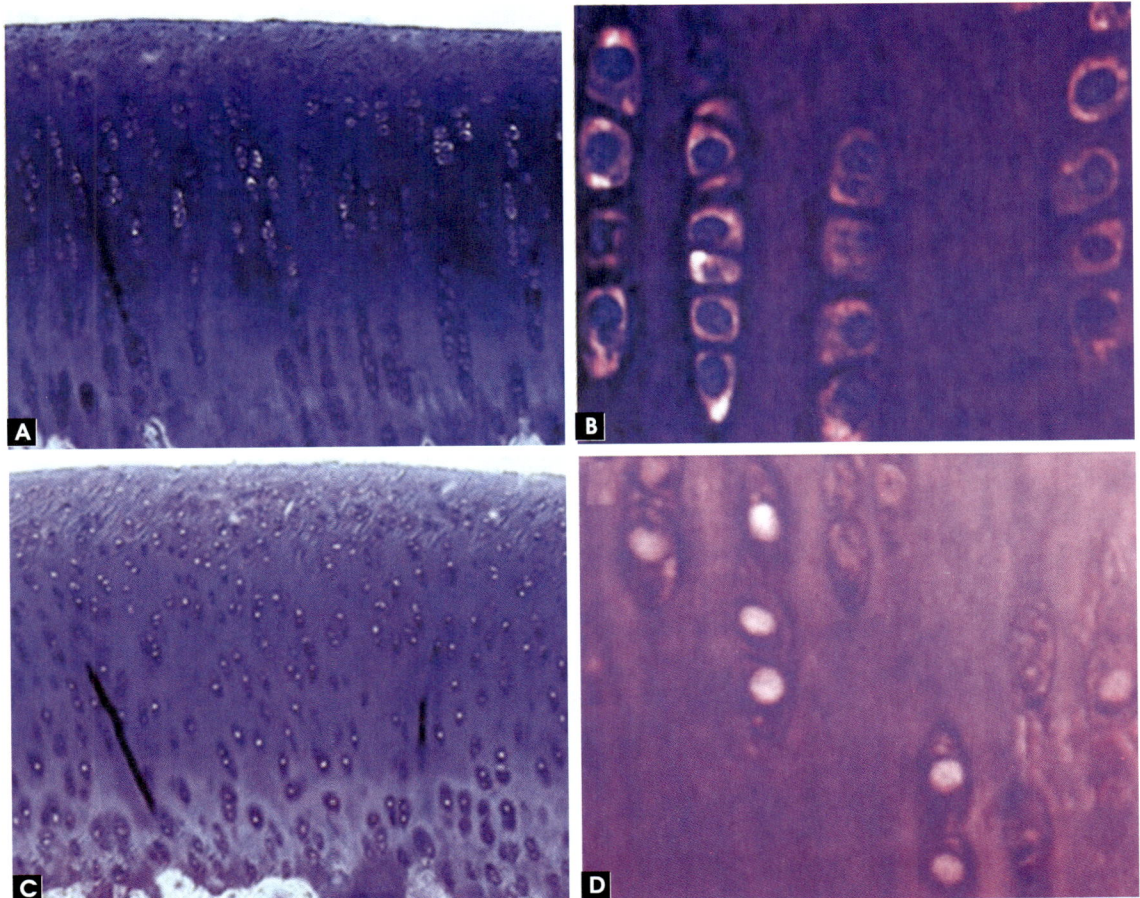

Fig. 4.2: Light photomicrographs of cartilage: (A) Decalcified with 10% EDTA and stained with thionin, (B) observe intense matrix staining and the presence of nuclei. The lower micrographs, (C and D) are of articular cartilage decalcified with the diluted TBD-1 solution and thionin staining. Note the weaker matrix staining and the absence of nuclei. Low-power and high-power micrographs in both staining are at (×50) and at (×400) magnifications.

4. It should have a reasonable speed for removal of calcium salts from within the tissue samples.

METHODS OF DECALCIFICATION

Decalcification methods include: (i) Acids, (ii) ion exchange resins, (iii) electrical ionization, (iv) histochemical methods using either buffer solutions or chelating agents, and (v) surface decalcification. Routinely, the process of decalcification may be done by using any of the following three methods. These depend upon the calcium contents of the given tissue and the facilities available in a laboratory.

1. Chemical Method

Simple aqueous solutions of decalcifying agents at least 30–40 times the volume of the tissue sample (to be decalcified) taken in a glass container are kept in a water-bath at room temperature (preferably at 25° C). Intermittent shaking of the tissue in decalcifying solution is recommended during the process for the specified time for that particular fluid. The sample is tested for the *end-point* of decalcification process, and is then transferred to be processed further for making the tissue-blocks.

The most commonly used decalcifying agents are shown in Table 4.3.

2. Ion-exchange Resin Method

An ion-exchange resin (ammonium form of a sulphonated polysterene resin) is layered on the bottom of a glass container to a depth of about half an inch (Fig. 4.3), and the specimen to be decalcified is allowed to gently rest on it. Thereafter, a decalcifying fluid, about ten times the volume of the specimen, is poured over the sample. The resin in the container helps to remove calcium ions from the fluids, thus ensuring a more rapid rate of solubility of the calcium from the tissues. After use, the resin is regenerated by washing twice with dilute (N/10) hydrochloric acid followed by the washes in distilled water. This allows the resin to be used over a very long period without renewal.

3. Electrophoretic Method

This method, first described in 1947, although still used in some laboratories, has been actually abandoned. However, it must be remembered that decalcification by electrophoresis is no doubt a slower procedure; it improves the staining quality of the tissue. By this method there is attraction of the calcium ions to a cathode (negative electrode) in addition to the solution of the calcium in the electrolyte. The electrolyte used when an electric current (6 volts) is passed generally is

Table 4.3	Some decalcifying agents' treatment time and strength used	
Decalcifying agents	**Strength used**	**Time required**
A. Acid reagents		
• Nitric acid (HNO_3)	5–10% aqueous solution	4 hours
• Formic acid (HCOOH)	Mixture containing	48 hours for small tissues
	Formic acid 8% 35 cc	15–20 days for larger tissues
	Sodium citrate 20% 65 cc	
• Trichloroacetic acid (CCl_3COOH)	5% aqueous solution (freshly prepared)	12–24 hours for small samples
B. Chelating agents		
• Ethylenediaminetetra-acetic (EDTA) acid	Solution (pH 7.0) of	4–40 days
	EDTA 10 g	Solution to be renewed every week
	Distilled water 70 cc	

Note: Decalcification carried out between 55° and 60°C speeds up the process, but should not be encouraged since it causes swelling of tissue components and impairment of subsequent staining.

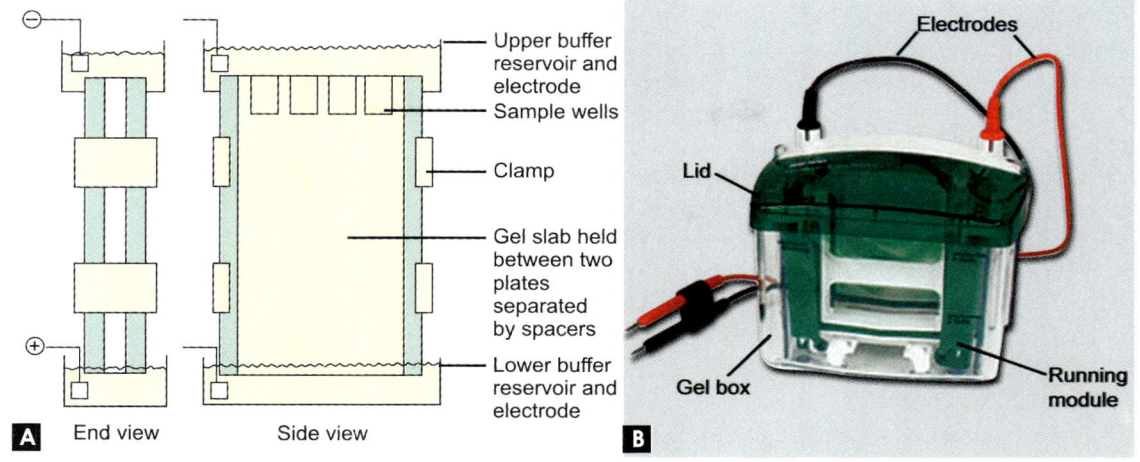

Upper buffer reservoir and electrode
Sample wells
Clamp
Gel slab held between two plates separated by spacers
Lower buffer reservoir and electrode
End view   Side view

Electrodes
Lid
Gel box
Running module
A
B

Fig. 4.3: (A) Schematic representation of end and side views of an electrophoresis apparatus; (B) Components of a vertical electrophoresis cell of the apparatus.

a mixture of equal parts of 8 per cent hydrochloric acid and 10 per cent formic acid. It is taken in a glass container and a perforated perplex sheet (screen) is placed between the anode (positive electrode) and cathode (Fig. 4.3), made up of coiled platinum wire (or stainless steel rod) and brass plate (or carbon rod) respectively. The sample of tissue to be decalcified is placed in the electrolyte near the anode. An electric current of 6–10 or even 12 volts (obtained generally from a car battery) is then passed through the electrolyte. This generates some amount of heat in the electrolyte solution and the acid ions, present in high concentration at anode, around the sample are produced. There is attraction of the calcium ions to a negative electrode in addition to the solution of the calcium in the electrolyte. The hydrochloric acid ionises as H^+ and Cl^-; and the formic acid as H^+ and $COOH^-$. The charge at the cathode is negative and this causes repulsion of the negative ions (Cl^- and $COOH^-$) to the anode (therefore the specimen) where they decalcify.

$$Ca(COOH) = Ca^{++} + 2COOH^-$$

$$CaCl_2 = Ca^{++} + 2Cl^-$$

The calcium ions being positive are repelled by the anode and are hence attracted by the cathode. The COOH$^-$ and Cl$^-$ ions being negative are attracted to the anode and repelled by the cathode.

The following are advantages of electrophoretic method:
• Shortened time for complete decalcification (Fig. 4.4)
• Better preservation of soft tissue details.

However, the method has a disadvantage that only a limited number of specimens can be processed at a time.

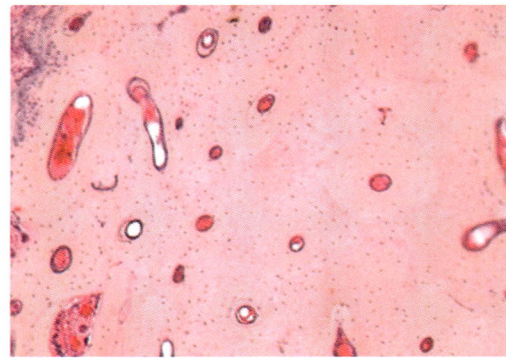

Fig. 4.4: Photomicrographs of hematoxylin and eosin (H&E) stained from a long bone optimally decalcified using formic acid. Notice several osteons with peripheral cement lines. Well-stained nuclei of osteocyte cells indicates that the decalcification endpoint was not exceeded.

KNOWING DECALCIFICATION END-POINT

If high quality results are to be obtained it is important to determine the point at which all the calcium has been removed, because from this point on, tissue damage seems to occur at an increasing rate. Over-decalcification, particularly with the strong acid decalcifiers, spoils the staining of basophilic elements such as cell nuclei and in some circumstances can cause maceration of the softer tissue elements. On the other hand, specimens that are incompletely decalcified may be difficult or impossible to section.

There are numerous tests which can be done to ensure that decalcification is complete. These tests are physical which involves simply bending the specimen, X-ray examination and chemical means. It is least important which decalcifying fluid has been selected and which method of decalcification has been utilized. The most important point to be considered (generally receiving least attention) is the removal of calcified specimens from the decalcifying fluid immediately after complete decalcification has occurred. If this is not done, the chances of subsequent good staining are reduced 10 per cent for every 2 hours the tissue sample remains in the decalcifying fluid.

For avoiding poor staining, one of the following procedures should be employed to determine the decalcification *end-point*.

1. Chemical Test

After the tissue has been in the decalcifying fluid for about 6 to 12 hours, take out approximately 5 ml of the fluid from the bottom of the container. Add 5 ml each of ammonium hydroxide and 5 per cent ammonium oxalate. Mix, and let the solution stand for 15 to 20 minutes. Note for any turbidity to occur. If turbidity appears, it indicates that the specimen is not thoroughly decalcified. Change the decalcifying fluid and repeat the test later. **Absence of turbidity is an indication for complete decalcification of the specimen.** In that case the specimen should be quickly removed from the decalcifying fluid without wasting any time (Figs 4.5A to C and 4.6).

This is the better method for determining complete decalcification. However, the method is not applicable to those tissues which have been fixed with fixatives containing mercuric chloride, because mercuric chloride is opaque to X-rays.

2. X-ray (Radiographic) Examination

Decalcification and point determination by X-ray is best method but very costly. Radiography is the most sensitive test for detecting calcium in bone or tissue calcification. The method is same as specimen radiography using Faxitrom with a manual exposure setting of approximately 1 minute, 30 kV, and Kodal X-OMAT.

Radiography and flame photometry have been compared as means of determining the end point of decalcification in relation to minimizing pulp-dentin separation in histological sections of teeth (Figs 4.7 and 4.8).

Radiography only indicates the presence of deeper foreign objects and care must be taken during microtomy not to damage the knife. This procedure is ideal for progress and end-point testing of decalcification, reveals the presence of some metal/amalgam fillings but not some implanted resin material.

The best method, particularly with large specimens such as femoral heads, is to X-ray the specimen. A good-quality X-ray will clearly reveal tiny residual calcium deposits and allow further treatment if required. It is an excellent method for following the process of decalcification of large specimens such as femoral heads (Fig. 4.9).

A simple chemical test can be applied when some acid decalcifiers are used (particularly formic acid). Ammonium oxalate solution is added to a sample of the final change of decalcified that has been neutralized with ammonium hydroxide. If calcium is present a precipitate of calcium oxalate will form indicating that decalcification is probably

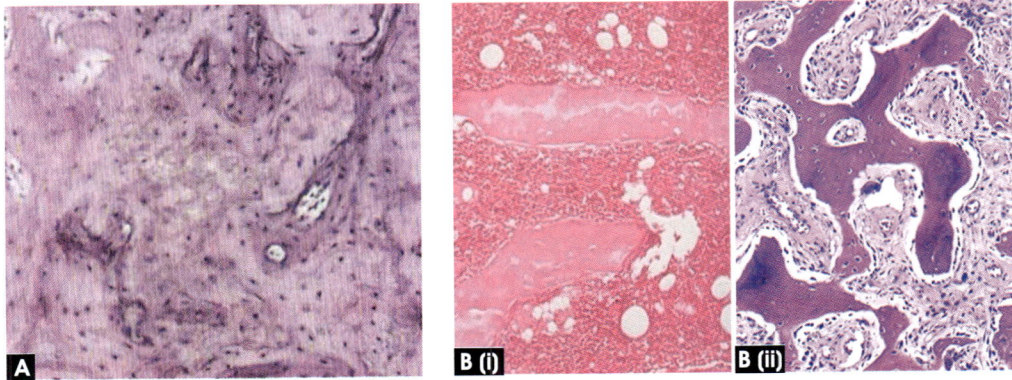

Figs 4.5A and B: Photomicrographs of hematoxylin and eosin (H&E) stained sections from: (A) Decalcified immature bone depicting relationship of cells to the extracellular matrix. Note that the bone has more cells, and the matrix is not layered in osteonal arrangement. (B) Light photomicrographs of the histological sections of bone showing: (i) Poor decalcification and (ii) proper decalcification.

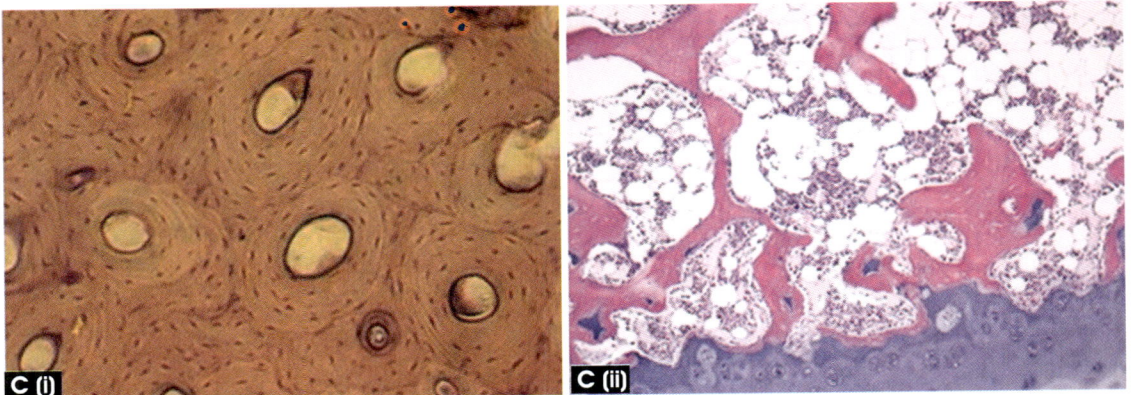

Fig. 4.5C: Photomicrographs of decalcified (i) compact bone and (ii) cancellous or spongy bone (pink) and hyaline cartilage (blue) from the epiphysis of a long bone (H&E stain). Notice that the delicate bone trabeculae are well preserved as in the fine structure of the bone marrow and associated fat cells (adipocytes). Remarkable feature is a well-maintained basophilic/acidophilic balance indicating that the formic acid decalcification was optimal.

incomplete and a longer time in decalcified is required. Of course this test is best done on a relatively recent change of decalcifier (exposed to the tissue for say, one hour only). Physical tests require manipulation, bending probing or trimming of the specimen to "feel" for remaining calcified areas. Whilst this method may be successful in experienced hands, it is generally considered to be unreliable. Mechanical damage can occur during bending or probing and small deposits of calcium can

easily be missed. A method of determining the end-point by carefully weighing the specimen after rinsing and blotting has also been described. This may be an effective method for large specimens.

FURTHER PROCESSING OF TISSUES AFTER DECALCIFICATION

After decalcification, it is not really necessary to neutralize the acid of the decalcifying fluids. The specimen, after 'end-point' determination,

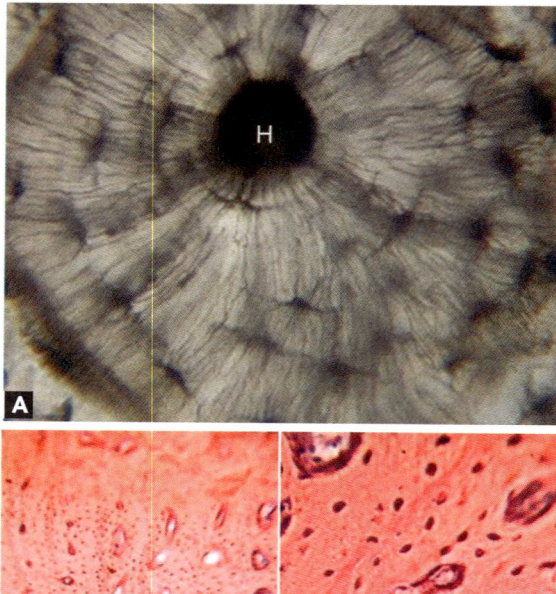

Fig. 4.6: Photomicrographs of compact bone: (A) Undecalcified showing osteocytes in lacunae interconnected through caliculi. Also, a prominent Haversian canal (H) is seen; (B) and (C) decalcified H&E cross sections viewed at 100 × and 400 × respectively.

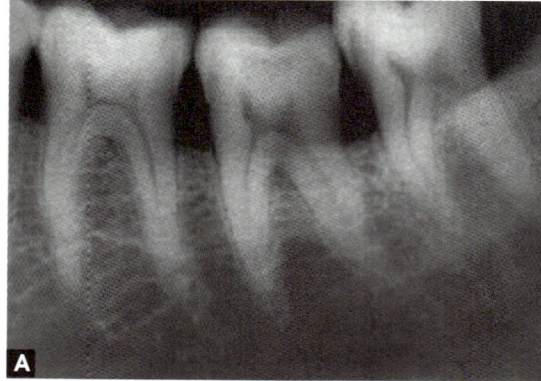

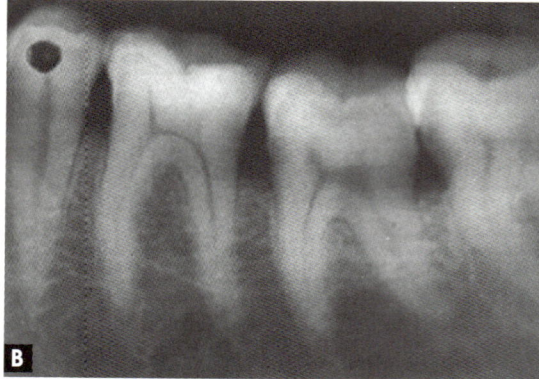

Fig. 4.7: Pre and postoperative digitized periapical radiographic images representing Gp I (second molar) and Gp II (first molar). (A) Preoperative view demonstrating the surface area of bone loss at the the furcation of the first and second mandibular molars, (B) postoperative view demonstrating marked surface area reduction and gray level gain.

is placed directly into 70 per cent alcohol with two changes of 12 to 18 hours each. Thereafter, the dehydration is completed in the routine manner and the tissue processed. Staining with Schmorl's picro-thionin method or Schmorl's thionin-phosphotungstic acid method is done.

SOFTENING OF NON-OSSEOUS HARD STRUCTURES

Although there are no highly satisfactory methods for keratin and/or chitin softening available, the following plans for these structures may be taken.

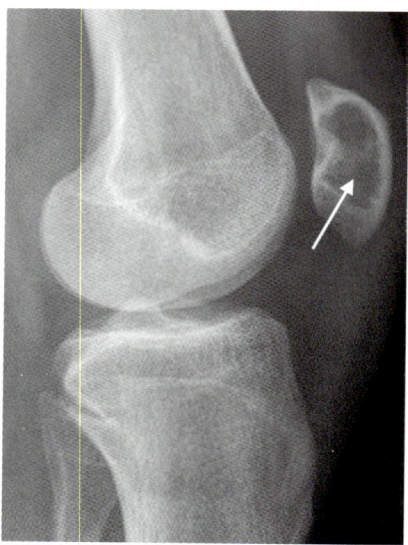

Fig. 4.8: A radiograph of the left knee joint showing a lytic lesion of the patella (knee-cap). Shown by an arrow.

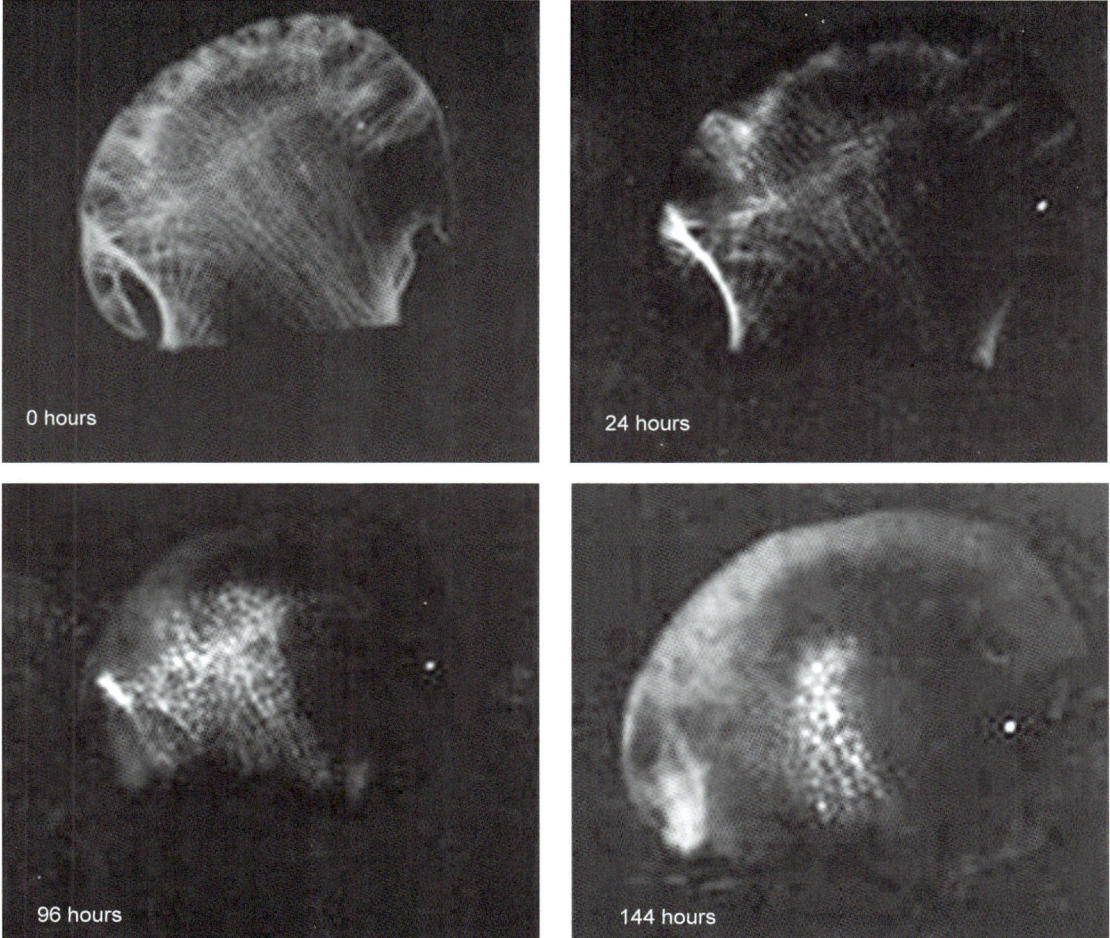

Fig. 4.9: An X-ray series following the process of decalcification of a femoral head with formic acid/citrate decalcifier. The radiographs were produced using the process to be accurately followed and the endpoint to be properly identified.

1. Keratin Softening

The use of concentrated sulfuric acid with the aid of heat is often followed. Although this completely dissolves the keratin from a tissue section, there is much destruction also.

2. Chitin Softening

After fixation the specimen is placed in the following solution

Mercuric chloride	4.0 g
Chromic acid	0.5 g
Nitric acid (conc.)	10.0 ml
Ethyl alcohol (95%)	50.0 ml
Distilled water	200.0 ml

Thereafter, the tissue is washed thoroughly in running tap-water for about three hours; dehydrated, cleared, and processed for impregnation with paraffin.

Factors Influencing the Rate of Decalcification

Concentration

The concentration of active agent will affect the rate at which calcium is removed.

Published formulations for decalcifying solutions strike a balance between speed and degree of tissue damage. It must be remembered that the concentration of active agent will be depleted as it combines with calcium and so it is wise to use a large volume of decalcifier and renew it several times during the decalcification process.

Temperature

Increased temperature will speed up the decalcification rate but will also increase the rate of tissue damage so must be employed with a great care.

Agitation

Gentle agitation may increase the rate slightly.

Fluid Access

As with fixation fresh decalcifier should have ready access to all surfaces of the specimen. This will enhance diffusion and penetration into the specimen and facilitate solution, ionization and removal of calcium.

Other Techniques for Increasing the Efficiency of Decalcification

Sonication used with EDTA has been successfully used to accelerate decalcification of trephine specimens for subsequent molecular analysis. During the process the temperature must be carefully controlled. Microwave treatment has been used with hydrochloric acid decalcifiers but the raised temperature may damage morphology and cause staining artefacts. Ion-exchange resins have been incorporated into some decalcification protocols. They are added to the container holding the decalcifier and take up the ionized calcium maintaining the effectiveness of the acid. If acid decalcifiers are used in adequate volumes and replaced regularly, the use of such resins is probably unnecessary. Electrolytic decalcification in which the bone is placed in acid decalcifier and attached to an electrode through which current is applied is a technique that has not found wide acceptance because of the potential to cause heat damage to the specimen.

If you believe the decalcification end-point is close and you wish to slow the process down so as to avoid over-decalcification and consequent tissue damage, as might be the case when your laboratory is unoccupied during a weekend, specimens can be removed from decalcifier, rinsed and placed back into formalin (important if hydrochloric acid is being used). Decalcification can then be resumed when convenient. An alternative is to refrigerate the specimen at 4°C in its decalcifier to slowdown the process.

SUMMARY

Decalcification

The term decalcification describes the technique for removing mineral from bone or other calcified tissue so that good-quality paraffin sections can be prepared that will preserve all the essential microscopic elements. Decalcification is carried out after the specimen has been thoroughly fixed and prior to routine processing to paraffin.

Decalcification is a straight forward process but to be successful requires:

- A careful preliminary assessment of the specimen
- Thorough fixation
- Preparation of slices of reasonable thickness for fixation and processing
- The choice of a suitable decalcifier with adequate volume, changed regularly
- A careful determination of the end-point
- Thorough processing using a suitable schedule.

Decalcifying Agents

Commonly bone and other calcified specimens are decalcified (demineralised) following fixation and processed using a standard method to produce paraffin sections. Two categories of decalcifying agents are:

i. **Acids**
   - *Weak acids* (picric, acetic, and formic acids)
   - *Strong acids* (nitric and hydrochloric acids)

   Nitric acid is used as 5% in solution in distilled water.

   It is rapid in action, exceeding end-point will impair staining. The hydrochloric acid (HCl) is used as 5 to 10% strength in distilled water. Formalin should be washed from specimen before placing in HCl to avoid the formation of **bis-chloromethyl ether** (a carcinogen). It is also rapid in action, and, like the nitric acid, exceeding end-point will impair staining.

ii. **Chelating agents:** Chelating agents such as ethylenediaminetetracetic acid (**EDTA**), work by capturing the calcium ions from the surface of the apatite crystal, slowly reducing its size. Decalcification process with use of EDTA is very slow but very gentle (weeks may be required). EDTA is *not suitable for urgent specimens* but more appropriate for research applications where very high quality morphology is required.

Qualities of a 'Good' Decalcifying Agent

i. should be capable of complete removal of calcium from a tissue
ii. does not cause any damage to tissue cells and fibres
iii. should not impair subsequent staining procedures
iv. may not be very slow in action
v. may provide moderate morphological details

Methods of Decalcification

i. Chemical method—simple aqueous solutions of agents (listed in Table 4.3)
ii. Ion-exchange resin method—using ammonium form of sulphonated polysterene resin
iii. Electrophoretic method—slower procedure (using equal parts of 8% HCl and 10% HCOOH; at 6 volts current), but improves staining quality

Determination of Decalcification End-point

It means to ensure optimum decalcification of tissue; and its *end-point* is determined by:

i. *Chemical test*—absence of turbidity of decalcifying solution
ii. Radiographic (X-ray)—flame photometry and radiography compared.

5

Dehydration and Clearing

DEHYDRATION

Processing—water is removed (dehydration) from the tissue and replaced by melted paraffin wax. The wax infiltrates the tissue and provides the necessary support when cutting the tissue into thin slices which will eventually be examined under a microscope.

Removal of water is referred to as "dehydration". Tissue processing is done to remove water from the biological tissues, replacing such water with a medium that solidifies, setting very hard and so allowing extremely thin sections to be sliced. This is important because biological tissue must be supported in an extremely hard solid matrix to enable sufficiently thin sections to be cut. Some typical values are:

5 μm thick for **light microscopy;** 5 μm (i.e. 5 micrometers) = 0.005 mm = 0.000005 meter 80–100 nm thick for **electron microscopy;** 80–100 nm (i.e. 80–100 nanometers) = 0.00008 mm to 0.0001 mm = 0.00000008 to 0.0000001 meter.

After a tissue has been suitably fixed, it has to undergo the next step termed *dehydration*, which literally means 'removal of water'.

However, the tissues generally contain inside them, two forms of water: *Free* and *bound*. The term dehydration, in its strict sense, is used for the removal of free or extractable water only. The process of dehydration is necessary for two reasons. Firstly, because water is not miscible with paraffin wax in which the tissue will be subsequently embedded for obtaining a paraffin 'block'. Secondly, the water content in the tissue will not be suitable, when its sections will be finally mounted over a slide with the commonly used mountant (Canada balsam) for making permanent preparations. Therefore, all the free water from a fixed tissue needs to be necessarily replaced by a dehydrating fluid during the process of dehydration.

Procedure of Dehydration

A piece of tissue which has been properly washed after fixation is generally transferred to a solution of weak strength of the dehydrating agent (commonly used agent is 30% alcohol). It is then placed successively in solutions of alcohol of gradually increasing strengths (50%, 70%, 90%), ending with two changes through absolute alcohol. The procedure replaces all the free water by alcohol.

Dehydration process can be started (after fixation) also directly with 90–95 per cent alcohol; but the diffusion-currents that are set up in the passage from water to alcohol may be minimised if the fixed tissue is treated with successive increasing strengths of alcohol.

In dehydrating extremely delicate objects injury to the tissue can be prevented by following special precautions. Such tissues may be placed with some of its liquid in a tube plugged at one end and closed at the other by a diaphragm of some suitable membrane. The tube is then immersed in a vessel containing the desired strength of alcohol. This is allowed to remain like this until, by diffusion through the diaphragm, the two liquids have become of equal density. However, to state that the last alcohol bath should consist of absolutely pure absolute alcohol is incorrect. A strength of 95 per cent (sometimes 90%) alcohol is sufficient in most of the cases. Usually the small amount of water that remains in the tissue after treatment with graded alcohol strengths is satisfactorily removed in the bath of clearing agent to be used after dehydration.

How to Make Different Strengths of Alcohol

In most of the laboratories only two strengths of alcohol are available ready for use. It therefore, often becomes necessary to know as to how various dilutions of alcohol may be prepared. The following schedule, the so-called *dilution method* is recommended.

By this method a particular percentage of alcohol is made by diluting a higher percentage solution of alcohol with the distilled water. In order to obtain a desired percentage from the percentage of alcohol available, the difference in strengths is the amount of water to be added. For example, if 50 per cent alcohol is to be prepared from 95 per cent alcohol; deduct 50 from 95, thus 45 ml will be the quantity of water to be added to the taken 50 ml of 95 per cent alcohol.

For one's own convenience, a dilution table may be prepared for a ready reference (Table 5.1).

Another method for making dilutions of alcohol is depicted in Table 5.2.

Time Required for Dehydration

This depends upon the thickness of the tissue sample. If the tissue is not more than 5 mm in thickness, dehydration may be satisfactorily achieved within 2 to 3 hours. The changes from one bath of alcohol to that of another should be made at an hourly intervals. It is also wise to keep in mind that too long keeping in higher dilutions of alcohol (above 80%) makes the tissue brittle and difficult to cut subsequently. Also, too long keeping in lower dilutions of alcohol (below 70%) should be avoided because this macerates the tissue. Generally a time of one to one and a half hour is sufficient for keeping a sample in the absolute alcohol. The tissue samples should not be left in absolute alcohol for longer time.

Placing the tissue in too much quantity of dehydrating agent causes considerable wastage. Routinely, the volume of dehydrating fluid should be approximately about 10 times the volume of the given tissue.

Substitute for Alcohol in Dehydration

The alcohol is far less satisfactory as a histological preservative, but it acts fairly well as a dehydrating agent. If tissues are placed in alcohol for a prolonged period (weeks and months), the minute structure of the tissues is considerably altered. Tissues get shrunken and too hard. They turn brittle. Their capacity for adequate staining becomes seriously diminished.

Acetone, as a substitute for alcohol is quite frequently used for dehydration purpose due to its very rapid action and relatively less cost. When haste is required, tissues can be well dehydrated with four changes of acetone. Fresh acetone is essential only for the last change.

Table 5.1 — Volumes of diluent to be added for making desired strengths of alcohol from alcohol of higher strength

	Percentage strength of original liquid																		
	100	96	95	90	85	80	75	70	60	50	40	30	20	15	10	8	5	4	3
95	5	1																	
90	10	6	5																
85	15	11	10	5															
80	20	16	15	10	5														
75	25	21	20	15	10	5													
70	30	26	25	20	15	10	5												
60	40	36	35	30	25	20	15	10											
50	50	46	45	40	35	30	25	20	10										
40	60	56	55	50	45	40	35	30	20	10									
30	70	66	65	60	55	50	45	40	30	20	10								
20	80	76	75	70	65	60	55	50	40	30	20	10							
15	85	81	80	75	70	65	60	55	45	35	25	15	5						
10	90	86	85	80	75	70	65	60	50	40	30	20	10	5					
8	92	88	87	82	77	72	67	62	52	42	32	22	12	7	2				
5	95	91	90	85	80	75	70	65	55	45	35	25	15	10	5	3			
4	96	92	91	86	81	76	71	66	56	46	36	26	16	11	6	4	1		
3	97	93	92	87	82	77	72	67	57	47	37	27	17	12	7	5	2	1	
1	99	95	94	89	84	79	74	69	59	49	39	29	19	14	9	7	4	3	2

Volumes of diluent fluids to be added

Note: a. **Heavy type** figures indicate strengths of alcohol most commonly required for use.
       b. Underline figures indicate most common strengths of formaldehyde. Formalin is 40% formaldehyde.

Table 5.2 — Dilution table for alcohol

To obtain alcohol at	80%	70%	60%	50%	40%
Add to 100 ml of 95% alcohol the following ml of water	21 ml	39 ml	63 ml	96 ml	144 ml

CLEARING OR DEALCOHOLISATION

After a tissue is dehydrated (usually with alcohol) the next step in the process of paraffin block-making is called as 'clearing'. The object of clearing is to remove alcohol before impregnation of tissue is started. The process of clearing is, therefore, sometimes called *dealcoholisation*.

The clearing agents (dealcoholisation agents or antemedia) are liquids whose function is to make tissue transparent. These agents are capable of raising the refractive index of any specimen soaked in them. This makes the tissue transparent (clear); and the agents are sometimes referred to as *clearer*. Majority of clearing agents are essential oils and themselves have high refractive index. Besides being able to dealcoholise, the clearing agents also facilitate the penetration, into tissues, of paraffin wax in which the tissue is to be embedded; and of Canada balsam in which final mounting is to be done. However, it is important to note that all clearing agents cannot be used as true clearers. For example, although glycerine is a clearing agent, it is neither suitable for making paraffin blocks, nor it can be satisfactorily used if sections are to be finally mounted in balsam. This is so because glycerine is not miscible with either of them. On the other hand, chloroform is an admirable clearing agent (precursor of paraffin and Canada balsam), but can hardly be utilized as a clearer because the desired transparency of tissue is not achieved.

Selection of a Clearing Agent

The selection of a clearing agent depends generally on the type of preparation of a histological section. Isotonic saline is quite useful for temporary preparations of animal tissues. For permanent preparations, however, xylene is the most frequently used clearing agent probably since it serves the purpose both in embedding and mounting. Toluene is relatively less harmful than xylene because it does not cause any hardening of the tissue; but at the same time clears less rapidly than xylene or chloroform. Chloroform also does not harden like xylene or toluene, but it has a very poor penetration in the tissue, hence not preferred as a clearing agent. For the best work on animal tissues the clearing agent of choice is cedar-wood oil. Table 5.3 depicts merits and demerits of some commonly used clearing agents.

Replacement of Water by Clearing Agents

Good quality clearing agents, specially oils, are capable of removing small quantities of free water that is left in the tissue even after the end of dehydration process. The cedar-wood oil will remove water from tissues saturated

Table 5.3	Merits and demerits of some commonly used clearing agents	
Name of the clearing agent	Merits	Demerits
Xylene	• Can be used as clearing agent both in embedding and mounting • Rapid in action • Does not dissolve celloidin • Staining with aniline is not affected	• Prolonged use over hardens and shrinks the tissue • Inadequate dehydration causes the formation of a whitish emulsion
Cedar-wood oil	• Can be used to clear both paraffin and celloidin sections • Penetration is well • No shrinkage caused • Immersion in cedar-wood oil improves section cutting • Tissues may be left in it indefinitely	• Clearing is slower than xylene • Sections tend to retain traces of oil till mounting, hence a quick dip in xylene becomes essential • Treatment with xylene necessary prior to paraffin impregnation • Quite expensive
Benzene and toluene	• Less hardening and shrinkage than xylene	• Emulsification occurs even with slightest traces of water
Chloroform	• Hardening almost nil as compared with xylene and toluene	• Penetration rate is very slow • Adequate transparency is not achievable
Dioxan	• Clearing possible even if tissues are not put into alcohol solutions (it is also capable of dehydrating tissues) • Prolonged immersion does not damage tissues	• Extremely toxic to nasal mucosa and conjunctiva

with 95% alcohol. Oils of bergamot and aniline oil are capable of such removal from the tissues saturated with 90 and 70% alcohols respectively.

With certain precautions, tissues may be made dehydrated satisfactorily even without the use of alcohol at all. For routine work, the usual commercial aniline will suffice. Even if it has turned brown through oxidation aniline (otherwise colorless) can still be utilized. However, for better clearing, perfectly anhydrous oil is required.

In routine histological procedures the term clearing may be used in two forms.

Clearing in Embedding

In this form of clearing, the tissue is transferred to a clearing agent after being adequately dehydrated. Since the tissue is to be embedded in paraffin wax, the clearing agent used here should be readily miscible with alcohol and also must be a solvent of paraffin. The examples of such clearing agents are: Xylene, dioxan, chloroform, and cedar-wood oil.

Clearing in Mounting

Although this form of clearing would be required after the sections have been stained and again dehydrated, one should be familiar with clearing in mounting. It also means dealcoholisation, not of the tissue but of the section. Since the stained sections are ready to be mounted, the clearing agents, employed for clearing in mounting must not only be capable of dealcoholisation but they must also be solvents of mounting media (usually Canada balsam). The examples of this category of clearing agents are: Xylene, toluene, and benzene.

The common practice of taking out the tissue out of the alcohol and then placing it on the surface of a cleaning agent is faulty. The reason for this is that the alcohol escapes from the surface of the tissue sample into the air quicker than the clearing agent can get into it. This causes tissue shrinkage. To prevent or at least to minimize the shrinkage, it is advisable that a small quantity of clearing agent must be placed under the alcohol containing the

Mixtures for clearing		Time periods up to which tissue is to be kept in the mixtures	
		Bulk tissue	Sections
Mixture—I (25: 75)			
Clearing agent	25 ml	30 minutes	2 minutes
Alcohol (absolute)	75 ml		
Mixture—II (50: 50)			
Clearing agent	50 ml	30 minutes	2 minutes
Alcohol (absolute)	50 ml		
Mixture—III (75: 25)			
Clearing agent	75 ml	30 minutes	2 minutes
Alcohol (absolute)	25 ml		
Pure clearing agent		Till transparent	Till transparent

Note: If clear-wood oil is used as a clearing agent, do not shake the oil with the alcohol. Gently add the alcohol to the oil and drop the tissue in it. As the tissue is penetrated with oil, it will sink. The time for clearing bulk tissues varies with different clearing agents, and also with the size of the specimen being processed. Generally a period of about half an hour is sufficient for clearing. However, 24 hours may be considered as the maximum time required for clearing.

tissue in it. The word 'under', used here, needs to be little clarified. The clearing agent may be put 'under' the alcohol by following one of the two methods.

1. Alcohol is taken in a tube; the tissue to be cleared is put in it; and then a sufficient quantity of clearing agent is introduced, at the bottom of alcohol, with the help of a pipette.

2. First the clearing agent is poured in a tube; then alcohol is gently added on the top of the agent. After this, the specimen to be cleared is carefully transferred into the supernatant alcohol. Gradually the specimen sinks and at the same time the two fluids (alcohol and clearing agent) mix together slowly. After the specimen gets sunk to the bottom of the tube, alcohol is gently drawn off with a pipette.

Technique and Stages of Clearing

When any tissue gets completely cleared, it appears transparent. In the process the tissue is passed through mixtures of alcohol and clearing agent as per the following schedules.

SUMMARY

Dehydration

Definition: Removal of water and fixative from tissue, and replaced them with dehydrating fluid.

- For this purpose alcohols of various strengths are used.
- As alcohols are hydrophilic in nature, it drags water from tissues by diffusion.
- So the increasing concentrations starting from 50%, 70%, 90% and two replacements of absolute alcohols are used.
- For delicate tissues like embryos, animal tissue start concentration with 30% dilution.

Clearing (Dealcoholization)

Clearing is the next step after dehydration. It removes alcohol from the specimen before the impregnation starts. The agents used for clearing are called *antemedia;* and these raise the refractive index of the tissue to make it transparent (clear)—hence the name of the process. Commonly used clearing agent is **xylene**. For merits and demerits of some other commonly used clearing agents refer to Table 5.3.

Usage of the Term Clearing

The term clearing is used in *two* forms:

i. *Clearing in Embedding*—after dehydration, the specimen is transferred to a clearing agent that should be clearly miscible both with alcohol and solvent of paraffin wax.

ii. *Clearing in Mounting*—means dealcoholization of a stained 'section'—not of specimen

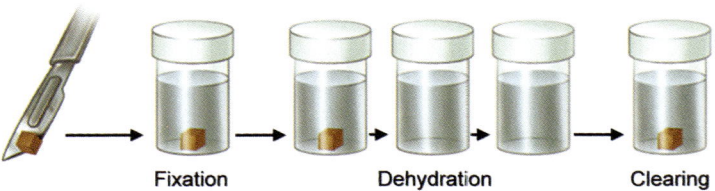

Fixation   Dehydration   Clearing

6

Impregnation (or Infiltration) and Embedding of Biological Samples

IMPREGNATION (OR INFILTRATION)

The early attempts to hold any biological specimen while it was sectioned with a sharp knife produced a variety of encasing methods. Although the methods initially gave external support only it was soon realized that support of specimens internally as well as externally was essential to obtain 'good' tissue sections.

Paraffin Infiltration

In this procedure, tissue is dehydrated through a series of graded ethanol baths to displace the water, and then infiltrated with wax. The infiltrated tissues are then embedded into wax blocks. Once the tissue is embedded, it is stable for many years.

The most commonly used waxes for infiltration are the commercial ***paraffin waxes***. A paraffin wax is usually a mixture of straight chain or n-alkanes with a carbon chain length of between 20 and 40; the wax is a solid at room temperature but melts at temperatures up to about 65°C or 70°C. Paraffin wax can be purchased with melting points at different temperatures, the most common for histological use

being about 56–58°C. At its melting point it tends to be slightly viscous, but this decreases as the temperature is increased. The traditional advice with paraffin wax is to use this about 2°C above its melting point. To decrease viscosity and improve infiltration of the tissue, technologists often increase the temperature to above 60°C or 65°C in practice to decrease viscosity.

In the schedule below, it is presumed that the working day is from 8:00 AM to 5:00 PM. If other than that, appropriate adjustments should be made (Table 6.1).

Trim fixed tissues and keep in neutral buffered formalin (NBF) until ready to proceed. Put tissues in a labeled (usually with pencil, as solvents dissolve the ink) cassette. Once fixed, tissue is processed as follows, using gentle agitation, usually on a tissue processor.

1. 70% ethanol for 1 hour.
2. 95% ethanol (95% ethanol/5% methanol) for 1 hour.
3. First absolute ethanol for 1 hour
4. Second absolute ethanol 1½ hours
5. Third absolute ethanol 1½ hours

Table 6.1	Main precautions in the tissue preparation schedule
Tissue preparation	
Thickness	No more than 3 mm thick.
Area	20 mm × 30 mm.
Fixed tissue	Cut large organs into 3 mm slices and store in neutral buffered formalin for 48 hours. Select tissue from fixed areas, trim to size and refix until the evening. If the trimmed sample is visibly unfixed, refix for a further 24 hours.
Unfixed tissue	Slices of tissue should be *thoroughly* fixed before processing.
Times	All times in processing fluids for this schedule are for tissues 3 mm thick or less. Tissues thicker than that will require longer times.
Clearing agent	Xylene or another clearing agent that will clear tissues in similar times should be used.
Processing time	This schedule takes 12 hours, and processes overnight. On weekends tissues should be left in fixative until Sunday evening with a 48 hours delay.

6. Fourth absolute ethanol 2 hours
7. First clearing agent (xylene or substitute) 1 hour
8. Second clearing agent (xylene or substitute) 1 hour
9. First wax (Paraplast X-tra) at 58°C for 1 hour
10. Second wax (Paraplast X-tra) at 58°C 1 hour.

Due to the viscosity of molten paraffin wax, some form of gentle agitation is highly desirable. If the processor is to be run overnight it should be programmed to hold on the first ethanol bath and not finish until the next morning so the specimens do not sit in hot paraffin longer than the time indicated. If specimens are fresh, they may incubate in formalin in the first stage on the machine. It is important not to keep the tissues in hot paraffin too long or else they become hard and brittle. Processed tissues can be stored in the cassettes at room temperature indefinitely.

EMBEDDING OF SAMPLES

After processing the specimens are placed in an embedding centre where they are removed from their cassettes and placed in wax-filled molds. At this stage specimens are carefully orientated (Fig. 6.1) because this will determine the plane through which the section will be

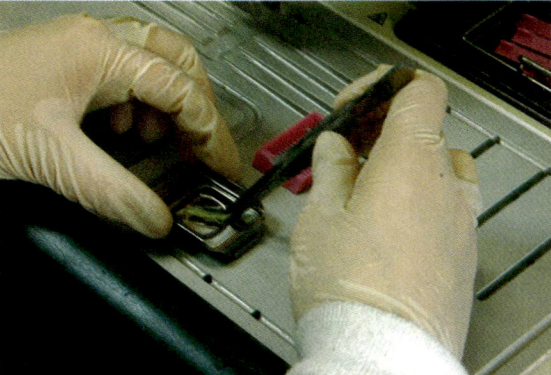

Fig. 6.1: A specimen being placed and oriented in an embedding mold on the hot plate.

cut and ultimately may decide whether an abnormal area will be visible under the microscope. The cassette in which the tissue has been processed carries the specimen identification details and it is now placed on top of the mold and is attached by adding further wax. The specimen "block" is now allowed to solidify on a cold surface and when set the mold is removed. The cassette, now filled with wax and forming part of the block, provides a stable base for clamping in the microtome. The block containing the specimen is now ready for section cutting.

In a general sense embedding of tissues begins with placing of a specimen in such a medium (*infiltration* or *impregnation*) that replaces the clearing agent from inside the specimen and terminates with the making of a *block*.

The process of embedding enables specimens too small and/or too delicate to be surrounded with some suitable material, for example, paraffin, that will support them on all sides with firmness but without producing any injurious effect on the specimen. The embedded tissue may then be sectioned into sufficiently thin slices (called sections) without distortion. Furthermore, embedding also helps to fill out, with the embedding material the natural cavities inside the specimen. In this way, their lining membranes or other structures contained in them may be properly cut *in situ*.

Embedding for Making Paraffin Blocks

Tissues processed into paraffin will have wax in the cassettes; in order to create smooth wax blocks, the wax first needs to be melted away placing the entire cassette in 58°C paraffin bath for 15 minutes. Turn the heat block on to melt the paraffin one hour before adding the tissue cassettes.

1. Open cassette to view tissue sample and choose a mold that best corresponds to the size of the tissue. A margin of at least 2 mm of paraffin surrounding all sides of the tissue gives best cutting support. Discard cassette lid.

2. Put small amount of molten paraffin in mold, dispensing from paraffin reservoir.

3. Using warm forceps, transfer tissue into mold, placing cut side down, as it was placed in the cassette.

4. Transfer mold to cold plate, and gently press tissue flat. Paraffin will solidify in a thin layer which holds the tissue in position.

5. When the tissue is in the desired orientation (Fig. 6.1) add the labeled tissue cassette on top of the mold as a backing. Press firmly.

6. Hot paraffin is added to the mold from the paraffin dispenser. Be sure there is enough paraffin to cover the face of the plastic cassette.

7. If necessary, fill cassette with paraffin while cooling, keeping the mold full until solid.

8. Paraffin should solidify in 30 minutes. When the wax is completely cooled and hardened (30 minutes), the paraffin block can be easily popped out of the mold; the wax blocks should not stick (Fig. 6.2). If the wax cracks or the tissues are not aligned well, simply melt them again and start over. The tissue and paraffin attached to the cassette has formed a block, which is ready for sectioning. Tissue blocks can be stored at room temperature for years.

After tissues have been dehydrated and before they can be "sectioned", i.e. sliced very thinly (see the thicknesses mentioned above) they must be secured in a **very hard solid block** in such a way that the hardened material used to secure all parts of the biological tissues in place is **transparent** to the optical method used for viewing the finished samples.

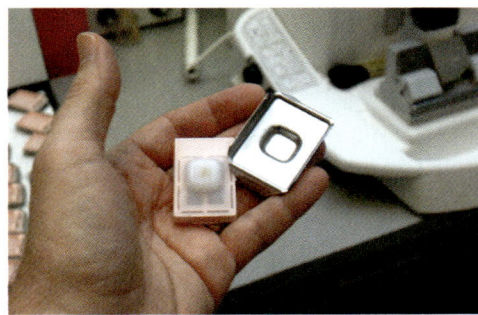

Fig. 6.2: Tissue processing—solidified paraffin block is popped out of the metal mould.

In general, tissue samples are placed in molds together with liquid embedding material which is then hardened. The result of this stage in the preparation of histology slides is hardened blocks containing the original biological samples together with other substances used so far in the preparation process.

Types of Embedding Media

Different types of embedding techniques and materials are used depending on the sample being prepared and the other types of processing involved in preparing that particular sample.

The substances used for infiltration or impregnation may be divided into two main types:

i. *Ribboning media*, e.g. soap, paraffin
ii. *Non-ribboning media*, e.g. sugar and gum solution, gelatin. The sugar and gum solution has a too long penetration time required for infiltration of tissues. The gelatin, due to its lower viscosity, requires less time.

Those materials which are to be sectioned with a microtome, can be embedded with two general categories of embedding media—*water soluble* and *water insoluble*.

Water Soluble Media

Such embedding media; for example, gelatin and gums and polyethylene glycol (carbowax); are required when shrinkage that occurs during dehydration has to be avoided or when lipids are to be stained. These media become fluid on warming and solidify on cooling. Generally, the specimens are cut as frozen-sections without embedding for preservation of fats. One of the shortcomings with water-soluble embedding media is that very thin sections cannot be cut.

Water Insoluble Media

The commonest examples of water insoluble media are paraffin-wax and nitrocellulose. The embedding of properly dehydrated samples in water insoluble media is the conventional method of embedding in most of the histology laboratories. Out of the two masses mentioned above, paraffin is the generally used material. The choice of embedding material depends whether or not the tissue is intended for sectioning with a microtome. If not, there is a need of preparing the whole mount. Such embedding media (e.g. clear plastic) mainly protect the specimen and enhance its transparency. Specimens so embedded are usually dehydrated and passed through a fluid that is miscible with plastic in its liquid form. By this method of embedding, the specimens get permeated and the plastic finally hardens.

Methods of Embedding

The embedding can be achieved in one of the *two* ways described below.

Evaporation Embedding Method

In this method the tissue is saturated with some material which whilst in solution is sufficiently fluid to penetrate into the tissue to be embedded, but at the same time, after the evaporation or removal by other means of its solvent, acquires and imparts the desired firmness for the section-cutting purposes. Celloidin embedding is based on this principle.

Celloidin Embedding

Procedure: The tissue may be embedded in celloidin after it has been subjected to suitable preliminary treatment already described for obtaining paraffin sections. The celloidin or cellulose nitrate is dissolved in equal parts of absolute alcohol and ether. Specimens after being dehydrated, are brought into 50:50 alcohol:ether, and then placed in celloidin solution (thin, medium, and thick); and could be cut from alcoholic solution or could be taken through chloroform into cedar-wood oil.

The following advantages and disadvantages may be noted.

Advantages:
1. Procedures for celloidin method do not require heat.

2. The processing is done at room temperature, hence tissue shrinkage is less.
3. Large specimens, like teeth and whole embryos, can be sectioned to the best advantage because celloidin holds all their parts firmly in place.

Disadvantages:
1. Celloidin embedding takes more time than paraffin embedding.
2. Sectioning requires a heavy microtome with a mounted microtome knife.
3. Sections cut are not as thin as those cut from a paraffin block.

However, it is noteworthy that celloidin embedding introduced by Schiefferdecker (1882), does not provide very thin sections specially from small specimens. It is mainly used to obtain satisfactorily thin sections from the samples of greater size (about an inch in diameter).

Fusion Embedding Method

In this method the tissue to be embedded is saturated by soaking with some material that is liquid while warm, but solidifies as it gets cooled. Paraffin embedding introduced by Klebs (1869) falls into this category of embedding; details are given in Chapter 7.

Plastic Embedding

Plastics are groups of substances, produced artificially and used mainly for varying purposes including molded objects of different kinds. The oldest commercial plastic, *celluloid*, a mixture of camphor and celloidin (nitrocellulose), is nowadays replaced by less inflammable and more durable materials which are classified into two types: *Thermoplastic* and *thermosetting*. Whereas the former types soften when heated and solidify on cooling; the latter ones harden on heating and remain hard at the setting temperature or even higher temperatures.

The thermoplastic plastic is generally used to embed tissues for sectioning with a microtome. Out of several forms of thermoplastics available in the market, the *methacrylates*

appear to be particularly well suited for the cutting of extremely thin (about 650 Å) sections needed for electron microscopy of animal tissues. Methacrylates exist as liquids (monomers) in their non-polymerized condition and solidify by addition of a catalyst. When subjected to heat, causing polymerization, the tissue embedded in plastic sets in a hardened form (block). As with other embedding media, thorough infiltration of methacrylate should occur before hardening starts. In routine practice, for cutting 1 μ or less thick sections of very small pieces (1–2 mm) of tissues, the time required for permeation of plastic is only for a few hours. The detailed technique of plastic embedding should, however, be consulted from any book on electron microscopic methods for biological samples.

Epoxy Embedding

Introduction of epoxy embedding media has greatly reduced artifacts due to shrinkage and also has allowed thinner sectioning than was possible with paraffin. The thinner sections (approximately 1 μ) may be viewed after staining with the light microscope or may be sectioned thinner and examined by electron microscopy.

Double Embedding

Sometimes, artifacts common to paraffin embedding are minimized if the tissues are first infiltrated by celloidin and later by paraffin. Such embedding process termed *double embedding* may be carried out as follows:
1. Dehydrate the fixed tissue sample in the usual manner up to 95 per cent alcohol.
2. Soak the tissue in an alcohol–ether mixture, prepared as follows, between the last step of dehydration and the step of placing them in a *l*ow *v*iscosity *n*itrocellulose (abbreviated as LVN).

LVN	200 g
Absolute alcohol	140 ml
Shake thoroughly, when LVN is dissolved mix	
Anhydrous ethyl alcohol	230 ml

3. Transfer to a 3–5 per cent solution of LVN for 1–2 days (for small pieces) or 3–5 days (for larger pieces).
4. Transfer specimens to chloroform, after blotting off the LVN solution, and let them remain in chloroform overnight.
5. Replace specimen in the second fresh change of chloroform for about 4–6 hours.
6. Remove the specimen from chloroform and blot off.
7. Transfer in molten paraffin in the oven for 3–24 hours.
8. Change in fresh paraffin, and cast blocks in the usual manner.

Vacuum Embedding

Wax embedding for embedding the specimens that contain numerous tiny cavities inside them (e.g. lungs), though not yet in common use, and is called *vacuum embedding*. It denotes the technique of *impregnating* the tissue with paraffin wax under *reduced pressure*. Hence, the term vacuum embedding is doubly wrong in its use:

i. The process is one of impregnation rather than embedding; and the reduction in pressure is nowhere near a total vacuum. Vacuum embedding is based on the principle that volatile clearing agents leave the tissue quicker when the surrounding pressure is low.

ii. Also, any air that might be dissolved in the clearing fluid or wax is withdrawn from the tissue and thus there are no chances of any air bubbles left within the specimen processed.

Paraffin embedding enables cutting very thin sections with fairly small specimens. With large blocks there is a tendency to split the section under the impact of the microtome knife. Hence, it is a general rule to keep in mind that *celloidin embedding should be preferred for larger specimens and paraffin embedding should be opted for small specimens.*

Since paraffin wax is the most commonly used material for making tissue blocks, it would be worthwhile to apprise a student of histology about the details of paraffins in the simplest form followed by procedures of wax embedding.

SUMMARY

Wax Impregnation/Infiltration

- It is the process by which tissues are surrounded by a medium such as *agar, gelatin*, or *wax* which when solidified will provide *sufficient external support* during sectioning.
- Tissue is impregnated with melted wax.
- Xylene + Paraffin   1 hour
- Paraffin wax I   2 hours
- Paraffin wax II   3 hours

Paraffin Infiltration (Embedding)

- Most commonly used waxes for infiltration are the commercial paraffin waxes
- It is solid at room temperature but melts at temperatures up to about 65°C or 70°C.
- Available in melting points at different temperatures.

- Dehydrated material is gradually infiltrated with wax.
- Liquid wax is recommended for the initial infiltration.

Types of Embedding Medium

- Ribboning media:
  e.g. paraffin, soap
- Non-ribboning media:
  e.g. sugar, gum solution, gelatin

Methods of Embedding

- Fusion embedding method:
  e.g. paraffin embedding
- Evaporation embedding method:
  e.g. Celloidin embedding

Paraffin infiltration

7

Paraffin Waxes and Paraffin Block Making

PARAFFIN WAXES

The term paraffin is derived from two Greek words: *parum*, which means little; and *affins*, which means affinity. In chemistry, *paraffin* is used synonymously with *alkane*, indicating hydrocarbons with the general formula C_nH_{2n+2}. The name is derived from Latin *parum* ("barely") + *affinis*, meaning "lacking affinity" or "lacking reactivity", referring to paraffin's unreactive nature.

Paraffin wax, first created in 1830 in Germany, is a white or colorless soft solid derivable from petroleum, coal or oil shale that consists of a mixture of hydrocarbon molecules containing between twenty and forty carbon atoms. It is solid at room temperature and begins to melt above approximately 37°C (99°F), its boiling point is >370°C (698°F).

Common applications for paraffin wax include lubrication, electrical insulation, and candles; dyed paraffin wax can be made into crayons. It is distinct from kerosene, another petroleum product that is sometimes called paraffin.

The paraffin waxes are saturated hydrocarbons (simplest aliphatic organic compounds) which are very stable and different reagents have a little effect on them. According to modern system, the paraffin waxes are called *alkanes*, and have variable numbers of carbon atoms. If they possess one to four carbon atoms, they are colorless and odorless *gases*. When the number of carbon atoms is between five and thirteen, paraffin waxes are *liquids*, for example, liquid paraffin. Paraffin possessing fourteen or more carbon atoms are *solids*.

Paraffin wax, a mixture of many hydrocarbons ranging from $C_{22}H_{46}$ to $C_{28}H_{58}$, is insoluble in water and alcohol; burns with smoky flame. It must be kept in mind that there is no sharp melting point of paraffin waxes. The melting point varies between 35°C and 65°C. It is the commonest crystalline material used in tissue block making as an infiltration medium to obtain routine histological sections of biological specimens. Paraffin wax is strongly hydrophobic and even traces of water in the specimen prevent complete infiltration of the wax resulting in a poor section-cutting which can sometimes become impossible.

The *plastic point* of a given paraffin wax determines its hardness and is usually a few, but variable, number of degrees below the melting point of paraffin. It is the lowest temperature at which permanent deformation may be made without fracture. Consequently, the plastic or melting point of paraffin has to be adapted to the room temperature where sectioning is being carried. A paraffin wax with low plastic point is more transluscent, is more brittle, and compresses more in sectioning.

Paraffin wax is a white or colorless soft solid derivable from petroleum, coal or oil shale, that consists of a mixture of hydrocarbon molecules containing between twenty and forty carbon atoms.

When heated, paraffin wax gradually softens until it becomes liquid. A low-melting point paraffin wax is recommended when thick sections are needed. On the other hand, paraffin wax with higher melting point is preferred for thinner sections.

Other Kinds of Waxes

Besides paraffin wax, there are some other kinds of waxes. Some of the examples are:

a. *Bees wax* or myricil palmitate $C_{15}H_{31}COO$ $C_{30}H_{61}$ is an ester of myricil alcohol ($C_{30}H_{61}$ OH) and palmitic acid ($C_{15}H_{31}$ COOH). The bees wax does not burn with smoky flame.

b. *Candle wax* contains stearic acid mixed with 9 per cent paraffin wax. Stearic acid is responsible for the hardness of the wax and also for the strength, thus reducing the tendency of paraffin wax to soften in very hot atmospheric conditions.

c. *Ester wax* is soluble in alcohol; and harder blocks can be made than those made with paraffin of the same melting point. The use of clearing agents is thus avoided.

d. *Carbowax* solid polyethylene glycol is water-soluble. It is also called HEM (*H*arleco *E*mbedding *M*edium). Generally, carbowax-4000 and carbowax-5000 strengths are avail-

able; the numbers suffixed indicating approximate molecular weights. The higher the number, harder is the carbowax. Because of carbowax's solubility in water, the fixed tissues can be directly transferred in an embedding mixture, containing 9 parts of carbowax-4000 and 1 part of carbowax-5000, melted down in a paraffin oven (Fig. 7.1) at 56–58°C. The tissues are kept in an incubator with three changes of one hour duration each.

However, following advantages and dis-advantages of carbowax must be considered (Table 7.1).

EMBEDDING OF SAMPLES IN PARAFFIN WAX

Technique of Wax Impregnation (Infiltration)

After a given specimen which has been fixed, dehydrated, and cleared, it is ready for infiltration or impregnation with paraffin. For this the tissue is transferred to specimen-tubes filled with paraffin wax to a depth sufficient to cover the specimen. It is then kept in an oven at a regulated temperature not more than 5°C

Fig. 7.1: Paraffin wax

Table 7.1	Advantages and disadvantages of carbowax	
Advantages	**Disadvantages**
Process is rapid because both dehydration and clearing are eliminated	More care is required to obtain good sections
Fat may be demonstrated	Tissues containing large amounts of gross fat are not penetrated
There is less shrinkage and distortion than in methods which require dehydration	Air-conditioned laboratories are required in warm and humid areas, because of hygroscopic nature of carbowax
With proper embedding, sections between 1 and 3 microns can be cut	

above the melting point of the wax used. It is advisable that paraffin wax of melting point 48–53°C (and not higher) should be used for infiltration. If the wax used is of too high a melting point, the specimen is likely to be *cooked* during impregnation, and difficulty may be found in getting a good ribbon of sections. Over-hot paraffin also causes shrinkage of the specimen. Two to three changes of paraffin wax are to be made for full infiltration of specimen with the wax. However, the wax used for the second and third time infiltration should be one which is used from time (for 12 months or more) as continued reheating improves cutting later on. Addition of 0.5 per cent petroleum resin to embedding wax is recommended for fine texture on cooling.

Merits and Demerits of Paraffin Embedding

Although paraffin embedding is most commonly used in the study of animal tissues subjected to subsequent sectioning, some of the advantageous characteristics as well as shortcomings should be recognized. The popularity of embedding in paraffin is based on the following merits.

Merits

1. The actual embedding process is rapid, since the pure paraffin can be melted in an oven, and permeation of specimens completed in a few hours or even less.
2. The subsequent sectioning and staining require minimum amount of time.

3. Ribboning is possible. A ribbon of any desired length can be obtained, because sections will adhere to each other. This facilitates keeping sections in serial order.
4. Very thin sections (even 2–3 μ) can be cut.
5. The paraffin blocks and sections cut may be stored dry for sufficiently long periods.

However, there are several shortcomings of the paraffin embedding which are equally important, and are summarized as under:

Demerits

1. A considerable distortion occurs when sections cut from the paraffin embedded specimen are affixed to the glass slides. The sections take up water and swell while placed over warm water (45°C) at the time of flattening. Thereafter, when the slides are dried, a lot of shrinkage causes *artefacts* to appear. Artefacts are usually at their worst after tissues are fixed in plain formalin. These can be minimized by fixation in Bouin's fixative or in Zenker's solution.
2. Paraffin embedding is less well suited for large specimens (about 2 cm) because of much time required in the paraffin oven for permeation.
3. Sectioning is rendered difficult when a temperature of about 30°C is reached in the laboratory. This may be overcome by mixing harder waxes with paraffin.
4. A substance or structure not naturally present in the matter being observed but

formed by artificial means, as during preparation of a microscope slide.

5. A spurious observation or result arising from preparatory or investigative procedures.

6. Any feature that is not naturally present but is a product of an extrinsic agent, method, or the like: *Statistical artifacts that make the inflation rate seems greater than it is.*

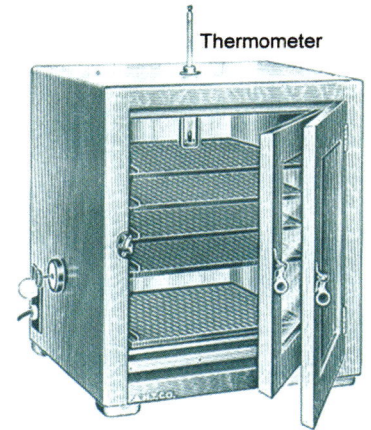

Note

There is often confusion in the meanings of two commonly used words: *artifact* and *artefact*. In realty these are two different words and, hence, a beginner should understand the meaning of both. Whereas, **artifact** is an object produced or shaped by human craft, especially a tool, weapon, or ornament of archaeological or historical interest; the word **artefact** means an artificial product or effect observed in a natural system, especially one introduced by the technology (such as histology and microscopy) used in scientific investigation or by experimental error.

FURTHER STEPS FOR MAKING PARAFFIN BLOCK

The wax-infiltrated specimen is now ready for block making suitable for placing in the microtome. The wax is molten by placing the solid paraffin in a paraffin embedding oven (Fig. 7.2) with adjustable temperature, about 5°C above the melting point of chosen wax.

The block is made by pouring molten paraffin wax into a mould, and at the same time the specimen is so placed and oriented in the mould as to enable it to be sectioned in the desired plane.

Types of Moulds

Three types of moulds are in common use:

1. *Brass Moulds*

These are also called Leuckhart's embedding boxes or 'L'-pieces. They are made of two pieces of brass about 5 mm thick. Each piece is 'L' shaped with long side 40 mm and short side 25 mm long. The mould may be constructed

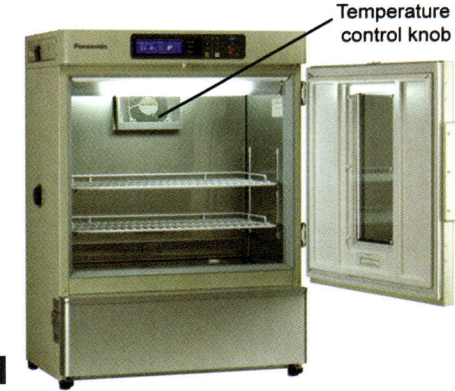

Fig. 7.2: Two common types of paraffin-embedding laboratory incubators with: (A) Two doors and an external thermometer, and (B) an automatic temperature control knob inside, and a single door.

by placing the two pieces together (Fig. 7.3) on a glass-plate smeared with glycerol. The steps for the block-making with 'L'-pieces are given as under:

• The inner surface of 'L'-pieces may be smeared thin with glycerol to prevent paraffin wax sticking to the metal.

• The two 'L'-pieces are placed on a sheet of glass which should also be smeared with glycerol.

• The 'L'-pieces should be cold, not warmed, so that when wax is poured, immediate chilling takes place. This results in a block with smaller crystal size and greater homogeneity.

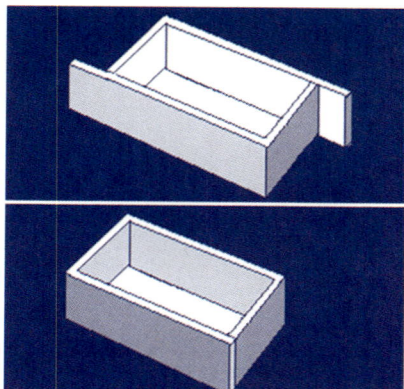

Fig. 7.3: Two L-shaped brass pieces for making a mould for paraffin block.

- The 'L'-pieces are kept to make a block appropriate to the size of the specimen.
- Pour molten wax into them so that 'L'-pieces are about 3/4 full.
- Wait until a thin skin of wax begins to form on the inside of the metal 'L'-pieces. This is called *congealing* of wax.
- Take the infiltrated specimen from its bath of liquid wax and place it in the block by gently pressing it against the skin of wax. The specimen will stay exactly where placed.
- Carefully place the block (on the sheet of glass) in a shallow dish with iced-water (water should reach just to the top of 'L'-pieces and should not cover the block).
- Place the block in a refrigerator.
- Allow the block to set and to harden before attempting to remove it from 'L'-pieces. Ordinarily the time varies from twenty minutes to two or three hours.
- Free the block by dropping briskly the 'L'-pieces on the bench.

Note

i. The infiltrated specimen should be placed eccentrically, because the central portion of the block solidifies the last, during the setting process.
ii. Block should not be cooled suddenly, as this causes split in the outside skin of the wax, due to production of partial vacuum inside.

2. *Paper-boat*

These are also called *paper-trays*. The paper-boat blocks have the advantage over the 'L' piece method that the chances of reduced pressure regions inside block are diminished because as the block cools, the paper-tray (Fig. 7.4) bends inwards (unlike 'L'-pieces).

Making Paper-boats

The paper-boats are made usually of 1" × 1.5" dimension. For this a thin cardboard or some stiff paper of 2" × 3.5" size is taken as shown in Fig. 7.3; and the following steps may be carefully followed for making a paper-boat.

- Fold the paper at pp' and qq'; and then at rr' and ss'.
- Then fold PP' by applying Pr against Pp and pinch out PP'.
- Repeat the steps for QQ', RR', and SS'.

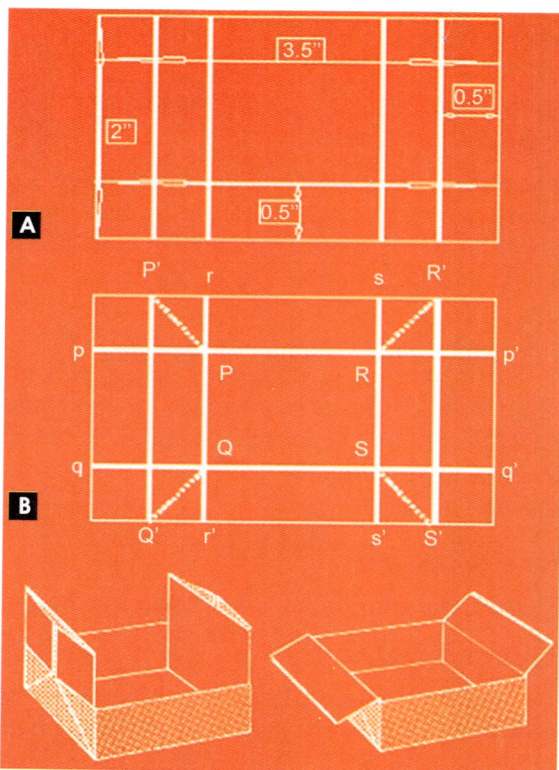

Fig. 7.4: Plan for making a paper mould for making paraffin block (*see text for steps*)

- It is important to remember that while folding the paper letters should always face inwards.
- Turn the 'dog-ears' (formed by foldings of PP', QQ', and SS') outwards until they are horizontal at the level of the low sides.

It is also possible to make paper-boats by twisting a strip of stiff paper around a soft wooden cork (Fig. 7.5). A pin piercing the paper is inserted in the cork. Such paper-boats are sometimes called *paper thimbles*.

Although some workers prefer to use paper-boat blocks, 'L'-piece method of block making is more widely used as it is easier and quicker to draw 'L'-pieces together to form a box.

3. *Watch-glass Blocks*

This method of block making makes the use of a clean watch-glass lightly smeared on its concave surface with glycerol. Molten wax is poured after keeping the watch-glass under a binocular microscope. This method is very convenient for embedding very small pieces of tissues.

Whatever type of mould is selected for making the block, the steps as described in the 'L'-piece method of block making must be followed. Furthermore, it is advisable to insert a slip of paper bearing a description of the tissue, so that it protrudes at one corner of the

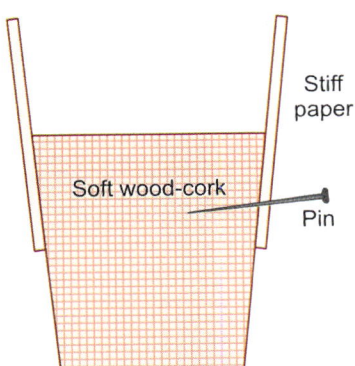

Fig. 7.5: Paper thimbles used for paraffin block making.

mould between the mould and wax. The piece of paper will stick to the wax on cooling. This avoids any confusion when wax blocks are unlabelled.

HARDENING OF PARAFFIN

The hardness of the infiltrated or impregnated specimens may be *natural* (bone, teeth, calcified chitin) or *acquired* (yellow elastic fibres in bulk, colloid of thyroid, albumen, white fibrous tissue). It should be kept in mind when infiltration is being performed. Nearer the hardness index is to 1, the better is the section likely to be.

$$\text{Hardness index} = \frac{\text{Specimen weight}}{\text{Wax weight}}$$

When the molten paraffin wax is poured in any type of mould, it first sets or congeals; and then it hardens. At the same time it also contracts. The hardening of paraffin wax bears important relationships with:

i. **Room Temperature**: If the temperature of the molten wax and that of the working place is nearly equal (less temperature difference), early crystallization of wax occurs. This is not desirable because block would be so crystalline that it is full of air bubbles. Hence, the temperature of the wax should be raised 5–10°C above its melting point before it is poured in the mould.

ii. **Age of the Block**: Faster the setting, more even the condition of the wax throughout the block, and hence the smaller the crystal size. However, it should be noted that although setting of the wax takes place within a few minutes, true hardening may take many hours or even days. The time which a block takes to congeal or to set will vary with the surrounding temperature. In a refrigerator the setting will be much faster than in a laboratory room with a temperature around 23°C.

SUMMARY

Paraffins are saturated hydrocarbons that are very stable. They allow very little effect of chemicals on them. Chemically, paraffins are called *alkanes*, and have variable number of carbon atoms:

i. If carbon atoms are 1–4: Paraffins are colorless and odorless **gases**
ii. Carbon atoms 5–13: Paraffins are **liquids**
iii. Carbon atoms 14 or more: Paraffins are **solids**

Paraffin wax is a white or colorless soft solid mixture of many hydrocarbons ranging from $C_{22}H_{46}$ to $C_{28}H_{58}$. Paraffin wax is strongly hydrophobic (hence, even traces of water in the specimen prevent complete infiltration).

Melting point, 35 to 65° C, a low-melting point wax is recommended for making blocks requiring thicker sections to be cut. Higher melting point paraffin wax suits for cutting thinner sections.

Plastic point determines the hardness of paraffin wax; and is a few degrees below the melting point of that wax. Waxes with low plastic point are more translucent, more brittle, and compress more in sectioning.

Waxes other than paraffin wax

1. *Bees wax*: Does not burn with smoky flame
2. *Candle wax*: Reduces the tendency of paraffin wax to soften in very hot atmospheric conditions
3. *Ester wax*: Because of its solubility in alcohol, the use of clearing agent may be avoided in tissue processing
4. *Carbowax*: It is water soluble (carbowax 4000 and 5000; the higher numbered carbowax is relatively harder). Because of carbowax's water-solubility the specimens can be directly transferred in an embedding fluid medium. For merits and demerits refer to Table 7.1.

Technique for Wax Impregnation (infiltration)

A dehydrated and cleared tissue is ready for impregnation with paraffin. The following *steps* are to be followed:

i. Specimen transferred to specimen tube is kept in an oven not more than 5°C above the melting point of the wax used
ii. Make 2 to 3 changes of paraffin wax (of melting point 48 to 53° and not higher)
iii. Addition of 0.5% petroleum resin to embedding wax is recommended for fine texture on cooling.

Merits and Demerits of Wax Impregnation

Reference to text on pages 74–75 is suggested/recommended.

Making Paraffin 'Block' of Specimen

Wax-impregnated specimen is now ready for making blocks suitable for use in a microtome (Chapter 8). Follow the steps given below:

i Melt the wax by keeping solid paraffin in paraffin embedding oven at 5 degrees above the melting point of the wax used
ii. Pour molten wax into a mould and at the same time place the specimen (with proper orientation to be sectioned later in the desired plane)
iii. Allow the 'block' to solidify without any disturbance
iv. Store in a cool place/or refrigerator

Types of Moulds

i. Brass moulds—Leuckhart's 'L'-pieces (*see* Fig. 7.3)
ii. Paper-boat—for plan *see* Fig. 7.4
iii. Watch-glass blocks (paper thimbles) (*see* Fig. 7.5).

8

Microtomes and Microtome Knives

INTRODUCTION TO MICROTOME

The word *microtome* is derived from a Greek word which means *to cut small*. The word was derived in 1839 by Chevalier to designate an instrument used for cutting 'sections' of properly embedded biological specimens, too large or not sufficiently transparent for direct examination with a microscope. Earlier, the microtomes were called *cutting engines*. The construction of these instruments is based upon *three* main features:

1. A *block-holder* for securing a tissue-block
2. A *knife-carrier* for firmly clamping the microtome knife
3. A *ratchets device* with a metal rod that can be turned an adjustable thickness for advancing the specimen past the cutting edge of microtome knife.

TYPES OF MICROTOMES

The basic features of microtome mentioned above are incorporated in a variety of designs. On the basis of fundamental mechanism involved in section-cutting, there are the following types of microtomes available (Table 8.1).

Hand Microtome

The hand microtomes were introduced in 1770 by Cumming as simple instruments suitable for coarse work where great accuracy in the thinness of section was not required. They are

81

Table 8.1	Types of microtomes and their applications		
Type of microtome	**Type of sample**	**Sample cut thickness**	**Application**
Saw microtome	Hard and brittle	>30 µm	Samples such as bone and teeth
	Embedded samples	1–60 µm	Samples sliced at an angle
Rotary microtome	Embedded samples	0.5–60 µm	Thin samples, manual control
Vibrating microtome	Soft, fresh/fixed samples	Fixed >10 µm Fresh >30 µm	Less pressure and sample disruption
Laser microtome	All samples	>1 µm	No sample contact and no sample preparation
Cryomicrotome	Frozen samples		Very specific thickness
Ultramicrotome		TEM 40–100 nm	Extremely thin cuts for analysis with specialty microscope

generally employed by botanists for cutting sections of roots and stems. The instrument (Fig. 8.1) consists of a microtome-well, a metal rod with a fine thread and a milled wheel for advancing the specimen. Its action is based on a differential screw which moves the table and the specimen fixed to it by 0.01 mm for each graduation on the screw drum.

A botanical specimen, such as a stem or root, is placed in the well of the microtome. The well is then filled with molten paraffin wax. When the wax becomes hardened, it is trimmed away from the front and sections are cut by sliding the knife across the specimen. After each cut, the milled wheel turned at the microtome base advances the specimen.

Bench Microtome

Introduced by Pritchard (1835), this type of microtome has the advantage that it may be fastened to a bench (Fig. 8.2). The instrument is basically a hand microtome used to cut celloidin blocks.

Fig. 8.1: Hand microtome.

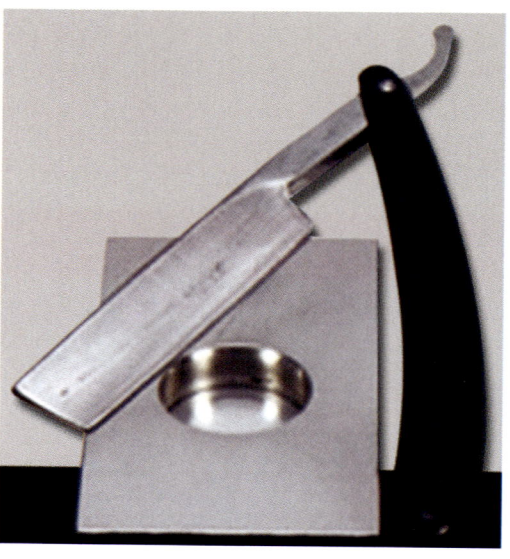

Fig. 8.2: Bench microtome.

Rocking Microtome

This variety of microtome (Fig. 8.3) was invented by Caldwell and Trefall (1881). Rocking microtome is oldest in design. It derives its name from the rocking motion of the upper (or cross) arm of the microtome as the sections are being cut. The movement of the specimen against the knife is an arc of a circle which is at right angle to the knife in case of small models of instruments; and parallel to the knife in large models. The rocking microtomes may be obtained in a variety of forms; some designed to obtain sections between 0.1 and 5 microns; and still another to cut ultrathin sections between 100 and 1000 Å. Generally, the small (standard) models of Cambridge rocking microtome are used for cutting sections from paraffin embedded material; but the large models can be adapted, by means of a special knife-holder, to cut celloidin sections also. To obtain good results with the rocking microtomes, the movement of object against the microtome knife needs to be rather abrupt. Steady backward and forward movement of the handle gives ribbons of good sections. Advantages and disadvantages of rocking microtome are summarized in Table 8.2.

Base-sledge Microtome

The sledge microtome (Fig. 8.4) consists of a knife supporting column with screw

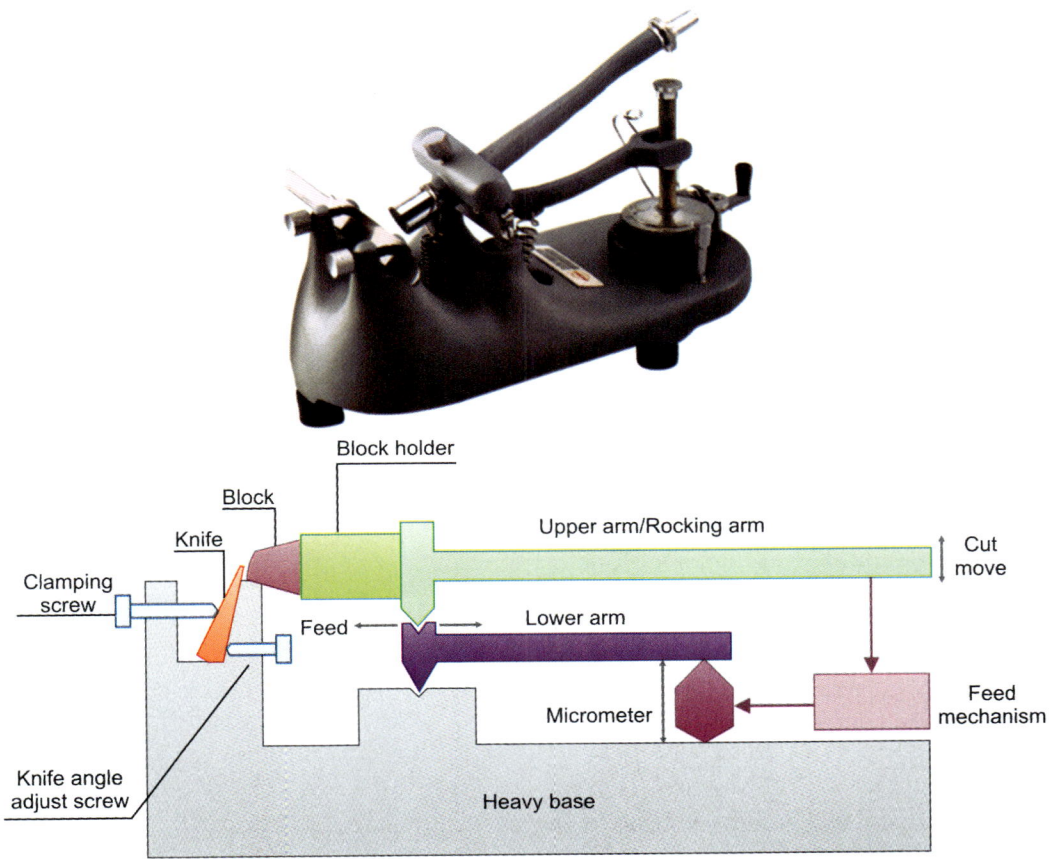

Fig. 8.3: A rocking microtome model RMT-10. A diagrammatic sectional plan of the rocking microtome depicts its mechanism of action.

Table 8.2	Advantages and disadvantages of rocking microtome	
Advantages	**Disadvantages**
• Cheap cost	• Limited size of the block can be cut
• Simple to use	• Sections are cut in a curved plane
• Extremely reliable	• Instrument is light, hence advisable to fit it into a tray which is screwed to the bench
• Very minimum maintenance	

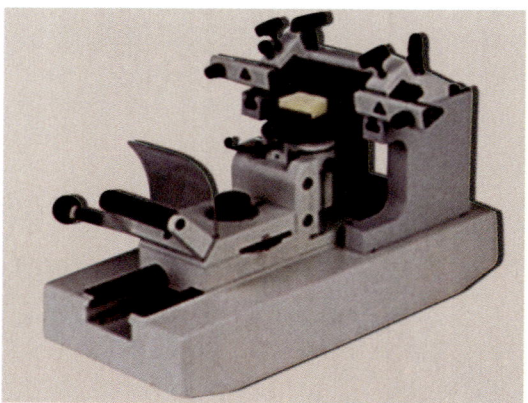

Fig. 8.4: Model of a base sledge microtome.

adjustments for altering tilt; a ball-jointed specimen holder; and a slide which brings specimen against knife and also carries mechanism controlling thickness of sections. The instrument is usually reserved for special purposes, like the cutting of refractory paraffin material and celloidin embedded objects. In some of the makes, the microtome knife moves against the specimen; but in the base-sledge variety of sliding microtome (introduced by Capenema in 1848) the knife is rigidly fixed and the object, mounted on a heavy sliding base, is slowly and steadily moved to cut sections. Perhaps, the main advantage of a sliding microtome lies in its capacity to deal with larger specimens than those cut on a rotary microtome. A freezing attachment can be adapted to several models, making a sledge or sliding microtome an 'all purpose' instrument. A sledge microtome was originally designed for cutting sections of very large blocks of samples (such as whole brain). In addition, this kind of microtome was used primarily for hard tissues, whole mounts, especially useful in neuropathology and ophthalmic pathology. The mechanism of action is based on the fact that the block holder is mounted on a steel carriage which slides backwards and forwards (hence the name) on the guides against a fixed horizontal knife. The sledge microtomes have heavier blades and cannot cut as thin as a regular microtome.

Sliding Microtome

The sliding microtome was first developed by Adams in 1798. It is similar to a sledge microtome (Fig. 8.5). This microtome is unusual because in this type the knife moves horizontally against a fixed block. This machine was developed for cutting celloidin embedded blocks of tissue. However, it can also be used for paraffin embedded samples.

Saw Microtome (or Saw)

The saw microtome is especially designed or hard materials such as teeth or bones. Its main feature is a recessed rotating saw (Fig. 8.6 arrow), which slices through the sample. The minimal cut thickness is approximately 30 μm and can be made for comparatively large specimens.

Freezing Microtome

This instrument was first used by Rutherford and Cathcart (1873) exclusively to cut unembedded tissues which have been frozen to the right consistency with carbon dioxide, Freon, or other refrigerants. In the freezing microtome (Figs 8.7 and 8.8) the knife moves against the vertical specimen-holder which has

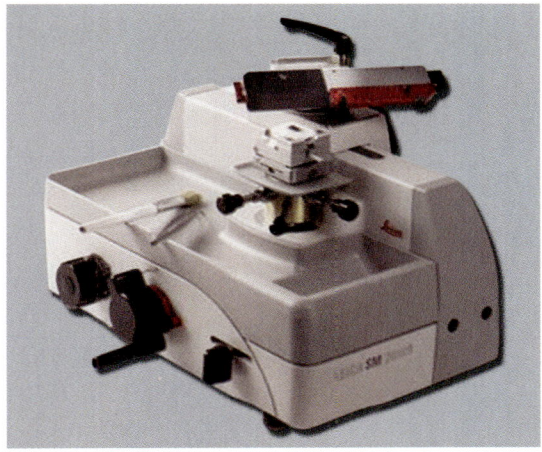

Fig. 8.5: Two models of sliding microtome.

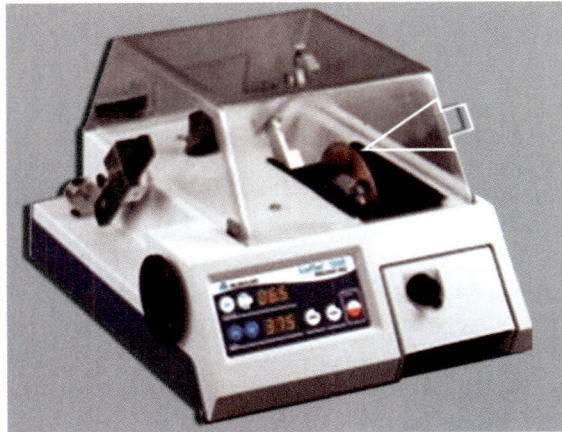

Fig. 8.6: Saw microtome for cutting hard materials. Note a circular sharp-toothed saw (arrow).

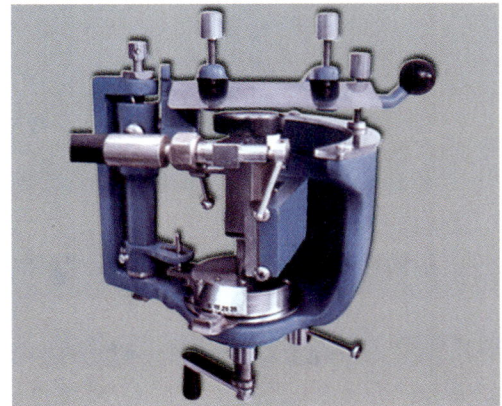

Fig. 8.7: Freezing microtomes.

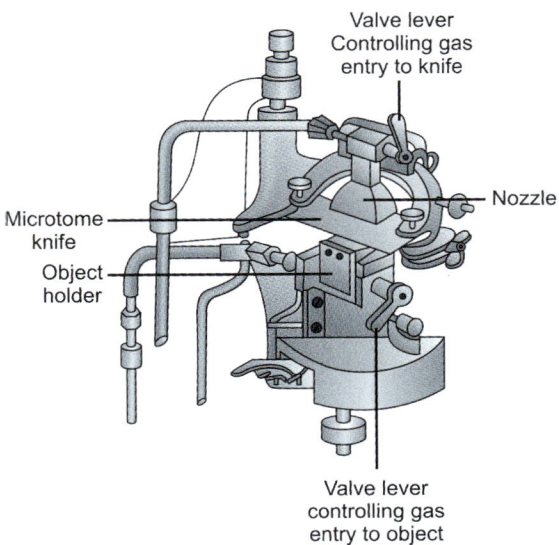

Valve lever Controlling gas entry to knife

Nozzle

Microtome knife

Object holder

Valve lever controlling gas entry to object

Fig. 8.8: Components of a freezing microtome.

holes in it through which carbon dioxide escapes and freezes the tissue. The block remains stationary. A nozzle directs another stream of the refrigerant against the knife. Separate valve levers control the admission of carbon dioxide to the specimen and knife, and these keep both the block and knife at the correct temperature. The advantage of frozen sections is that they can be prepared very quickly. Thus a histologist may use this procedure for 'quick-section' diagnosis. The freezing microtome gives best results for cutting frozen sections. Machine is clamped to the edge of a bench and connected to a cylinder of CO_2 by means of a especially strengthened flexible metal tube. Separately controlled flow of CO_2 delays the thawing of sections on the knife and makes it possible to transfer them directly from knife to slides. Knife freezing attachment is supplied with most machines. Sections thickness gauge is graduated in units of 5 μm instead of 1 μm.

Vibrating Microtome

The vibrating microtomes (**vibratomes**) are usually used for fresh biological samples that can be accurately cut without freezing or embedding. The cut thickness is usually around 30–500 μm for live tissue and 10–500 μm. The sections are, therefore, thicker. Such precise instruments (Fig. 8.9) maintain cell morphology as well as enzyme and cell activity to provide more viable cells per slice.

The vibrating microtome operates by cutting using a high speed vibrating safety razor blade (hence the name), allowing the resultant cut to be made with less pressure than would be required for a stationary blade. The amplitude of vibration is adjusted by altering electrical voltage applied to the 'knife'. To prevent tearing, soft material is cut whilst immersed in a fluid which also aids in dissipating heat produced at the vibrating edge of the razor as it cuts. Two main advantages using vibrating microtome are: (i) Greatest application in enzyme histochemistry and

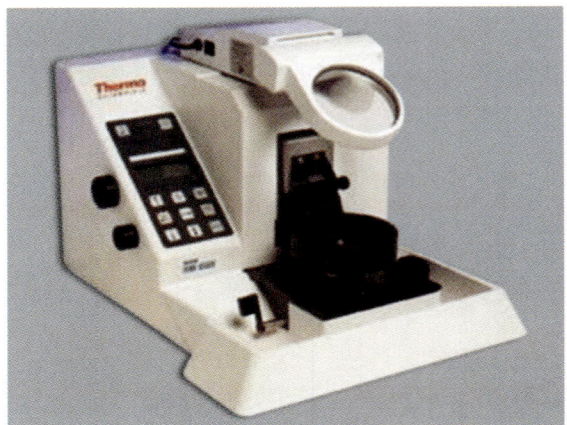

Fig. 8.9: A vibrating microtome consists of a vibrating blade allowing the resultant cut with comparatively less pressure.

ultra structure histochemistry, and (ii) disintegration is avoided because of cutting tissues at very low speed.

Rotary Microtome

This type of microtome is so named because the fly-wheel of the instrument (Fig. 8.10) rotates as sections are being cut. Invented independently by Pfeifer (1883) and Minot (1886), the rotary microtome is sometimes referred to as 'Minot' microtome. It is very suitable for general work and is used in majority of histology laboratories. A paraffin block can be firmly attached to it. The microtome operates so that when the fly-wheel is turned, the paraffin block is swept past the edge of the microtome knife and then back again. The feed mechanism is independent of the up-and-down movement of the object. The device in which the object clamp is carried forward on a horizontal slide-way is carried up and down on a broad and very rigid, perpendicular slide-way. The feed screw is quite heavy, being 12 mm in diameter, and has two threads to the millimeter. There are 250 (or 500) teeth in the wheel at the end of this screw. Each tooth, therefore, represents a section of 2 microns (or 1 micron) in thickness. Any thickness can be cut in multiples of

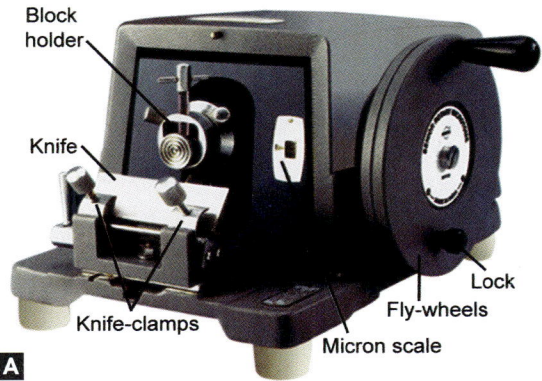

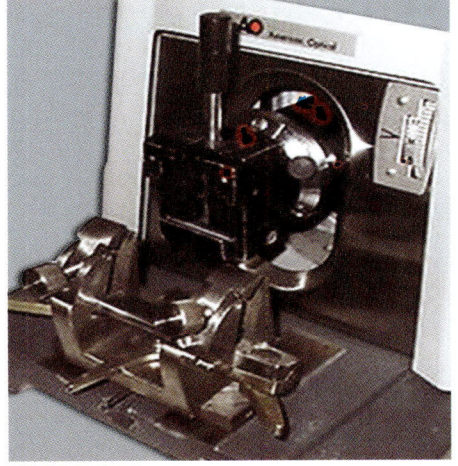

Block holder

Knife

Knife-clamps

A

Lock

Fly-wheels

Micron scale

B

Fig. 8.10: A rotary microtome: (A) Viewed from front and side, (B) enlarged portion of the front part shows the part for holding the knife (knife is not shown here).

the face of paraffin block. Each slice, after it is cut, tends to adhere to the cutting edge of the knife until it is displaced by the next slice, and then it adheres to the free edge of the slice that displaces it. This results in a paraffin 'ribbon' of sections, consisting of individual slices adhering to one another.

Though primarily designed for sectioning paraffin-embedded material, the rotary microtome can be used for cutting celloidin blocks, provided that the microtome knife is fixed diagonally in a special knife-holder. A steady motion of the fly-wheel is needed to obtain good ribbons of sections. Many workers regard rotary microtome as the best instrument for cutting large paraffin sections, particularly when serial sections are wanted.

Cold Microtome or Cryostat

A **cryostat** (from *cryo* meaning cold and *stat* meaning stable) is a device used to maintain low *cryogenic* temperatures of samples or devices mounted within the cryostat. The cryostat is usually used in a process called frozen section histology. Basically, a cryostat (Fig. 8.11) is usually a stationary upright freezer, with an external wheel for rotating the microtome mounted outside the chamber. The external wheel minimizes unnecessary warming all necessary mechanical movements of the microtome.

Cryostats have numerous applications within science, engineering, and medicine. In medicine cryostats are used to cut histological slides through essentially an ultrafine *deli slicer*, a *microtome placed in a freezer*. The freezer is either powered by electricity, or by a refrigerant like liquid nitrogen. Small portable cryostats are available and can run off generators or vehicle inverters. The temperature can be varied, depending on the tissue being cut—usually from − 20 to − 30° C. Low temperatures may be maintained within a cryostat by using various refrigeration methods, most commonly using cryogenic fluid bath such as *liquid helium*. Hence it is usually assembled into

2 microns up to 20 to 40 microns. The total excursion is 22 mm and the feed screw automatically ceases to operate at the end of this excursion. At the end of each stroke, the pawl actuating the toothed wheel is *not lifted* out of the tooth but *recedes* from it, returning on the back stroke entirely free from the teeth. The small crank at the end of the feed screw provides a means for quickly returning the screw to any desired position, or for quickly and definitely feeding the object up to the knife.

Between each full stroke the specimen is moved a few microns (adjustable) nearer the knife edge and a thin slice of tissue is cut from

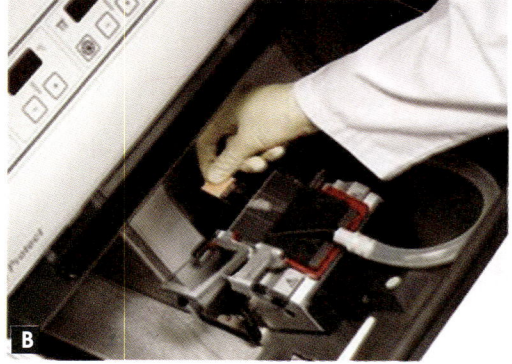

Fig. 8.11: Cold microtome (cryostat or cryomicrotome) is: (A) An instrument for cutting semi-thin sections from frozen samples, (B) shows a diagnostic section being prepared is being picked up onto a warm slide where it will be immediately fixed and stained.

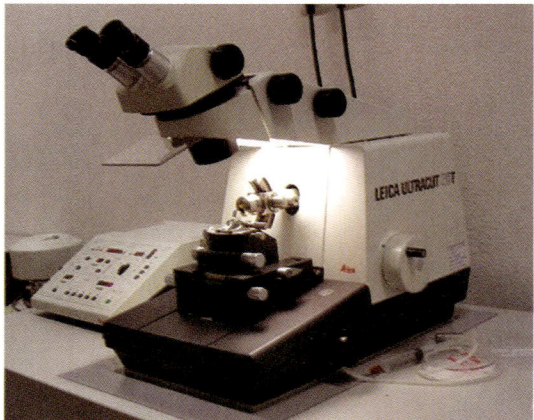

Fig. 8.12: An ultramicrotome is a device for cutting extremely thin sections exclusively for transmission electron microscopy.

a vessel, similar in construction to a *vacuum flask*.

The cold microtomes are different from the freezing microtomes; being rotary microtome mounted within a refrigerated cabinet. The microtome can be operated by various 'controls' fitted outside the cabinet. For preventing the sections to roll up as they are cut, most of the cryostat makes are fitted with an anti-roll plate. The sections cut in a cryostat are much better than those cut with a freezing microtome; hence they are ideal for certain kinds of enzymatic and histochemical studies.

Ultramicrotome

An *ultramicrotome* is a device to cut extremely thin sections, functioning in the same manner as a rotary microtome, but with very tight tolerances on the mechanical construction. The linear thermal expansion of the mounting is used to provide very fine control of the thickness. These extremely thin sections between 40 and 100 nm (40 and 1000 Å) are important to use exclusively with transmission electron microscope (TEM) [*see* Part 2, Chapter 18 for further detail]. Thicker sections up to 500 nm are used for TEM applications or for light microscopy *survey sections* to select an area for the final thin sections (Fig. 8.12).

Diamond knives (preferably) and **glass knives** (Figs 8.15 and 8.16), and **sapphire knives** (Fig. 8.17) are used with ultramicrotomes. To collect the sections, they are floated of the top of a liquid as they are cut. Thereafter, the sections are carefully picked up onto grids suitable for TEM specimen viewing. The thickness of the section can be estimated by colors of reflected light as seen as a result of extremely low sample thickness.

Laser Microtome

The laser microtome is an instrument in which prior preparation of the sample through normal steps is not required and, therefore, there is contact-free slicing. Obviously, the preparation artefacts are minimized.

Alternately the laser microtome can also be used for very hard materials such as bones or teeth. The thickness achievable is between 10 and 100 µm.

Principle of Working: The laser microtome operates using cutting action of infrared laser. As the laser emits a radiation in the near infrared, in this wavelength regime, the laser can interact with biological materials. The sharp focusing of the laser probe within the sample (Fig. 8.13), a focal point of very high intensity can be achieved. Through the non-linear interaction of the optical penetration in the focal region, a material separation in a process termed photo-disruption is introduced. By limiting the pulse durations to the femtoseconds range (Fig. 8.13), the energy expended at the target region is precisely controlled, thereby limiting the interaction zone of the cut to under 1 µm. External to this zone the ultra-short beam application time introduces minimal to no thermal damage to the remainder of the specimen. The laser radiation is directed onto a fast scanning mirror-based optical system, which allows 3-D positioning of the beam crossover, whilst allowing beams traversal to the desired region of interest.

PROPER CARE OF MICROTOME

The microtomes are precision instruments and require proper and regular care for perfect functioning. A long useful life of precision section cutting may be expected when the

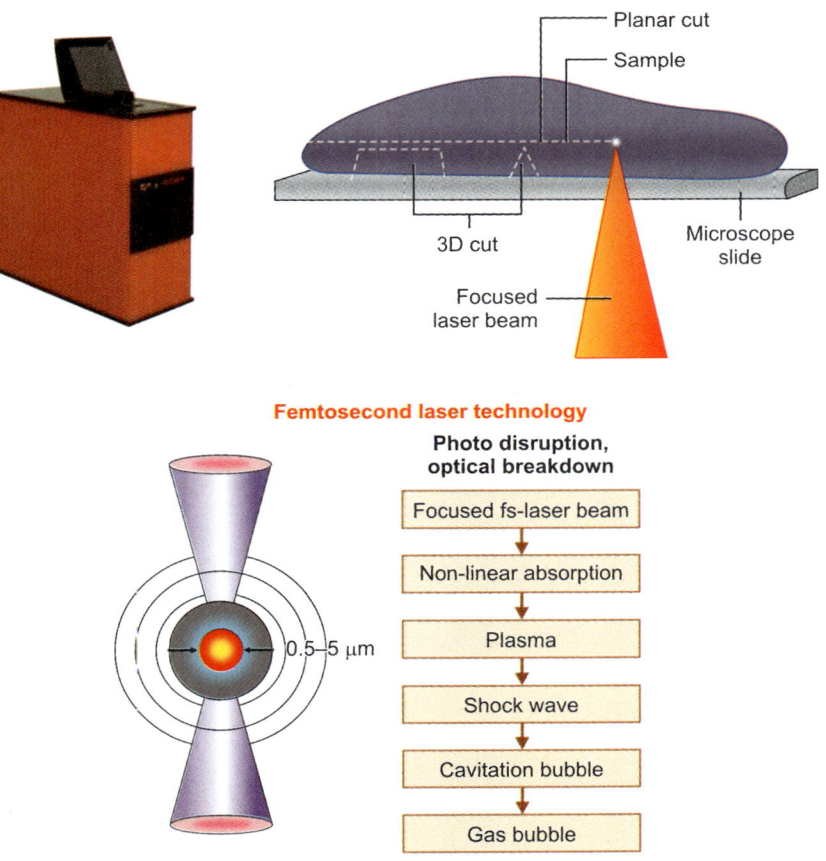

Fig. 8.13: The laser microtome and principle of laser microscopy.

instrument is given reasonable care. Following are important and noteworthy points:

- When not in use, microtomes must be, as a rule, covered properly with a plastic cover (usually provided by the instrument's supplier) or some other cloth.
- All moving parts should be applied every month with a light machine oil to avoid any rust.
- After paraffin-sectioning, the metallic parts of the microtome should be cleaned with xylene, and the knife-holder replaced back (after complete drying) into the knife-case.
- Paraffin residue may be removed from the knife holder with a cloth moistened with xylene.
- The slides of the knife holder should be kept oiled with Pike Oil or light grease (Fig. 8.14). Relubrication should be done when cleaning removes the oil or grease and as necessary for free moving use of the knife clamp parts.

- A lighter oil (used by watch repairing mechanics) when the microtome is used in cold chamber or air-conditioned laboratories.
- After celloidin and/or frozen sections have been cut, all moving parts should be oiled with any light machine oil (sewing machine oil etc.)
- Any paraffin wax adhered to the cutting edge of the microtome knife should be removed with a soft camel-hair brush.
- After use clean the microtome by brushing or wiping off any ribbon fragments.
- To avoid damage to the teeth of the micron-scale of the microtome, the setting of the section thickness must be accurately done with care.
- After considerable use the friction on the feed screw decreases and **must** be tightened for continued regular section cutting. When the hand feed crank does not offer a slight resistance, screw may be tightened.

MICROTOME KNIVES

The microtome knives are greatest single factor for cutting good tissue sections. There are many shapes, sizes and materials for microtome knives, which were developed to fit certain microtomes and to cope with different degrees of hardness of tissues and embedding media.

1. Steel Knives

The steel knives are in most common use for cutting tissue sections from some sample either unembedded, but hardened, or embedded in some suitable embedding medium. A microtome knife is made up of very high grade of good quality steel are used to prepare sections of animal or plant tissues for light microscopy. A steel knife consists of a blade, about 15 cm in length, with a broad base fitting in a knife-carrier or 'honing-guide'. It has a cutting edge formed by meeting of two facets. The *facet* or *bevel* is the flat surface at the extreme edge of a

Fig. 8.14: Microtome knife: Its case and parts. Notice that heel is the angle formed by the cutting edge and end of the knife nearest the handle; while toe is the angle formed by the cutting edge and end of the knife farthest from the handle. At the bottom of the figure is shown a **Heiffor knife** (used on rocking microtomes with a fixed handle).

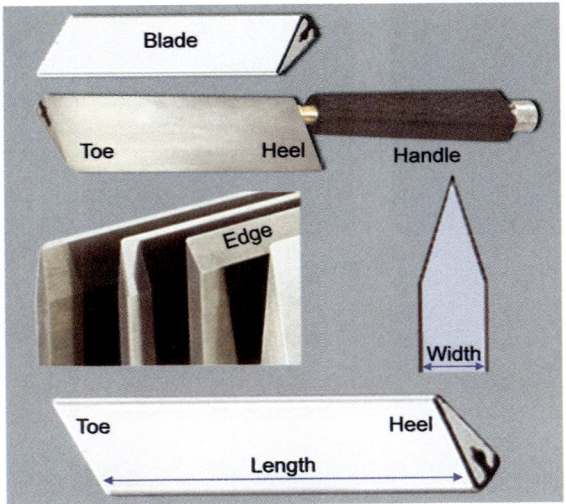

Fig. 8.15: Microtome knife blade with a hole for attaching the knife-handle (not shown here). The *heel is the angle formed by* the cutting edge and end of the knife nearest the handle; while toe is the angle formed by the cutting edge and end of the knife farthest from the handle. At the bottom is shown a **Heiffor knife (used on rocking** microtomes with a fixed handle).

microtome knife. The *handle* of the knife can be screwed into a hole at the posterior or *heel* end; while the opposite end is termed the *toe* of the knife (Fig. 8.15). There are several varieties of steel knives (vide infra).

2. Disposable Steel Razor Blades

The disposable blades, made up of high quality stainless steel (Fig. 8.16), replace conventional steel knives in microtomy. These blades produce consistently high quality sections; the sharp edge of the blade may cut 2–4 μm thick sections.

3. Glass Knives

These knives offer a valid alternative for ultra-thin sectioning of plastic-embedded tissues for transmission electron microscopy. In most of the electron microscope laboratories, the glass knives are routinely used. The glass knives are generally made from a 12 inches strip of hardened good quality glass, 1.5 inches wide

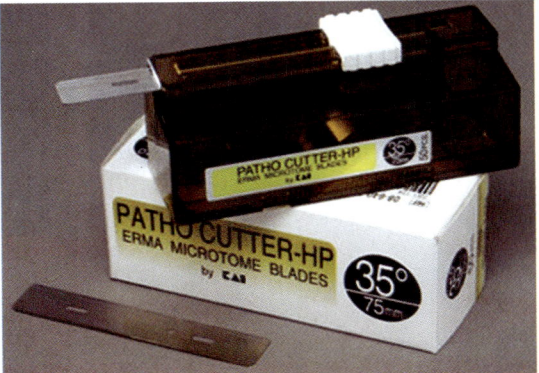

Fig. 8.16: Disposable microtome steel blades.

and ⅜ inches thick. The glass strip is scored, with the help of a special knife making device (Fig. 8.17B), to obtain perfect "square" with orthogonal edges. Each square is then cut diagonally into two equilateral right-angled pieces by breaking the glass as close as possible through the corners. Each piece has two cutting edges ⅜ inches wide which meet at 45°. The closer the break goes to the two vertices better become the chances of getting 45° angled sharper "good" glass knives (Fig. 8.17C). All the knives made by this method do not possess an ideal cutting edge; it is fully adequate for getting desired ribbons from the specimens used for electron microscopy. For sectioning, the glass knife is fitted in a metallic trough (filled with distilled water) and set in the ultra-tome in such a way so as to have a clearance angle of nearly 4° to 6°. A glass knife is hard but brittle, and great care is required while handling. The glass knives **deteriorate with storage** due to changes in the 'flow' or 'strain' of the glass after fracture and from oxidation impurities remaining in the hardened glass (and thus in the glass plates, and the 'square' by 'fracture' with the glass breaking machine) after manufacture. Hence, the **glass knives should be prepared immediately before use.**

4. Diamond Knives

The diamond knives (Fig. 8.18) are manufac-tured from gem quality diamonds. These

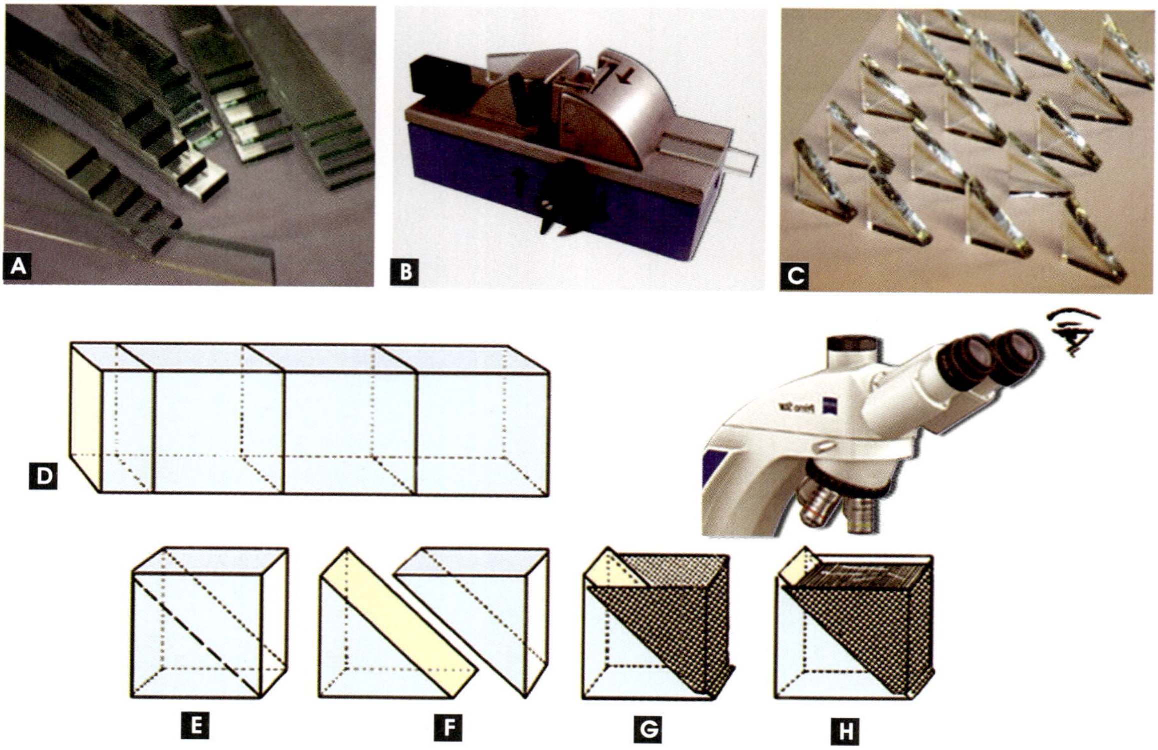

Fig. 8.17: Glass knives are prepared from hardened good quality glass plates (A) with a device (B) preparing several knives (C). Sketches show (D) glass squares. Each square is broken diagonally, (E) a glass square with a diagonal cut-line, ready to be broken; (F) into two pieces; (G) a glass knife fitted with a metallic clamp making a boat-like trough which is filled with distilled water to receive ultrathin ribbons obtained by using an ultratome (H) ribbons viewed under a microscope.

Fig. 8.18: A diamond knife.

knives are used less commonly because of their high prices and limited use in cutting *cryosections* of fresh or fixed samples. Despite they are being very expensive the knives are extremely durable, because of the hardness factor of the diamond, and are used primarily for cutting ultrathin, resin sections. These knives are used to slice hard materials such as bone, teeth and botanical specimens for both light microscopy and electron microscopy.

5. Sapphire Knives

Sapphire is harder than tungsten carbide or glass. The sapphire knives (Fig. 8.19), therefore, ensure high durability of the cutting edge for all types of material to be sectioned. They are manufactured from one piece of solid sapphire

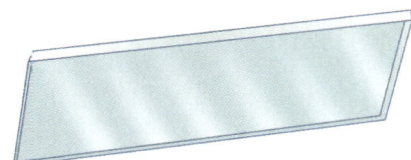

Fig. 8.19: A sapphire knife.

artificially produced from an alumina mono-crystal under computer controlled thermal conditions. The only restriction when using sapphire knife is block size as the knife edge is limited to 11 mm. A special knife holder is required.

STEEL KNIVES: VARIETIES AND POSITIONS

Varieties of Steel Microtome Knives

The chief kinds of steel microtome knives (Fig. 8.20 and Table 8.3) are given below:

i. *Plane wedge knife*—also called *wedge knife*, used for cutting frozen sections.

ii. *Plano-concave* or *hollow-ground knife* used for cutting celloidin-embedded sections. If the concavity is less, the knife may be used for cutting paraffin embedded tissues with a sliding microtome.

iii. *Biconcave* or *Minot-type* routinely used for cutting paraffin sections with a rotary microtome.

iv. Chisel-shaped knife

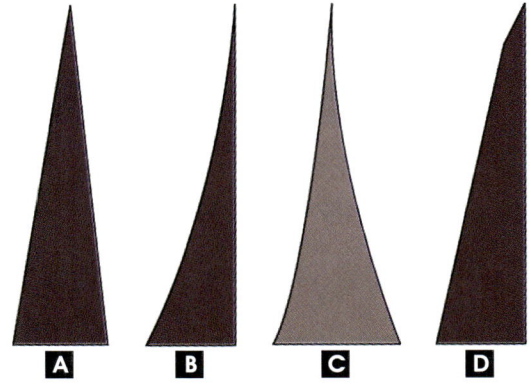

Fig. 8.20: Sectional views of different knives used for microtomy: (A) wedge knife; (B) plano-concave (may have mild or moderate concavity); (C) biconcave knife; and (D) chisel-shaped or tool knife. For features of these knives refer to Table 8.3.

Square and Slant Positions of Knives

There exists a twofold relationship between the microtome steel knife and the block surface since the process of section-cutting involves movements both in the horizontal as well as

Table 8.3	Knife profiles and their characteristics			
Profile	**Diagram**	**Used**		**Remarks**
Planar concave (wedge/standard knife profile)		• Used in all types of microtome to section all type of material • For cutting moderately hard specimens (epoxy or cryogenic samples) • For ordinary paraffin section-cutting		• Both cutting surfaces are plane • Due to extra thick wedge-insufficient grounding at the tip • Plane-wedge knives have good rigidity
Plano-concave (hollow ground)		• For sledge microtomes and some rotary ones • Knives with very concave surface for celloidin work • Used for very soft tissues (nitrocellulose embedded)		• Extremely sharp but very delicate • Plain surface of knife towards block surface
Bi-concave (Minot type knife)		• Used rocking and base-sledge		• Position of knife to be oblique (slanting) to the block to be cut • Easy to sharpen but less rigid and prone to vibration
Chisel-shaped (tool-edge knives)		• Used for cutting hard tissues		• Blunt edge increases the knife's stability. • Significantly more power required in section-cutting

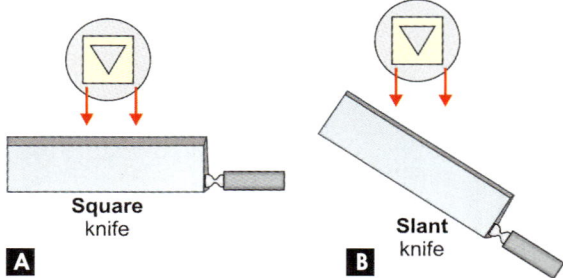

Fig. 8.21: Two positions of knife used in section-cutting: (A) *square knife*—for paraffin wax block cutting by a rotary microtome, and (B) *slant* or *oblique knife* used for celloidin section cutting. Remember that a slant knife is **never** used paraffin sectioning.

Square knife

Slant knife

A

B

the vertical planes. The knife may be used either *square* or *slanting* (Fig. 8.21).

The term *square* knife denotes that the knife edge is parallel to the cutting surface of the block, and at a right angle to the side of the block. This position of the microtome knife is used when paraffin-embedded tissue is to be sectioned, and ribboning is required. Square knife is never used for cutting celloidin sections.

Slanting or *oblique* knife is set when sections of either hard material embedded in paraffin are to be sectioned or celloidin sections are needed. The *slant* refers to the relationship of the knife edge to the edge of the tissue block. The greater the slant, the farther the knife will cut into one corner of the block before it even enters the other front corner. With a slanting knife, the surface of the object with the knife is lessened and the resistance during section cutting is consequently diminished. Celloidin has a certain elasticity; and unless celloidin embedded sample is cut obliquely (slanting), the celloidin is likely to get compressed during sectioning. Hence all celloidin material must be cut with a slant knife only. One corner of the block will be cut first and 'crumbled' sections would be avoided.

A slant knife is entirely different from what is described as the *tilt* of the knife. The tilt of the knife is an angle that a plane passing through its back edge makes with the plane of the section. This angle is generally between 0° and 5°. A biconcave knife needs only a slight tilt for producing good sections. On the contrary, microtome knives with a plane surface on both sides usually require a greater tilt of about 20° to prevent the tissue from being pressed against the knife while section-cutting. The angle of tilt is entirely different than a slanting knife. Various angles in relation to a microtome knife are described later.

KNIFE ANGLES AND RELATIONSHIP WITH PARAFFIN BLOCK

Various angles play important roles in obtaining the perfect tissue sections. The following *terms* and *angles* in relation with the parts of the microtome knife (Fig. 8.22) and their relationship with the paraffin embedded tissue block need careful attention for understanding the principles involved in carrying out good microtomy.

Types of Knife Angles

1. Bevel (or Facet) Angle

It is the angle between the two facets that meet to form the cutting edge. It usually varies between 27° and 32° (reported by some authorities to vary between 18° and 35°). **Smaller the bevel angle sharper is the knife**. However, too small bevel angle permits elastic distortion of the knife edge. The width of the two facets of a knife makes the cutting edge of knife and is recommended from 0.1 to about 0.6 mm. Longer facets and smaller bevel would give rise to a keener edge. Sharpness or acuity of the knife is verified by the reflection of light by the knife edge when viewed under the microscope. Figures of 0.3 and 0.1 µm are necessary for maximal acuity or sharpness.

If the bevel angle is less, the breadth of the bevel will be large (2–4 mm). But if the bevel angle is more, it results in a narrow bevel (0.5–1 mm). A smaller bevel angle with broader bevel gives a weaker cutting edge to the knife

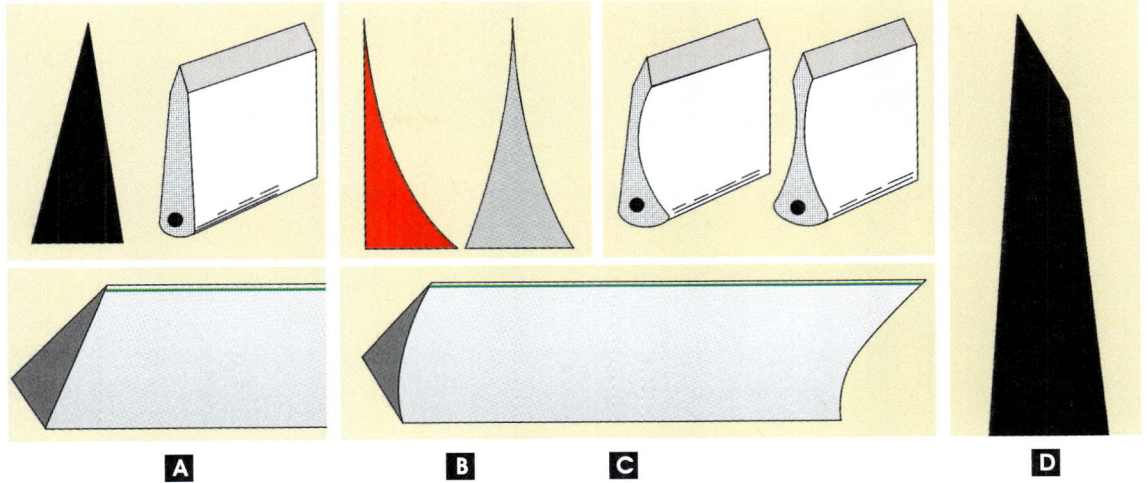

Fig. 8.22: Profiles of steel knives: (A) Wedge knife; (B) plano-concave; (C) biconcave; and (D) tool wedge or chisel-shaped.

which is more likely to vibrate. This results in sections showing a number of thick and thin stripes across them. On the other hand, a microtome knife which has a larger bevel angle (about 40°–50°) and possesses a narrower bevel (0.5–1 mm) is said to have a stronger edge. Although such cutting edge lasts long, curved sections are likely to be produced. However, a narrow bevel enables sections to escape the frictional-drag imposed by a large bevel. Thus *ribboning* is facilitated. Also the intensity of the *charge* is less. It is desirable to have a microtome knife with a bevel angle of 25°–30°, and a bevel breadth (Fig. 8.22) of about 0.7 mm for easy sectioning and ribboning.

2. Rake (or Cutting) Angle

It is the angle between the upper bevel of the knife and a line at 90° to the block surface. High rake angles suitable for soft tissues and need to be reduced for harder tissues. This is an angle that cannot be adjusted on a microtome. It is the result of the clearance angle and the upper bevel angle of the knife or blade.

The rake angle and cutting angles are actually not the same. There is slight difference between the two terms. What is the difference between the rake angle and cutting angle?

If the file is sectioned perpendicular to its long axis, the rake angle is the angle formed by the leading edge and the radius of the file. If the angle formed by the leading edge and the surface to be cut (its tangent) is obtuse, the rake angle is said to be positive or cutting. If the angle formed by the leading edge and the surface to be cut is acute, the rake angle is said to be negative or scraping. However, the rake angle may not be the same as the cutting angle. The cutting angle, effective rake angle, is a better indication of the cutting ability of a file and is obtained by measuring the angle formed by the cutting (leading) edge and the radius when the file is sectioned perpendicular to the cutting edge. In some instances, as with some Quantec files, a file may have a blade with a negative rake angle and a positive cutting angle. If the flutes of the file are symmetrical, the rake angle and cutting angle will be essentially the same.

Negative angles result in a "scraping" action and positive angles result in a cutting action (Fig. 8.23A). Although cutting actions can be more efficient and require less force for enlarging a canal, a scraping action may have a smoother feel. The operator may erroneously confuse smoothness with efficiency. However,

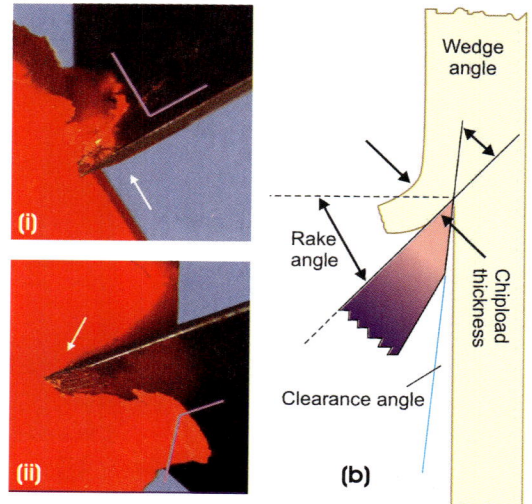

3. Clearance Angle

It is the angle by a line drawn along the block surface and lower bevel of the knife. It is dependent upon tilt. **The clearance angle is the only knife angle that can be adjusted on a microtome using the knife tilt.**

a. Set the clearance angle between 3° and 8°. This is the angle that we are most concerned with, which can be adjusted easily to suit the blade and block for better sectioning. Once the proper clearance angle is found for a particular blade, it rarely needs to be changed. Most microtomes using low profile blades cut best when the clearance angle is set at 5°.

Fig. 8.23A: (a) Diagram showing direction and action of the leading edge (angle of incidence) of a knife cutting tissue blocks (in red): (i) Negative cutting angle is an acute angle, while (ii) positive cutting angle is an obtuse angle. (b) Shows a diagrammatic sketch with a partially sectioned block (in yellow) and associated important angles.

Practical Hint

It should be noted that a **high rake angle and low clearance angle gives less compression** to the tissue block and produces a smooth plastic flow type during sectioning. A clearance angle of 2 to 4° is used for paraffin and 5 to 7° for frozen sections

if excessive pressure is applied to a cutting file excessive torsion could be the result (arrows indicate the direction of the blade motion).

b. When the clearance angle is too wide, the tip of the blade will scrape the block and **chatter** will result.

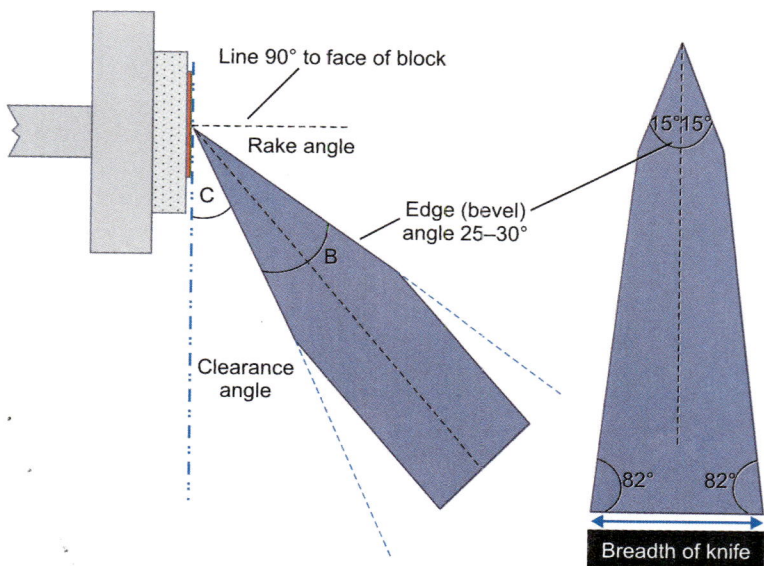

Fig. 8.23B: Diagram showing the concept of bevel angle, clearance angle, rake angle and breadth of microtome knife (see text for definitions of the terms used). Notice that if bevel angle is less, breadth of the knife will be large; likewise, a greater bevel angle will be with ncrrow bevel (see also Fig. 8.22).

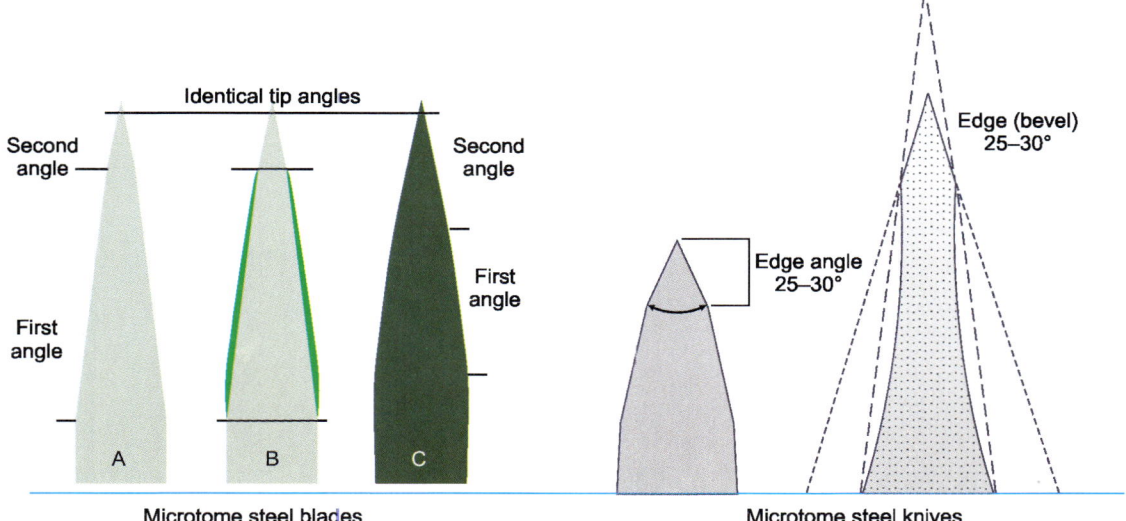

Fig. 8.24: Some angles related with microtome blades are shown on the left side of the diagram: A: Accu-Edge blade; B: The difference; C: Encore blade. Notice that the microtome blade has a 32% sharper blade edge and has 48% longer blade life over AccuEdge. On the right is shown a schematic depiction of breath and edge (or bevel) angle in a steel knife.

c. When the clearance angle is too small, the body of the blade will scrape the block and skipped sections or poor ribboning will result.

4. Wedge Angle

This is the angle (Fig. 8.24) between the edges of a wedge knife (normal value 15°)

Relationship between Knife Angles and Cutting Face of Block

The relationship between the knife angles and the cutting face of the paraffin block being cut is shown in Figs 8.25 to 8.27.

Rake angle is a parameter used in various cutting and machining processes, describing the **angle** of the cutting face relative to the

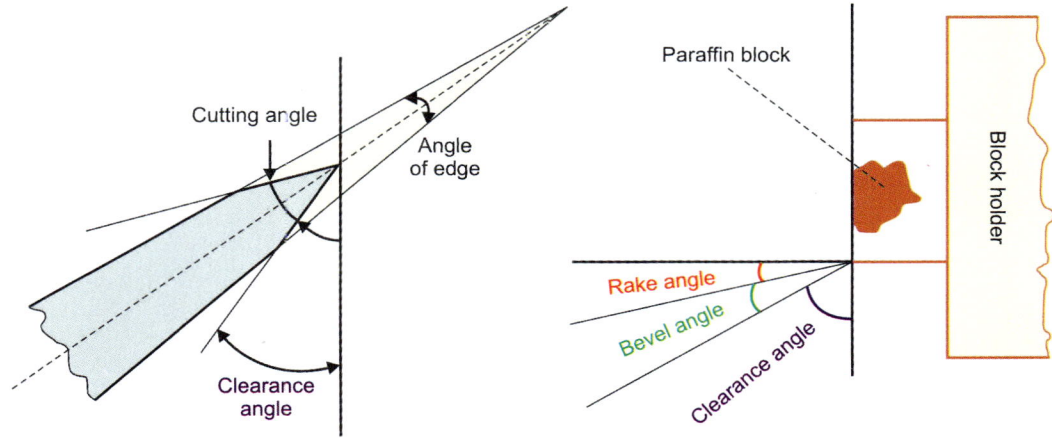

Fig. 8.25: Knife angles associated with the cutting surface of paraffin block (see also Fig. 8.26).

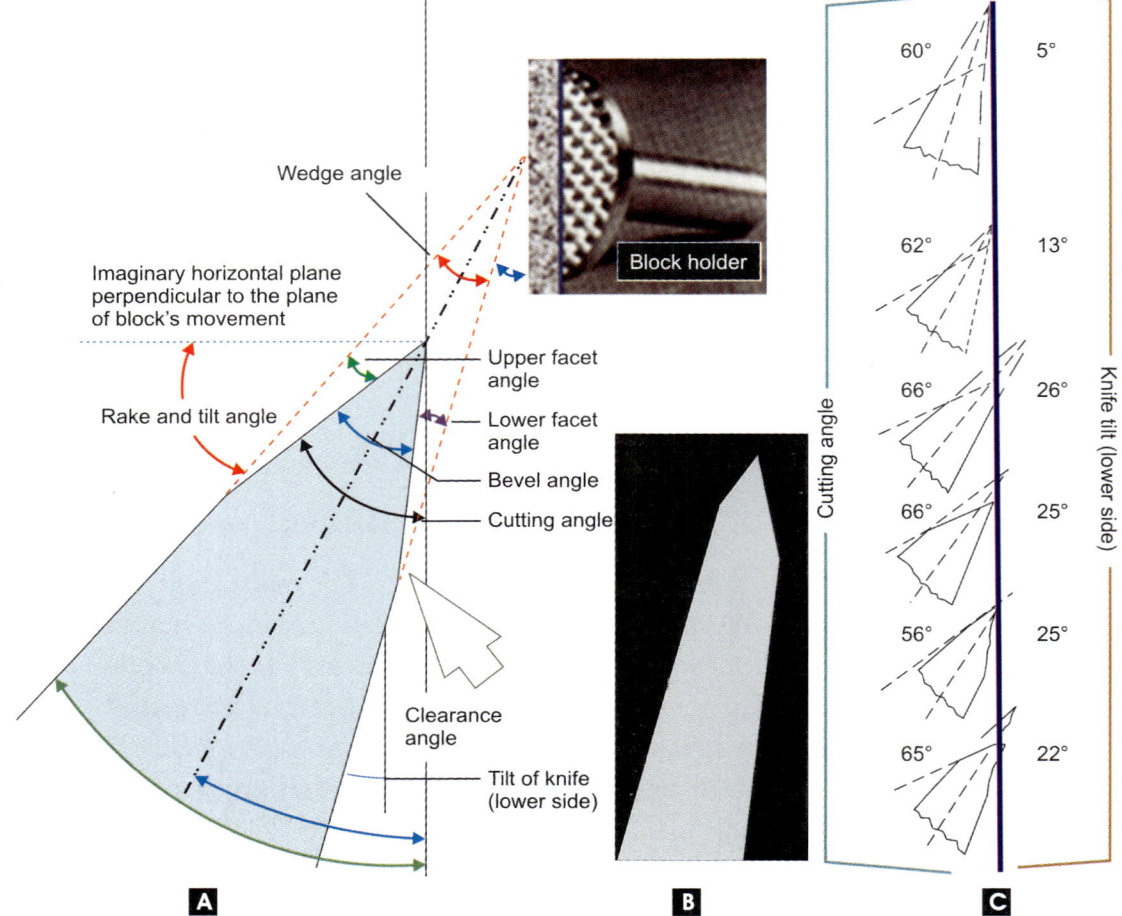

Fig. 8.26: (A) Geometry of the microtome knife edge angles; (B) is the profile of a knife cutting edge showing unequal sharpening; and (C) rake and tilt angles for a clearance angle of 5° for proper placing of knives with unequal facets. Other values are also shown for an advance set-up of the knife (not purposely described in the present book to avoid confusion).

work. There are two **rake angles**, namely the back **rake angle** and side **rake angle**, both of which help to guide chip flow. There are three types of **rake angles**: Positive, negative, and zero.

Generally, positive rake angles:
- Make the tool more sharp and pointed. This reduces the strength of the tool, as the small included angle in the tip may cause it to chip away.
- Reduce cutting forces and power requirements.

- Helps in the formation of continuous chips in ductile materials.
- Can help avoid the formation of a built-up edge.

Negative rake angles, by contrast:
- Make the tool more blunt, increasing the strength of the cutting edge.
- Increase the cutting forces.
- Can increase friction, resulting in higher temperatures.
- Can improve surface finish.

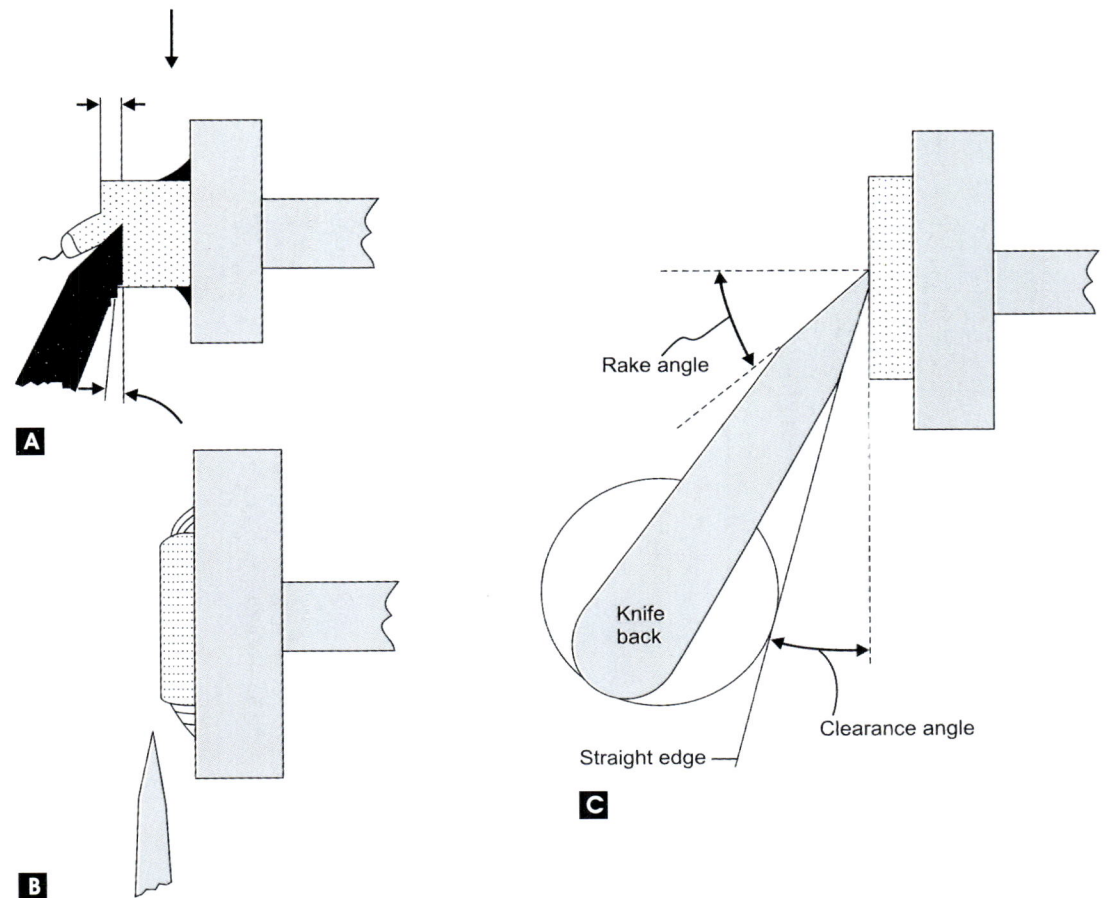

Fig. 8.27: Relationship of microtome knife with a paraffin wax block during microtomy: (A) Clearance angle and increase in section thickness from compression (*see* Chapter 9);(B) shows wedging when there is no clearance angle. (C) shows how to set the optimum (correct) clearance angle for obtaining good ribbons of sections by a skilled microtome operator.

A zero rake angle is the easiest to manufacture, but has a larger crater wear when compared to positive rake angle as the chip slides over the rake face.

Types of Sectioning

The sectioning is of the following three types:
- **Vertical sectioning**, the most common method of sectioning is perpendicular (i.e. at right angles) to the surface of the tissue.
- **Horizontal sectioning** is often done for the study of hair follicles and structures that include hairs (such as erector pili muscles, and sebaceous glands). Such structures are sometimes called "pilosebaceous units".
- **Tangential** to **horizontal sectioning** is done in chemosurgery (also called "Mohs surgery") which is a form of microscopically controlled surgery used to treat certain types of skin cancers.

CARE AND MAINTENANCE OF STEEL KNIFE

There is no much need to emphasize that proper care and maintenance of every item in any histology laboratory is of great utility. Maintenance of microtome has already been

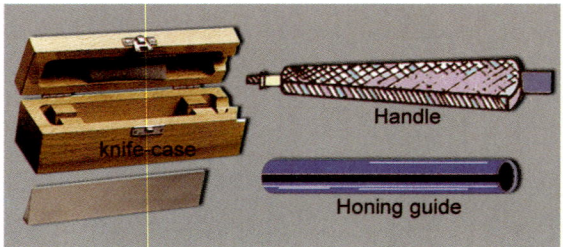

Fig. 8.28: Microtome knife: Its case and parts. Notice that heel is the angle formed by the cutting edge and end of the knife nearest the handle; while toe is the angle formed by the cutting edge and end of the knife farthest from the handle.

described in the early part of this chapter. The microtome steel knives must be kept while not in use so that proper section-cutting does not become a problem for any one working on microtomy.

Precautions in Handling of Microtome Knife

The following *instructions* must be followed for the proper care of the microtome knife:

a. Always carry a knife in its case.

b. Never leave a knife to lie on the working bench, as the edges are likely to suffer damage as well as being a danger for the users.

c. Avoid cutting in the center of the knife edge.

d. Retain knives with their handles until clamped on to a microtome or clamped on to a knife-sharpening machine or kept in their cases.

e. Sharpen with automatic knife-sharpeners only after careful reading the instructions.

Methods for Examining Sharpness of Knife

A sharp microtome knife is an essential component of any microtome to obtain thin sections of properly embedded tissue. The sharpness of knife must be tested (Fig. 8.29) as it depends primarily upon its composition and secondarily upon the methods selected to obtain its cutting edge. Any granular material (like steel) could never produce a fine and smooth type of edge like that in a glass-knife. In the same way, the quality of the edge of a microtome steel-knife is dependent for its final sharpness on the skill of the sharpener. A good

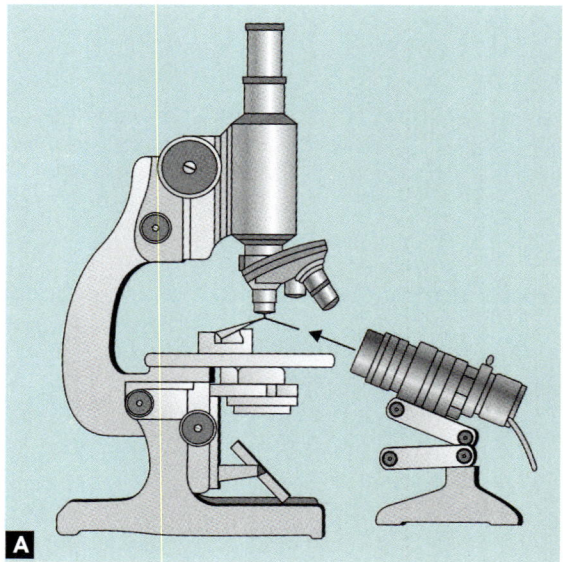
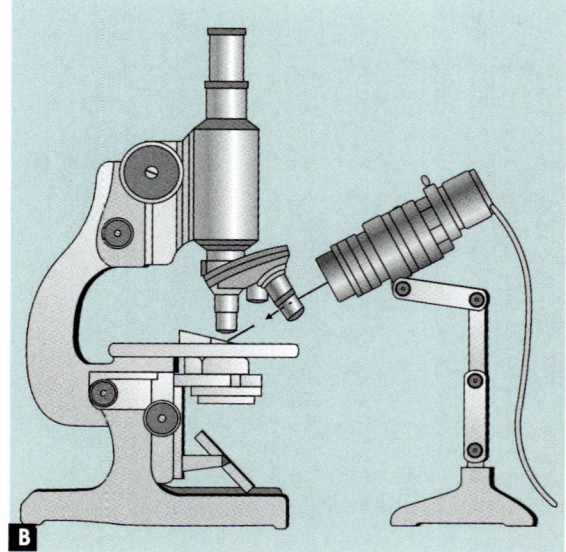

Fig. 8.29: Testing the status of a microtome steel knife to be used for good quality section-cutting are essential prerequisites for a skilled histology technical operator. (A) Shows the method for examining sharpness of the knife; (B) method for examining knife's facet polish (see text and Chapter 9). Refer to Fig. 8.30 for 'burred edge' of the knife requiring instant sharpening.

quality steel knife finished on a coarse hone would lack smoothness of a poorer knife finished on a fine hone.

After a knife has been sharpened, its edge is to be tested for fitness. For doing this set up a block of filtered tissue-free paraffin in the microtome and cut five microns thick ribbons. Check the ribbon for *striations*. If present, they indicate serrated or feather edge of the knife. On the other hand, if the ribbon is *compressed*, a dull or round edge is the cause. It must be remembered that a *burred edge-knife* (Fig. 8.30) results when one side of the knife has been sharpened more than the other. The burred edge is turned towards the least sharpened side. It is, therefore, as a general rule that the sharpening is repeated equal number of times for the two surfaces of the knife. Vibration of the knife edge is often referred to as *flutter*.

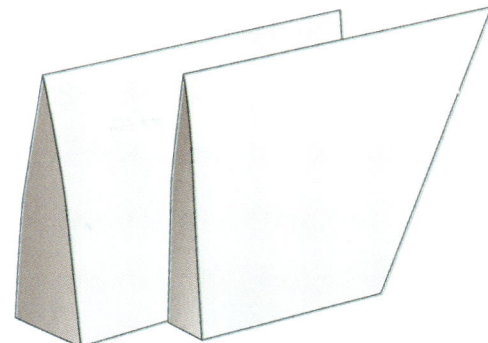

Fig. 8.30: Testing of knife-edge of microtome knives under a scanning objective using a low-power eyepiece. a damaged burred-edge seen under low magnification is an indication for honing and stropping of the knife (refer to Chapter 9).

For knife sharpening, reader may refer to Chapter 9.

SUMMARY

Microtome

Microtomes are machines that cut extremely thin sections from a processed sample for applications in histology or pathology.

- Uses special metal, glass, or diamond blades, depending on the type of specimen and the desired thickness.
- Consists of a blade holding unit with a blade holder and clamped blade, the object clamp, an advancing mechanism, and a mechanism for adjusting section thickness.

Types of Microtomes

There are numerous types of microtomes; the main ones are:

- Freezing microtomes
- Cold microtome (cryostat)
- Rotary microtome

Advantages of Rotary Microtome

- Ability to cut thin 2–3 mm sections.
- Heavier, so more stable
- Cutting angle of knife is adjustable
- Can cut celloidin-embedded sections by using a special holder to set the knife obliquely.
- Ideal for cutting serial sections

Types of Microtome Knives

The microtome knives are basically classified into two categories:

  i. Metal Knives
- Standard steel knives
- Razor blades including disposable blades

  ii. Non-metal Knives
- Glass knives
- Diamond knives
- Sapphire knives

Steel Knives

- Most common knife used for microtomy with rotary microtomy
- Made up of high grade steel
- Length about 15 cm; posterior end called **heel**, opposite end **toe**, **facet** or **bevel** is the flat surface at the extreme edge, broad **base** fits in a honing-guide; **handle** may be screwed into a hole at the heel.

Glass Knives

- Made up from hardened good quality glass strip (size $12 \times 1.5 \times 3/8$ inches)
- Cleaned glass strips are scored with diamond knife into 1.5×1.5 'squares'
- Each square cut precisely into 2 equilateral right-angled pieces by a diagonal cut-mark with a special knife-making device; knives thus obtained have two cutting edges 3/8 inches wide that meet at 45°.

Important Knife Angle and Relationship between Knife and Block

- **Rake angle** is an angle between the upper facet of knife and an imaginary horizontal line perpendicular to the block surface at the level of cutting edge of the block
  - High cutting angle–suitable for softer specimens
  - Reduced angle–suitable for harder tissue
- **Clearance angle** is an angle between a line drawn along the block surface and the lower bevel of the microtome knife. It is usually set between 3° and 8° to suit the knife for obtaining better ribbons.
- **Wedge angle** is an angle between the upper and lower bevel of a microtome knife (normal 15°).

Care and Maintenance of Microtome and Steel Knife

- When not in use—always keep the microtome with a cover provided by the supplier.
- After paraffin-sectioning, remove all wax adhered to the cutting edge of the knife used. If celloidin and/or frozen sections have been cut all moving parts should be oiled with any light machine oil
- Always use xylene for cleaning the metallic parts.
- Keep the knife-holder in the wooden case only after complete drying.
- Set the micron-scale very carefully (affixing the correct markings) to prevent damage to the 'teeth'. A careless setting will provide sections of unequal thickness if any damage has occurred.
- During microtomy, use the lock to avoid any unwanted movement whenever at pause.
- Never hesitate to hold the microtome knife without its handle.
- Avoid any thing coming in touch with the knife's edge or else nicks might result due to burred edge.

9

Microtomy: Technique of Sectioning

MICROTOMY—THE ART OF SECTIONING TISSUE BLOCKS

Microtomy is a method for the preparation of thin sections (thickness between 50 nm and 100 mm) mounted onto a delicate instrument called a **microtome**. The working of microtome is based on *two* basic principles. *First*, that the object to be sectioned is elevated by the shifting of the block holder up an inclined plane. The *second* principle is to elevate the object in a vertical plane by a micrometer screw. An extremely sharp knife is used to cut sections of the tissue embedded in the wax block. Since the sections studied in normal histology laboratories represent profiles through the body organs cut in different anatomical planes, a student of histology should familiarize himself with outlines obtained from cutting domestic items (Fig. 9.1) for a better orientation. To prepare biological tissue for observation under a microscope, the tissue is usually cut in thin slices.

Most biological tissue is too soft to cut; the knife would push into it and compress it, even if the cutting edge was very sharp. Therefore, the tissue is either frozen and sectioned in a cryostat or embedded in a hardening material like paraffin or resin, or cut while still soft with a vibrating blade microtome. The correct knife angle is the subject of much misunderstanding, misleading experience, and incorrect information passed between microtomists, but in fact can be logically derived.

Remember following points for easy and good sectioning:

1. Selection of appropriate microtome
2. Well-prepared (sharpened) microtome knife
3. Perfect setting of the material and equipment
   a. Trimming of the block
   b. Relation of the knife to the properly trimmed specimen block
   c. Tilt of inclination of the knife

Contd...

103

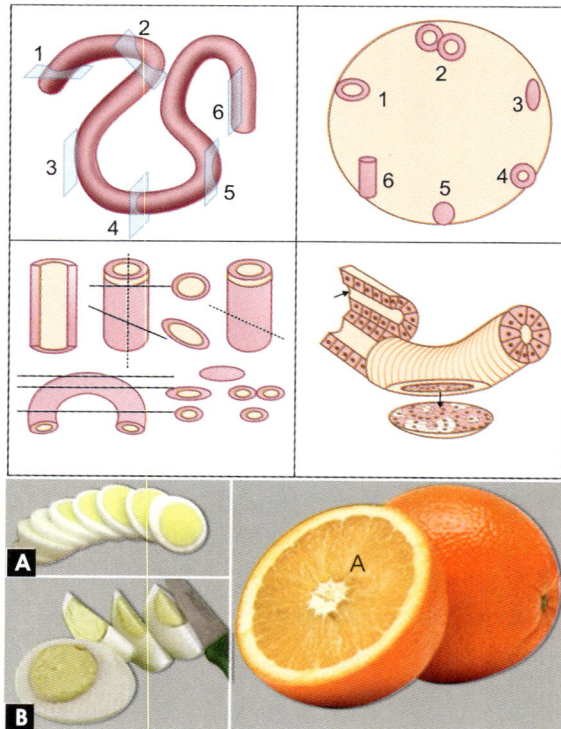

Fig. 9.1: Schematic appearances of sections cut in different planes from tubular structures in the body (at various sites numbered 1 to 6) in a 3-D model of an imaginary tubule, lined by low columnar epithelium cut in different planes to reveal the various shapes that may be seen in a microscopic section. Slices are shown from familiar objects like boiled egg and orange cut in transverse (A) and oblique (B) for better orientation and comparison with the sections likely to be encountered in given histological slides.

Contd...

    d. Clearance angle of the knife – it governs the angle of slant

4. Proper processed specimen
5. The skill of the technologist
6. The speed (rate) of section-cutting
7. The temperature and humidity of the atmosphere around the microtome in use.

Selection of Appropriate Microtome

The method used to actually cut sections from the hardened block of tissue depends on the type of microscopy that will be used to observe it and hence the thickness of sample required. Generally, a rotary microtome (Fig. 9.2) is of choice for paraffin-embedded specimens

Rotary Microtome

Mechanism of rotary microtome: In a rotary microtome (Fig. 9.2) the hand wheel rotates through 360° moving the specimen *vertically* past the cutting surface and *returning it to the starting position*. Block holder is mounted on a steel carriage which moves up and down in grooves and is *advanced* by micrometer screw-cutting perfectly flat sections. These sections are cut one after another to form a ribbon, which is floated on warm water to soften and flatten tissue sections. These sections are then placed on microscopic slides and stored for future procedures.

Advantages of rotary microtome: A rotary microtome has the following merits
1. Ability to cut thin 2–3 μm sections.

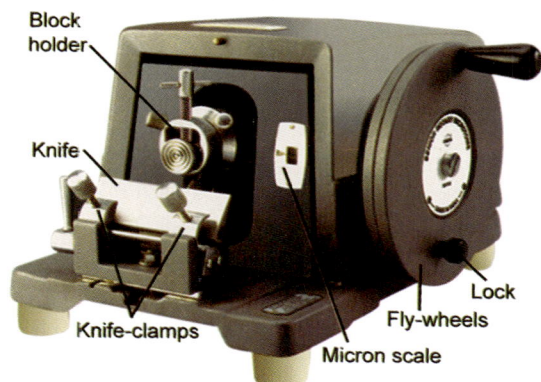

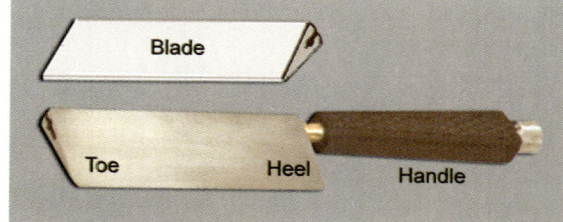

Fig. 9.2: The rotary microtome and steel knife make choices for routine microtomy (sectioning) of paraffin blocks of biological specimens.

2. Easy adaptation to sectioning of all types of tissues (hard, fragile, or fatty).
3. Ideal for cutting serial sections: Large number of sections from each block (may be large).
4. Cutting angle of knife (Chapter 8) is adjustable.
5. Large and heavier knife causes fewer vibrations while cutting hard tissue.
6. Microtome is heavier and more stable.

Well-prepared (Sharpened) Microtome Knife

In the case of samples to be studied using light microscopy, a **steel knife** (Fig. 9.2) mounted in a microtome may be used to cut 10 μm tissue sections which are then mounted on a glass microscope slide. In the case of samples to be studied using transmission electron microscopy, a **diamond knife** mounted in an ultra-microtome may be used to cut 50 nm tissue sections which are then mounted on a 3 millimeter diameter copper grid. The microtome knives can be sharpened by two methods: *Manual* and *automatic*. Also, there are two steps for sharpening: *Honing* (against a special stone called hone), and *stropping*. Honing refers to grinding the cutting edge of the knife on a hard abrasive surface to sharpen the knife.

Manual Knife Sharpening

These consist of a vertical support of stainless steel mounted on a fiber bar and two rods of stainless steel, with wooden handle, one rod threaded for insertion either in the back of knife, or in the back of knife-clamp for knives not provided with hole in back. The knife (or knife-clamp) is attached to the rod, which is placed in the vertical support at an angle determined by insertion in one of a series of eight holes. The support is moved by hand, in a diagonal direction, on the polished glass base plate. After each stroke, the knife is lifted and reversed.

Removal of gross nicks in the cutting edge: Gross nicks in the microtome knife can be removed on a fine carborundum stone or on a glass plate with emery powder on it. The edge of the knife is moved forward and the blade is drawn towards the operator in a diagonal stroke.

Honing

The purpose of honing is to remove fine nicks in the cutting edge of the microtome knife. It can be done either on a fine water-hone (hone = a stone for sharpening), or on a glass plate with diamantine powder. The purpose of using any liquid on a hone is about 10 percent for lubrication and 90 percent to remove the cuttings of metal as they accumulate. Using a viscous liquid such as oil on the hone tends to form 'metal-cuttings' into an abrasive paste capable of additional injury to the knife edge. It is, therefore, advisable to hone a microtome knife in a running stream of water. In hard honing, the hone should be kept lubricated with a fine grade of oil. The motion in the process of honing (Fig. 9.3) must always be from heel (H) to toe (T). The edge of the knife is then turned on its back with the edge of the knife upwards, and the blade is drawn towards the operator in a diagonal stroke. It is always preferable to turn the knife on its back without lifting it from the hone.

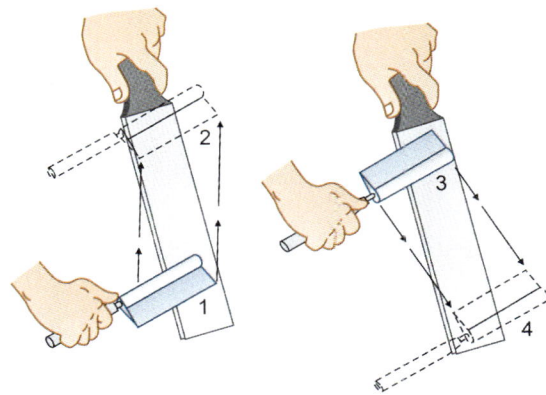

Fig. 9.3: Direction of strokes (1 to 4) recommended for honing of a microtome knife.

Types of Hone

1. Belgian black vein is a stone for fast coarse grinding and finishing
2. Arkansas used to finish a microtome knife after coarse honing on a coarse hone such as carborundum
3. Aloxite like Belgian black, aloxite is also fast and coarse, but is not suitable for finish.
4. Carborundum is only for coarse work, i.e. for removing large nicks in a badly-damaged microtome knife
5. Glass plate used as all types hone by applying an abrasive powder or paste (for example, aluminium oxide).

Stropping

It is a process of polishing an already fairly sharp edge. It is done to sharpen a knife which is free of nicks. The procedure is generally done either on a flexible (hanging) leather strop or a rigid strop mounted on a wooden board of about $12 \times 2 \times 2$ inches, or a glass plate with rough powder. Before use and regularly (usually once a year), strops must be oiled with some vegetable oil and dressed with fine carborundum powder.

Technique

The knife is laid on the near end of the strop with cutting edge of the knife towards the operator. Movement in stropping is exact reverse to that of honing (Fig. 9.4). The knife is held with forefinger and thumb. It is not the cutting edge but the back of the knife that is drawn forward towards the operator with the edge following. It is, however, important to note that over-stropping spoils the best edge of a knife. About 20 to 30 double strokes are usually adequate. Leather strops mounted on solid blocks give the best results; a loose strop should never be used since it will only round a well-sharpened knife. Before use and regularly (once in a year), strops must be oiled with vegetable oil and dressed with fine carborundum powder. There are two kinds of strops: (i) **Flexible**—which normally hangs

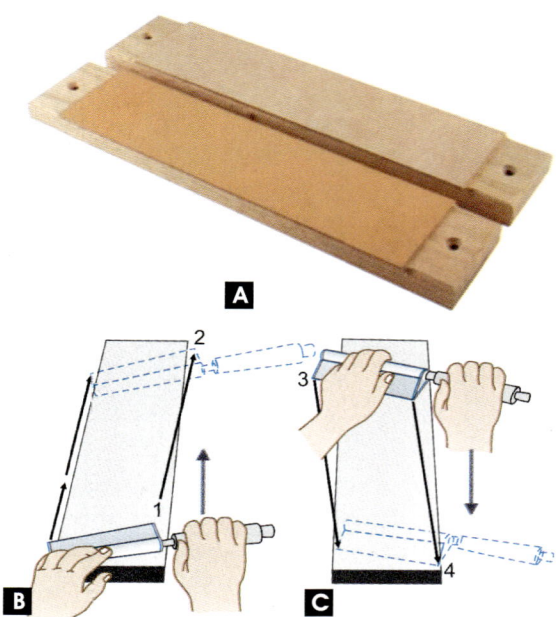

Fig. 9.4: Movements during stropping of a honed microtome knife are shown. (A) A leather strop; pasted on a plane wood base. (B and C) are the direction of strokes in stropping. Notice that the reversal of numbers 1–4 of Fig. 9.3.

with a fixed strong support, (ii) **Rigid type**—a single leather strop stretched over a wooden frame about $12 \times 2 \times 2$ inches.

Automatic Knife Sharpening

Automatic microtome knife sharpeners are of *two* basic types: (1) Hand models, and (2) motor-driven

1. *Hand model sharpeners:* These consist of a vertical support of stainless steel mounted on a fiber bar and two rods of stainless steel (Fig. 9.5), with wooden handle, one rod threaded for insertion either in the back of knife, or in the back of knife-clamp for knives not provided with hole in back. The knife (or knife-clamp) is attached to the rod, which is placed in the vertical support at an angle determined by insertion in one of a series of eight holes. The support is moved by hand, in a diagonal direction, on the polished glass base plate. After each stroke, the knife is lifted and reversed.

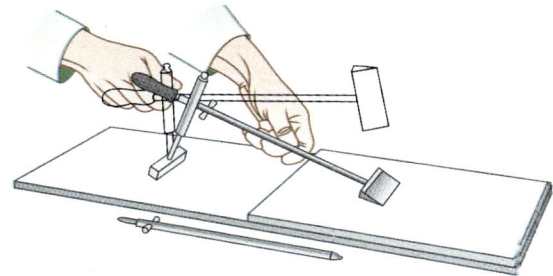

Fig. 9.5: A hand model of microtome knife sharpener.

2. *Motor-driven model sharpeners:* These instruments are provided with automatic knife lifting and reversing devices by means of a cam shaft gears, levers and some other finer adjustments (Fig. 9.6). These sharpeners give a more nearly perfect cutting edge than the most skillful hand honing and stropping, and permits to precise, reproducible setting on the angle of the bevel on the cutting edge as per the individual requirements. A glass disc is rotated mechanically at a constant speed (about 27 rpm). There is provided adjustable mechanism of holding, inclining, counter balancing and swinging a microtome knife to and fro in an arc through the center of the surface of the rotating disc. A sector-type metal scale, graduated from 0 to 25° in 1° divisions, indicates the exact angle between the bevel and the center line of the knife. Generally, a 15 inclination is satisfactory for routine purposes. The automatic lifting and reversing device holds the knife in contact with the rotating glass disc for approximately 27 seconds, during which period, it makes 4 to 5 horizontal reciprocating cycles. It is then lifted and turned over in approximately 8 seconds, and the process repeated for other side of the microtome knife. Despite high cost of automatic sharpeners, these are popular because less time consuming.

Mechanism

The knife is held in the feeding mechanism and is sharpened by revolving cast iron wheels with both edges sharpened alternating. A lubricant is used in the process. Coarse lapping compound consists of alumina, suspension fluid and water is used first, followed by a compound containing a finer grade of alumina.

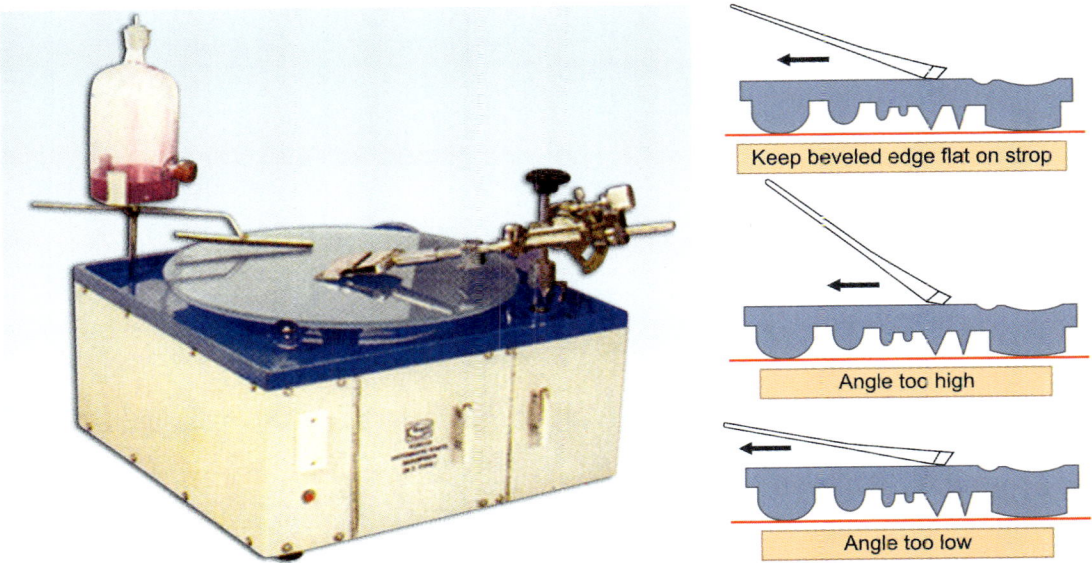

Keep beveled edge flat on strop

Angle too high

Angle too low

Fig. 9.6: An automatic knife sharpener. Notice the different angles which may be suited for stropping of microtome knife. Stroke direction is represented by an arrow in each diagram.

Lastly the suspension fluid is used alone to polish the edge. The time taken to traverse the whole of the cutting edge of knife should be about 25 seconds. In routine, 30 strokes in each direction should suffice with each grade of lapping compound.

STEPS BEFORE START OF MICROTOMY

Prior to microtomy, one has to ensure that the stored block is taken from the refrigerator, and kept at sufficiently cool temperature to get ready for the next step—**trimming** followed by **sectioning** after setting of the correct clearance angle.

Trimming of the Paraffin Block

Proper steps for trimming of a block (Fig. 9.7) must be followed for 'good' paraffin ribbons.

The following **steps** are useful in trimming of a paraffin block of tissue:

• No wax should be left on the sides of the speci-men. Usually, wax is removed with a sharp knife until ⅛ inch (3 to 4 mm) remains on all sides of tissue.

• Remember that only little flakes of wax are to be trimmed at a time. Attempts to trim large pieces can lead to splitting and exposure of the tissue

• Leave very little wax at the top and bottom

• Upper and lower surfaces of the block must be parallel

• Trim the upper or cutting surface of the block parallel to the knife edge until the surface of the tissue gets exposed

• Storing trimmed paraffin blocks of embedded tissue in water for a week or more before cutting of sections makes sectioning easier and prevents brittleness and cracking of the tissue.

Setting of Correct Clearance Angle of Knife

What is the Correct Knife Angle?

As a general principle, excessive bending of the section at the cut line is never advantageous in histology. Errors of too shallow an angle are more damaging to the specimen than errors of too steep an angle (Fig 9.8). Hence, the

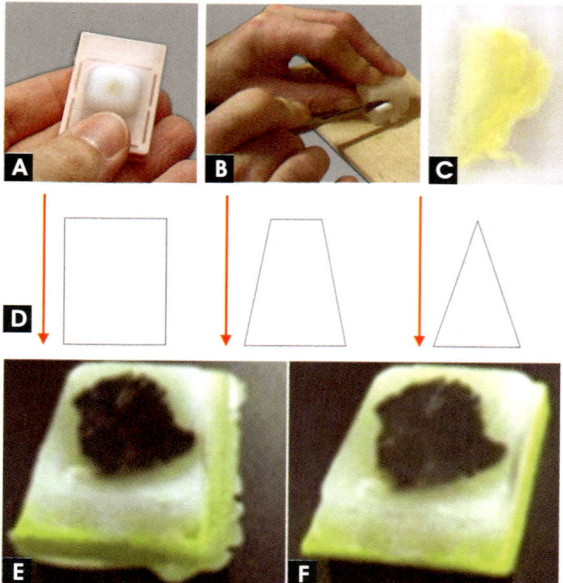

Fig. 9.7: Steps in preparation of material ready for microtomy: (A) Paraffin embedded tissue block; cooled with ice, (B) block is trimmed square with a sharp knife, and (C) front face of the trimmed block with only minimal paraffin wax around the specimen. The block is now ready for mounting before starting microtomy. (D) three common block front surfaces after trimming: The red arrows represent cutting direction and the cutting edge in each block. E and F are photos of paraffin block respectively before and after trimming.

conclusion is that the **correct knife angle should always be set** in the following manner:

• Position the lower bevel face of the micro-tome knife parallel to the specimen block and the plane of motion.

• Then raise the angle slightly (ideally ½ degree if can be managed) above the bevel to avoid having the lower bevel face slide.

Steps and Precautions in Actual Microtomy

Sectioning an embedded tissue sample is the step necessary to produce sufficiently thin slices of sample that the detail of the micro-structure of the cells/tissue can be clearly observed using microscopy techniques (either light microscopy or electron microscopy).

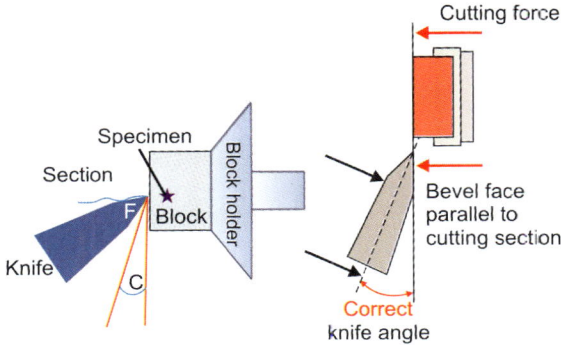

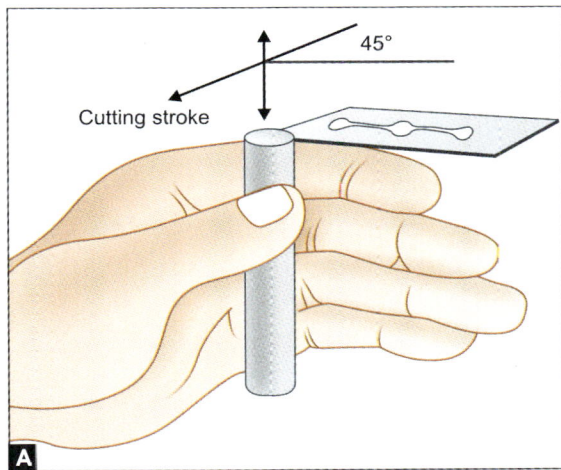

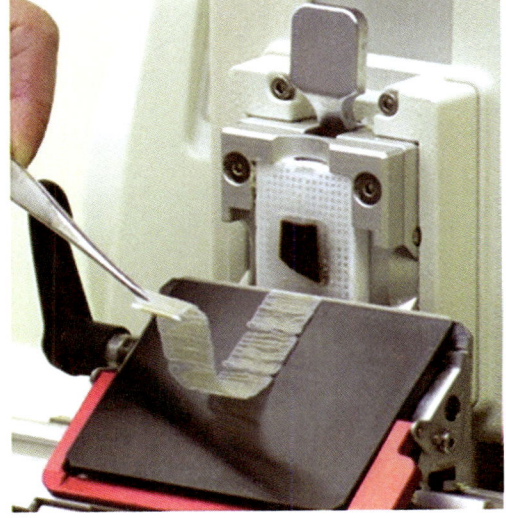

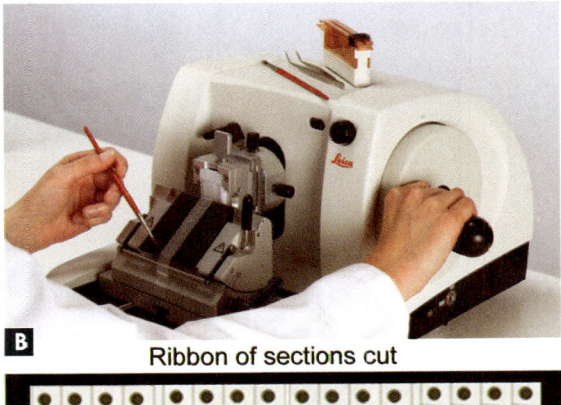

Ribbon of sections cut

Fig. 9.8: Blade clearance angle is adjustable and is set optimally for optimum performance in section cutting. On extreme right an extremely 'good' ribbon due to a correct setting of the clearance angle of a sharp knife is shown.

Fig. 9.9: Two basic procedures for cutting sections: (A) Manual or free hand cutting, with a sharp razor blade and (B) mechanical with the help of a microtome. In the upper picture the cutting stroke is approximately 45° with the vertical plane.

Paraffin sections are cut using extremely fine steel blades at a thickness of 3–5 μm ensuring that only a single layer of cells makes up the section (a red blood cell has a diameter of about 7 mm).

The sections may be of two kinds (Fig. 9.9):

i. **Free hand sections** can be done if the material is hard, 10 mm sections can be taken, or sectioning with razor is done; and

ii. **Serial sections**: For serial sections, paraffin infiltrated materials are affixed on wooden blocks and specimen is cut into a series of sections which are placed on adhesive smeared glass slides. Serial sections enable the reconstruction of structure of organ.

One of the advantages of paraffin wax as an embedding agent is that as sections are cut they will stick together edge-to-edge, forming a "ribbon" of sections. This makes handling easier. Sections are now "floated out" on the surface of warm water in a flotation bath to flatten them and then picked up onto microscope slides. After thorough drying they are ready for staining. Sectioning allows light pass through the material.

Technique of Microtomy [Section-cutting]

Following steps are recommended:

- Insert the microtome knife in the knife-holder and screw tightly.
- Fix the trimmed block in the block-holder and secure it properly
- Set an optimum knife angle.
- Adjust the feed mechanism until the wax block is almost touching the knife.
- All screws should be tight to avoid faulty sectioning
- Use a sharp knife for sectioning.
- Reset the thickness gauze to required thickness [4 to 5 mm sections are routinely cut]
- Performing a gentle, smooth regulated movement of fly-wheel with also the steady movement of fly-wheel of the rotary microtome to obtain a sufficiently long ribbon of section (Fig. 9.11) of desired thickness is both an art and skill of an operating histology expert.

Limitations for Cutting Very Thin Sections

In majority of rotary microtomes (even of good makes) two factors limit the minimum thickness of sections: These *factors* are:

i. The *first factor*—**degree of perfection of the cutting edge** of the microtome knife—is a difficult problem, but may be achieved satisfactory by superfine polishing of a steel edge, and the use of glass knives (which are used routinely for cutting ultra-thin sections for electron microscopy).

The glass knives may, however, be used by making suitable clamps for adapting them to use in a regular rotary microtome. Sections between 100Å and 1000Å, depending upon the final magnification required, may be cut. The glass knife has an advantage over the steel knife that it is amorphous, the edge showing no granularity. The disadvantage of a glass knife is that the knife edge produced cannot be altered. The diamond knives are also

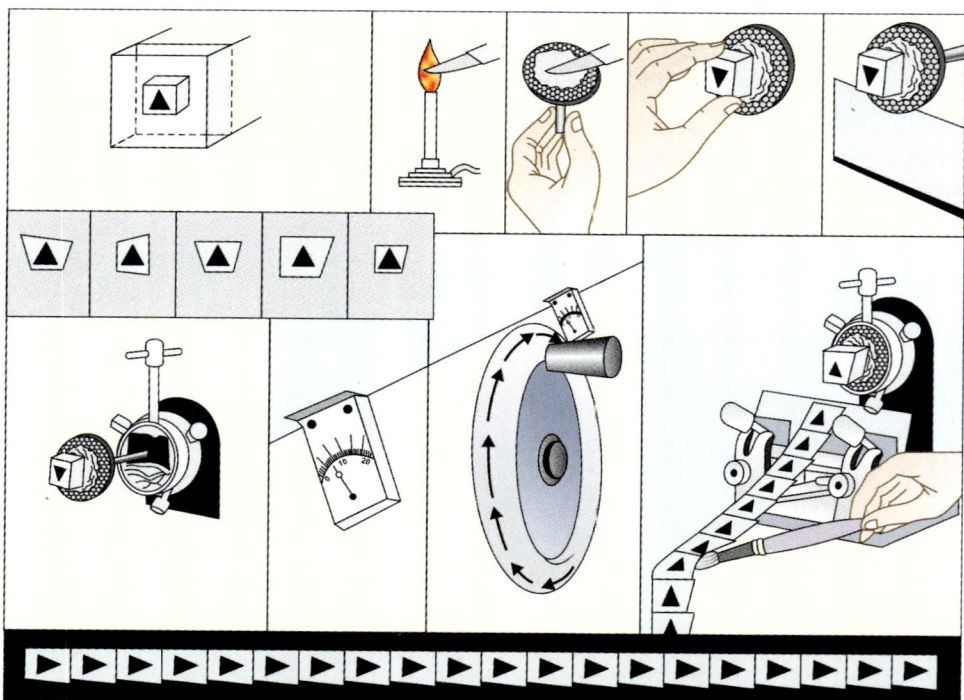

Fig. 9.10: Steps in section cutting of a paraffin block are shown in sequential manner (*see text*).

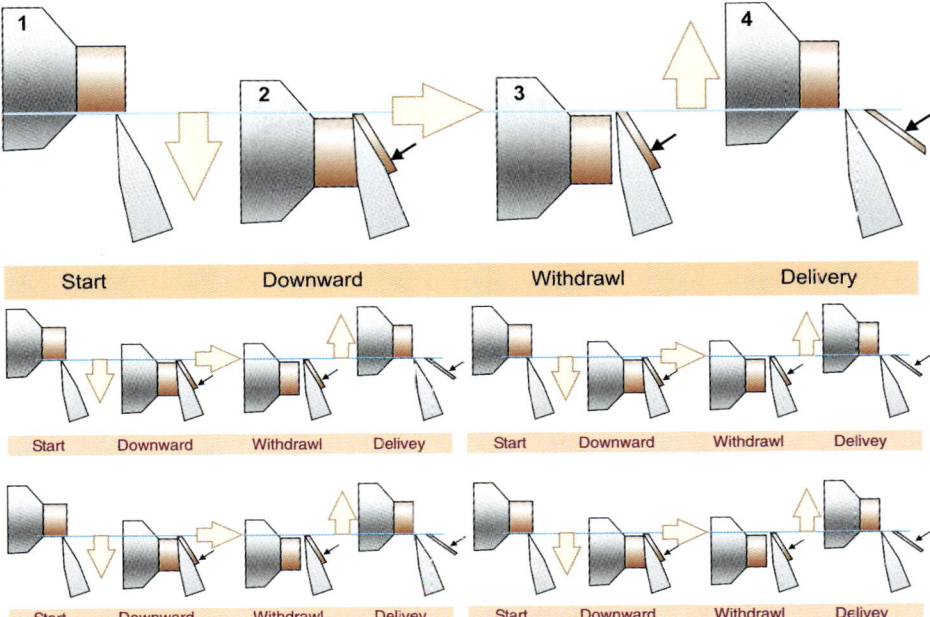

Fig. 9.11: Principle of section 'delivery' from a paraffin embedded tissue block and production of a long ribbon each section containing a thin section of desired thickness during microtomy with a rotary microtome is both an art and skill of a operating histology expert. Smooth regulated movement of the microtome fly-wheel is the key information of a 'good ribbon'.

available for obtaining ultrathin sections, but are quite expensive. A detailed description and use of glass and diamond knives is beyond the scope of the present textbook.

ii. The *second factor*—**precision for advancing movement of the tissue block**–may be achieved by a supplementary attachment, for the rotary microtome, with a slope of only one-tenth of that of a regular inclined plane. This provides a section in one-tenth of a micron for setting of micron-scale thickness control. Thus, a 0.5 mm thin section is obtained by setting the micro-scale adjusted to the 5 mm reading.

RIBBONS OF SECTIONS OBTAINED AFTER MICROTOMY

Keeping Ribbons on Black Paper Sheet

The following precautions should be noted to keep the ribbons on paper.

1. The first ribbon should be kept nearest to the operator.
2. A soft, rough-surfaced non-glossy paper (preferably black and not white) must be used for this purpose to avoid adhesion of sections to black paper surface that shows up the sections better.
3. Never cut more ribbons than can be mounted that day, because they may collect dust and are likely to stick to the paper over which they were kept.
4. If a block is ribboning satisfactorily, it is advisable to keep on cutting as long as possible. A long ribbon may then be divided into lengths to fit the size of the paper sheet (Fig. 9.12).
5. Always leave a few sections in ribbon form on the edge of the microtome knife. These will help to promote good ribboning when sectioning is re-started.
6. Use a curved scalpel blade for cutting ribbons into shorter lengths. The curved

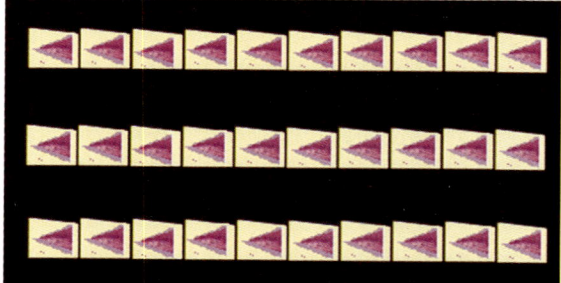

Fig. 9.12: Three ribbons of sections placed parallel to each other on a sheet of black paper for clear visibility.

blade makes a clearer cut and there is less likelihood of picking up the ribbon when removing the knife.

Affixing Wax-impregnated Sections on Uncoated Glass Slides

After a ribbon of paraffin sections has been obtained, the next step would be to attach section(s) to a glass slide. Some workers are of the opinion that the wax-impregnated tissue of a section will always make a little contact directly with the microscope slide when the slide is dried. According to these workers this contact alone is sufficient to make the section stick to the slide, and that no adhesive is necessary to retain the section on the slide, provided the slide is grease-free. The use of adhesive (usually protein) is considered detrimental to the subsequent staining step. When no adhesives are used, the following steps may be tried:

1. Floating on to the grease-free slides sections or ribbons, using distilled water.
2. Flattening the sections.
3. Arranging the sections after partly draining the slide.
4. Completely draining the slide.
5. Drying the slide overnight at a temperature about 8 to 10° C below the melting point of the paraffin wax used.

However, this method of adhesion of sections without adhesives cannot be guaranteed always to give satisfactory results. If 10 per cent formalin substitutes the distilled water when sections are flattened, better results are obtained. This is so because formalin hardens the proteins of the sections making them less liable to detach from the slides during staining.

AFFIXING OF SECTIONS ON ADHESIVE-COATED GLASS SLIDES

Selection of Adhesive

The adhesive used for the purpose must be sufficiently strong to retain the section(s) attached to slide. It must not dissolve off the slide during subsequent steps. Mayer's egg albumen, used either alone or with glycerin, is most commonly used adhesive in majority of histology laboratories. The human saliva is undoubtedly cheapest and quite effective as an adhesive for the sections. However, a few other preparations are also in common use and the following formulas may be noted.

1. Mayer's Adhesive Albumen

Fresh hen egg's white	50 ml
Beat and stir thoroughly	
Add glycerine	50 ml

Filter in a damp atmosphere (Fig. 9.13) and add a few crystals of thymol as preservative. The adhesive should be kept in a well-stopper glass bottle in a refrigerator.

If sections are coming off the microscope still, an improvement in adhesion is achieved by smearing the slide with Mayer's albumen and then floating the sections on to the slide using 10 per cent formalin instead of water.

2. Baker's Adhesive Albumen

Sodium chloride (0.9% aqueous)	100 ml
Sodium p-hydroxybenzoate	0.2 g
Dissolve and add	
Fresh egg albumen	100 ml

The mixture should be shaken thoroughly, and then centrifuged until the supernatant fluid is absolutely clear. Decant the

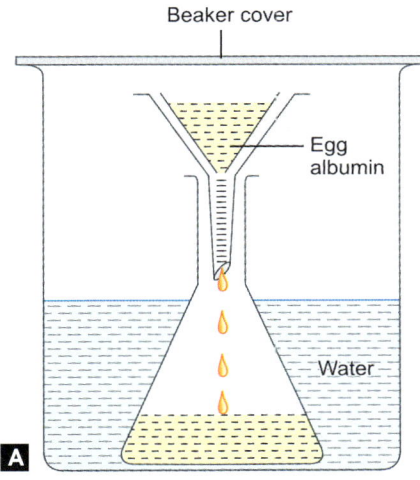

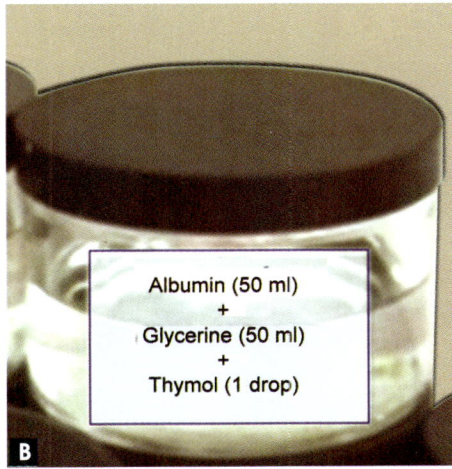

Fig. 9.13: Adhesive for section(s): (A) Method of filtration of most commonly used adhesive—the Mayer's egg albumin in a damp atmosphere; (B) finally the albumin solution is mixed with equal parts of glycerine, in distilled water; then filtered through coarse filter paper and a crystal of thymol is added.

supernatant. This may be used, like Mayer's albumen, as an adhesive.

3. Carbohydrate Adhesives

The carbohydrate adhesives should be used only when carbohydrate staining methods are not to be used subsequently. They have exceptional adhesive power, and when thoroughly dried these are strongly resistant to subsequent procedures.

Starch	3 g
Cold water	30 ml
Mix to a paste and add to:	
Boiling water	60 ml
Then add:	
Hydrochloric acid (conc.)	0.5 ml
Boil for five minutes.	0.1 g thymol as
Cool and add	a preservative.

Attachment of Sections to Glass Slides

1. Smear a dry clean grease-free slide very sparingly and evenly with adhesive (egg albumen or saliva) by applying it to one end of the slide and then spreading it with the edge of the palm lengthwise on the slide. It must be remembered that the adhesive drop, after being spread, should be practically invisible, otherwise it will take-up some of the stains and produce artefacts.

2. Throw the ribbon of section on to the surface of warm water inside a dish (Fig. 9.14). This would remove all wrinkles in the tissue and wax both. However, with very thin sections it is often difficult to get rid of one or two persistent folds. Removal of wrinkles from the tissue section and paraffin wax is called *flattening*.

   Another way of flattening of sections is by heating a mixture of adhesive and distilled water on a clean glass slide (coated with adhesive) on which a ribbon of section is placed. As the sections flatten, the ribbon will stretch (Fig. 9.15).

3. Remove excess water from the slide by either tilting it to one side or by covering the slide with a filter paper and gently pressing it over the sections.

4. Place the slide on a hot plate to finish drying for at least two hours. The temperature of the hot plate is to be maintained a little lower than that for flattening the sections. If possible, the slides may be left overnight in an incubator, supported on a pair of glass tubes or rods.

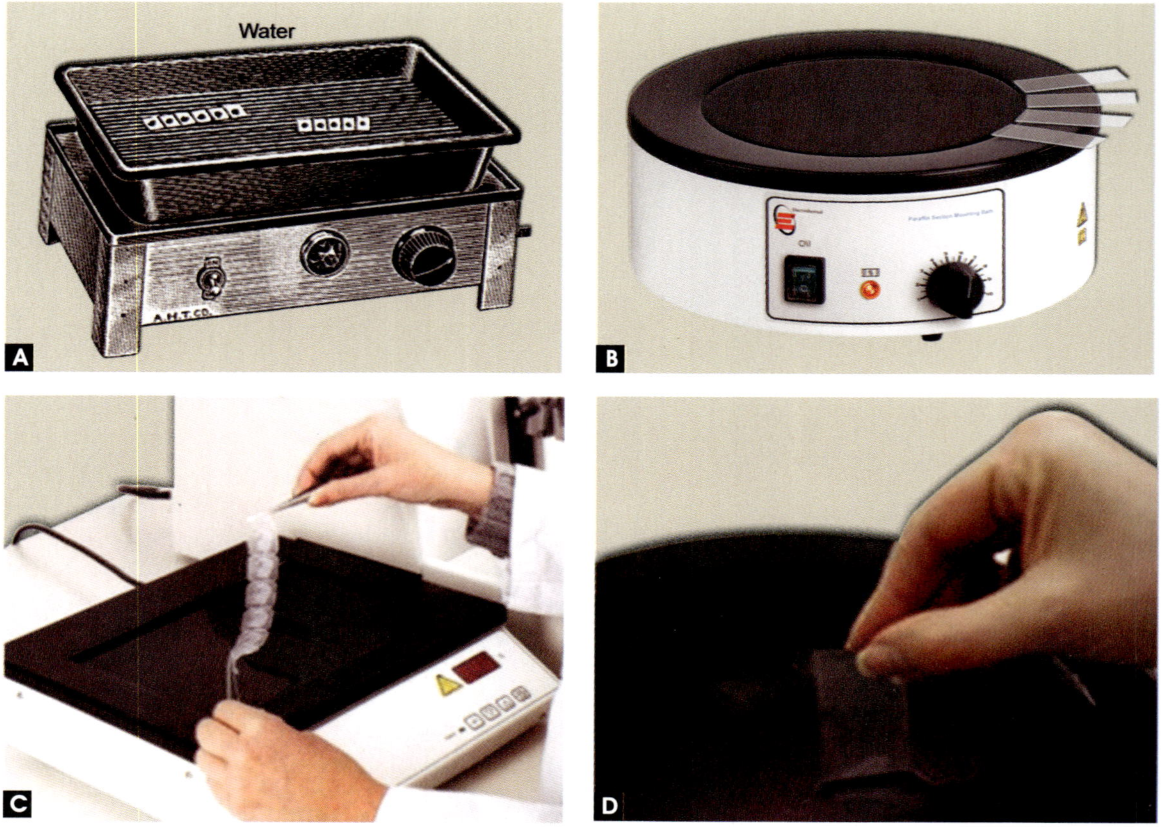

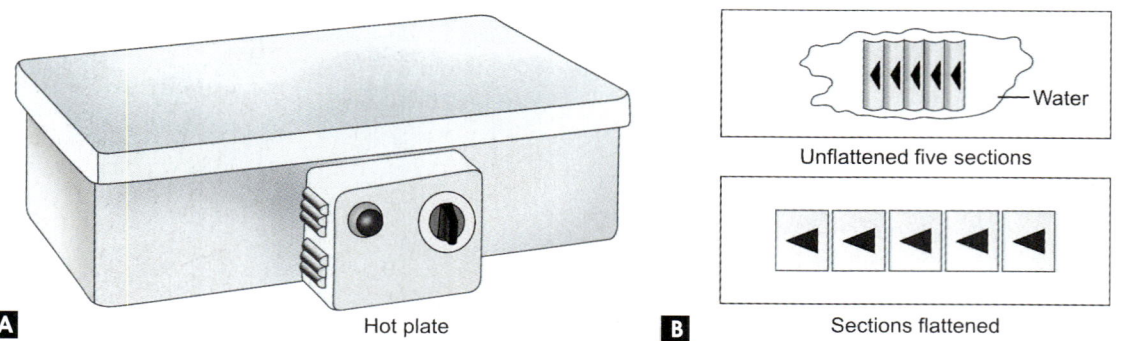

Fig. 9.14: (A) A tissue floatation bath with two ribbons of paraffin sections; (B) tissue floatation bath; (C) a ribbon being picked up gently; and (D) lifting up section(s) on a glass slide smeared with an adhesive.

Flattening of Sections

Fig. 9.15: A paraffin section being mounted on a microscope slide after being floated out on warm water to flatten it. (A) Unflattened section is placed with a little water and kept for a few minutes over a hot plate for flattening. (B) The same ribbon with sections flattened and water dried up by evaporation.

SECTIONING WITH FREEZING MICROTOME AND ULTRATOME

As described earlier in Chapter 8, often sections are required to be cut immediately during the progress of a surgical operation where histological diagnosis is the final deciding factor for the extent of surgery. This facility is achievable with the freezing microtomy (Fig. 9.16).

It is still preferable to use a cryostat (if the facility is available). Sometimes, for early diagnosis of certain pathological an ultratome (using diamond knives) is required for viewing under transmission electron microscope (Fig. 9.17).

COMMON FAULTS IN MICROTOMY: REASONS AND REMEDIES

At times some troubles occur which may be so great that section cutting has to be abandoned for the day. Should the reasons be known the remedies may prove useful. The common faults may be summarised as follows. Reasons and remedies are given in Table 9.1 and numberwise depicted in Fig. 9.18.

 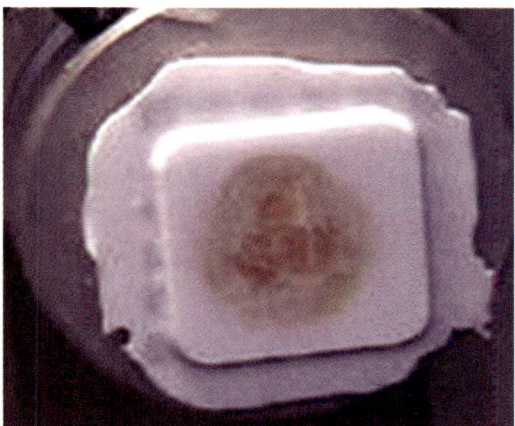

Fig. 9.16: On the left is a freezing microtome clamped to the edge of a bench and connected to a cylinder of CO_2 by means of a specially strengthened flexible metal tube. Knife freezing attachment is supplied with most freezing microtomes. On the right is the cut-face of a frozen section surrounded by an ice-block.

 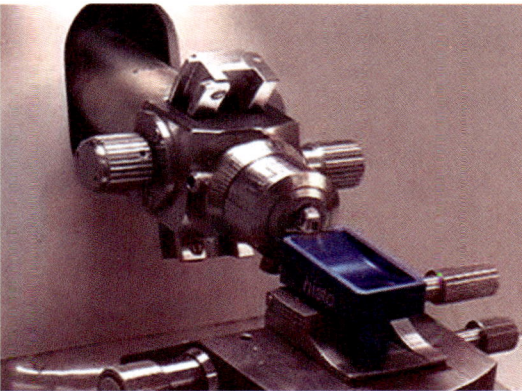

Fig. 9.17: On an ultratome is mounted a set-up for diamond knife for ultrathin ribbons received in a 'water-boat' and seen under a microscope.

Table 9.1	Common faults in microtomy: Reasons and remedies
Reasons	**Remedies**

1. The block does not make contact with the microtome knife even after repeated operations

• Advanced mechanism of the microtome is defective	• Consult the manufacturer

2. The block strikes the knife but sections are not produced

• There is insufficient tilt of the knife. Almost no clearance angle causes the block strike the bevel rather than the knife edge	• Re-adjust the microtome knife
• Insecure clamping of knife	• Tighten the knife in clamp
• Insecure clamping of block	• Tighten the block holder

3. The specimen crumbles and drops out the wax as the block strikes the knife

• Imperfect infiltration does not support specimen	
• Blunt knife	
• Specimen has been overheated or is "cooked"	• Discard the block. Start afresh. In case this is not possible, because of the non-availability of the specimen take the specimen back into xylene–90% alcohol. Leave the specimen in this strength of alcohol to soften. Bring the specimen again into absolute alcohol—molten paraffin. Re-block and try again for microtomy

4. The ribbon curves to the right or left side

• The top and bottom edges of the block are not parallel with the knife edge	• Rotate the block-holder until edges become parallel with the knife edge
• The top and bottom edges of block are not parallel with each other	• Re-trim the block
• The specimen is eccentrically placed in the block	• Re-trim the block
• The knife is sharp and blunt alternately	• Either shift the knife to a better position or re-hone it
• The paraffin-wax of the block is not even because of the traces of clearing agent (xylene) either in specimen or in paraffin	• Re-block or re-infiltrate the specimen

5. Lines and/or grooves appear along the ribbon

• The knife is blunt (not sharp), either before the start of microtomy or by un-intentional touch of the knife by some object such as forceps or needle etc.	• Hone and strop the knife

6. The ribbon of sections splits into two or more separate strips

• Grit in the specimen	• Examine the face of the block either with a powerful magnifying glass or under a very low-power (scanning) objective of microscope
• Catching the hairs of a sable brush between the block and knife edge during lifting ribbon	• Examine the face of the block either with a powerful magnifying glass or under a very low-power (scanning) objective of microscope
• Calcium salts, crystals, or some hard material in the specimen	• Nothing much can be done. However, celloidin method may be tried

Contd...

Table 9.1	Common faults in microtomy: Reasons and remedies (Contd...)
Reasons	**Remedies**

7. Sections and/or paraffin wax collapse and show excessive compression

Reasons	Remedies
• Dirty or very blunt knife	• Clean the knife with xylene or re-hone the knife
• Imperfect inclination of knife (too large or too small)	• Reset the knife angles
• Only sections (not the paraffin wax) collapse	• Re-infiltrate the specimen
• Both sections and the paraffin wax collapse	• Re-block the specimen in a purer and harder wax. In case of non-air-conditioned surrounding place some ice cubes near the block being sectioned

8. Irregularities in the thickness of consecutive sections or thick and thin sections follow each other

Reasons	Remedies
• The wax is too soft	• Re-block of the specimen in a purer and harder wax, or place the knife and block in a refrigerator
• Blunt knife	• Hone the knife
• Knife is loose	• Check the knife as well as other microtome parts
• Block is too large relative to the strength	• Reduce the size of block, if possible. If not, change to a bigger stronger knife
• Too variable speed of sectioning	• Regulate speed of cutting (18–20 strokes/ minute)

9. Holes appear in the sections

Reasons	Remedies
• Air bubble(s) present in the specimen	• Re-block if the holes are large in size
• Hard particles in the specimen and/or wax	• Remove the grit under low power of microscope
• Paraffin wax contains clearing agent's traces	• Re-block the specimen carefully
• Water droplets sucked during cooling of block	• Re-block the specimen carefully
• Some soft patch in the specimen	• Re-infiltrate the specimen

10. Sections show cracks parallel with the knife edge

Reasons	Remedies
• Blunt knife	• Hone the knife
• Paraffin wax too hard for thickness of section	• Warm the block slightly, or cut thinner sections
• Knife tilt is too small	• Increase the tilt angle of the knife
• Knife bevel is too great	• Replace with a knife having smaller bevel angle

11. Corrugations appear across the ribbon

Reasons	Remedies
• Knife is loose in clamps	• Tighten the knife clamps
• Knife-holder is loose on the microtome	• Tighten the block holder
• Block is loose on block-holder	• Remove block and reaffix it again on block holder
• Blunt knife	• Re-hone the knife, or select a sharper area of edge
• Vibrating knife edge	• Replace with a thicker edge knife

12. Failure of ribbon formation—sections lifted off the knife during knife's up-stroke

Reasons	Remedies
• Blunt knife	• Hone the knife
• Incorrect tilt of the knife (may be too horizontal)	• Adjust the knife more perpendicular
• Knife bevel angle may be too great	• Replace the knife having a smaller bevel angle
• Paraffin wax is of too small crystal size	• Reduce the section thickness slightly

Contd...

Table 9.1	Common faults in microtomy: Reasons and remedies (Contd...)
Reasons	Remedies
13. Failure of ribbon formation—sections lifted off the knife during knife's up-stroke	
• Longer stalk of block, hence bends as cut	• Shorten the block stalk/ cut in a cooler room
• Knife is greasy (oily)	• Clean the knife thoroughly
• Block is loose on the block holder	• Refasten the block to its holder
• Paraffin wax is too soft	• Re-block in fresh clean paraffin wax
• Too small clearance angle—block not 'cleared' during up-stroke of knife	• Adjust the clearance angle
• Blunt knife	• Hone the knife
• Static electricity generated at knife edge	• No other method except to wait to restart; if still not proper ribbon formed–abandon the sectioning for next day
• Uneven surface of block due to built-up deposits on the back of knife edge	• Keep cleaning the back of knife intermittently
14. Pin-cushion distortion of sections	
• Specimen collapses because it is softer than the surrounding wax—due to insufficient infiltration and hence the presence of water, alcohol, air, or clearing agent in the specimen	• Re-infiltrate or make a fresh block if more sample is available
15. Barrel distortion of sections	
• Paraffin wax collapses away from the section, but not near the center of the block due to the support of the specimen. This distortion is due to soft wax (high temperature)	• Re-block the specimen using new pure wax of higher melting point. Alternatively cut sections in a cooler room
16. Wrinkles appear in sections even after flattening because only wax flattens (not sections) on slide	
• Incomplete infiltration or the presence still of some clearing agent in the specimen	• Re-infiltrate the specimen
• Too soft paraffin wax	• Re-infiltrate the specimen or partly freeze the specimen block in the refrigerator for 1 hour

The problems associated with producing 'substandard-quality' tissue slides are very dissatisfying for a histology technologist because essential microtomy skills are wasted. Hence, focus on improving the quality and integrity of the tissue slide becomes the burden of a specialist of the subject. Successful microtomy is dependent on tissue fixation, processing, and embedding.

Troubleshooting skills for commonly encountered microtomy problems and artefacts are:

• The most common factors that directly affect the quality of microtomy

• The process of sectioning tissue blocks that focuses on improving the quality and integrity of the tissue slide

• Various cutting protocols for different tissue types and sizes particularly proper slide drying and preparation for deparaffinization

• The various hazards associated with microtomy

• Problems and artefacts associated with the flotation bath, paraffin block, or microtome

Causes and prevention are two very common problems occurring during microtomy (Figs 9.19 and 9.20).

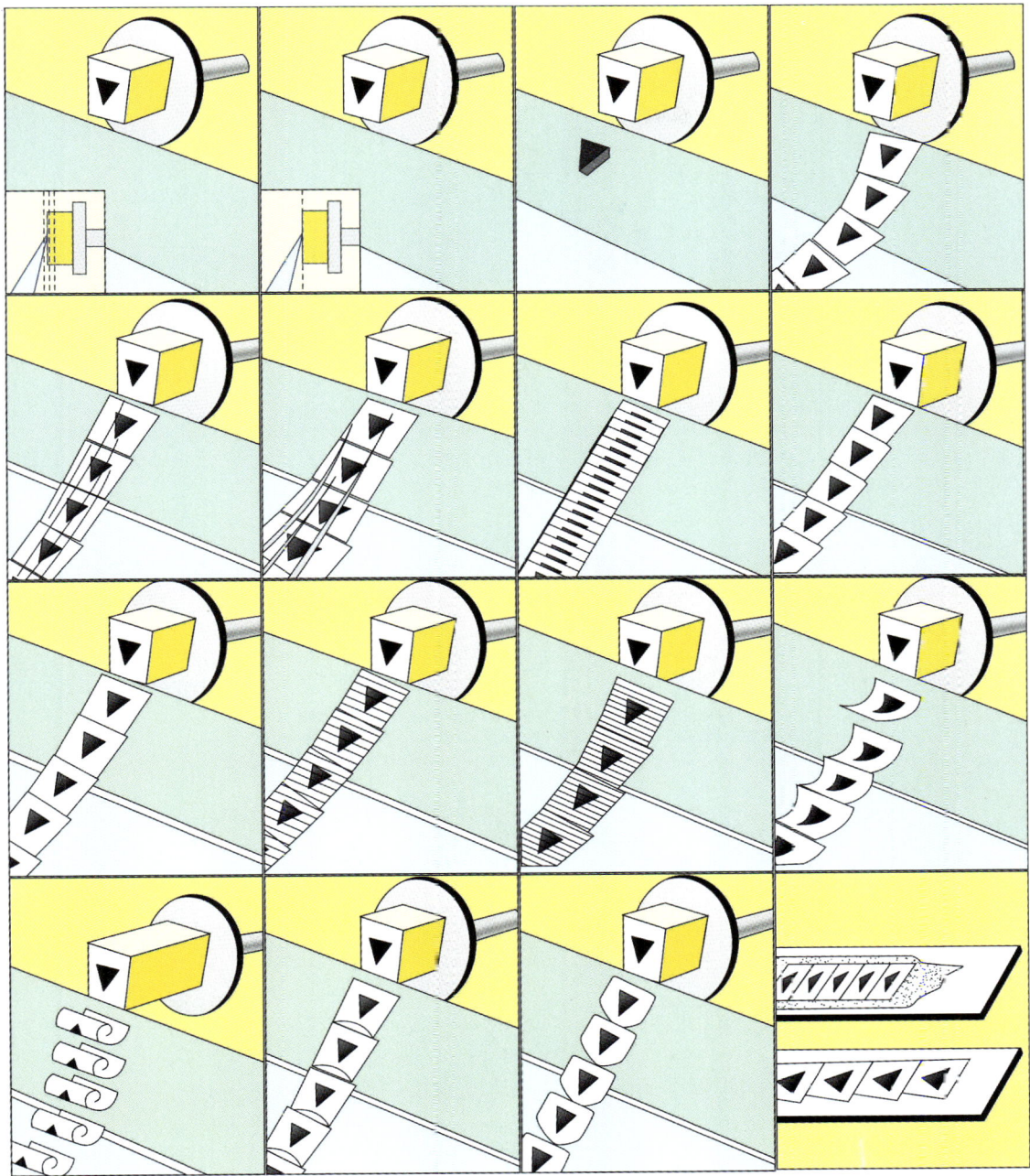

Fig. 9.18: Common problems encountered during microtomy. The number details are explained in Table 9.1, which shows reasons for the common faults and their remedies.

Precautions in Microtomy

• The safety lock, most often located on or near the advancement wheel, should be used whenever the microtomist is NOT actively sectioning paraffin block.

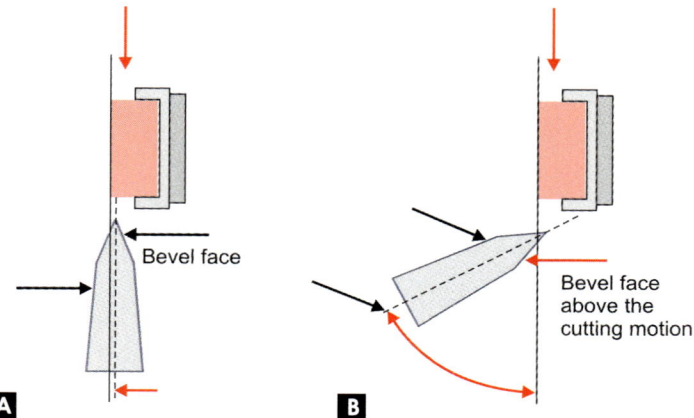

Fig. 9.19: Defects caused due to improper setting of the knife angle (see text): (A) The center line of the knife if parallel to the cutting motion causes pressure to the specimen causing improper sectioning, and (B) too steep a knife angle produces bevel face of the microtome knife above the cutting motion resulting into deformation of the specimen.

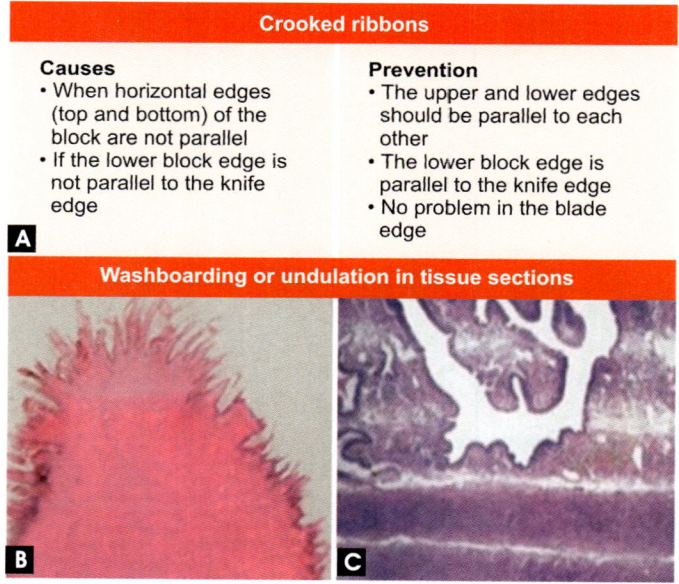

Fig. 9.20: (A) Crooked ribbons are due to faulty microtomy when the top and bottom edges of the block being sectioned are either not parallel to each other or when the knife edge is not parallel to the lower edge of the block; (B) and (C) are both photomicrographs of sections showing wash-boarding (or undulations) introduced into tissue sections is MOST often caused by an unstable work surface or microtome parts that are not properly tightened. These occur in very hard tissue such as uterus or in over-fixed tissue. The problem of undulations is the macroscopic type of charter caused by loose clamping of blade or block.

- The bevel angle can be adjusted on a microtome by moving the knife tilt/ knife holder to an angle between 3 degree and 8 degree.

SUMMARY

Steps Involved in Tissue Processing

1. Sacrifice of animal by perfusion fixation; 2. Dehydration; 3. Clearing; 4. Paraffin infiltration; 5. Wax impregnation; 6. Casting into paraffin block; 7. Storing the wax blocks at low temperature; 8. Attachment of the block into the block holder of the microtome; 9. Microtomy; 10. Affixing paraffin ribbon on glass slides smeared with adhesive; 11. Removal of wax; 12. Staining and mounting; 13. Viewing under microscope.

Paraffin Processing of Tissue

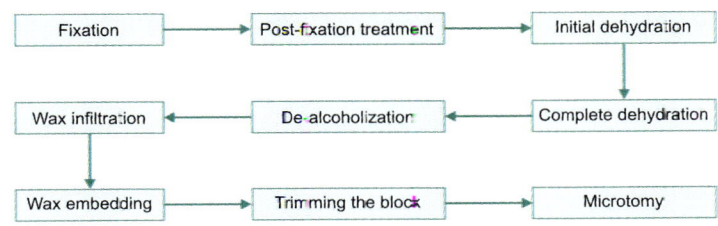

Faults in Paraffin Section Cutting (detected before staining): Their Reasons and Remedies

Fault	Reason	Remedy
Sections scored or cut vertically	a. Knife edge has small nicks	a. Sharpen
	b. Knife is dirty	b. Clean
	c. Calcium salts in tissue	c. Use softening solutions
Sections curl or roll up	a. Knife is blunt	a. Sharpen
	b. Tilt of knife is too great	b. Adjust
Sections crumble on cutting	a. Knife is blunt	a. Sharpen
	b. Wax too soft	b. Cool with ice
	c. Wax crystallized due to slow cooling or contamination with water or clearing agent	c. Repeat processing
Ribbons of section curved	a. Block edges are not parallel to each other	a. Remove block to re-trim
	b. Block edges not parallel to the knife	b. Adjust again
Excessive compression of sections	a. Wax too soft	a. Cool with ice
	b. Blunt knife	b. Use different part or hone

Faults Detected after Staining and Microscope Viewing: Their Reasons and Remedies

Fault	Appearance	Reasons/Remedies
Section **too thick**		• Wrong micrometer setting • Warm breath applied to cold block to facilitate sectioning • First section in ribbon chosen • Sectioning at too great a speed • Poor processing • Microtome needs recalibration
Section shows **too many holes**		• Block roughly trimmed (too quickly?) • Block surface not polished by cutting some thin sections after roughing • Inappropriate section thickness used when trimming • Block brittle (over-processed?) or too cold when trimmed
Section shows vertical striations (**knife lines**)		• Damaged knife or blade used • Poor processing • Block contains hard material such as calcium • Debris in unfiltered wax • Buffer salts precipitated in specimens
Section shows **Disruption**		• Rough handling of specimen during grossing • Poor processing (incomplete dehydration, clearing or infiltration) • Vigorous treatment to dislodge wrinkles during floatation • Floating out for too long or using water that is too hot
Section shows fine cracks or **micro-chatter**		• Tissue over-processed • Block too cold • Cutting too fast • Clumping mechanism not securely locked • Clearance angle needs adjustment

Contd...

Faults Detected after Staining and Microscope Viewing: Their Reasons and Remedies

Fault	Appearance	Reasons/ Remedies
Section shows **coarse chatter**		• Clumping mechanism not securely locked • Very hard or large specimen • Poor processing • Insufficient clearance angle • Sectioning too rapidly • Worn microtome
Section shows **folds**		• Poor flotation technique • Poor fixation and/or processing • Insufficient support • Warm block • Section too thin • Clearance angle too great • Water bath too hot
Section shows **excessive compression**		• Poor processing (insufficient support) • Warm block • Cutting too fast • Dull cutting edge • Clearance angle too great • Poor quality paraffin wax
Bubbles appear under the section		• Bubbles adhering to base and sides of flotation bath • Poor flotation technique trapping bubbles under section
Overexpansion during flotation		• Temperature of bath too high • Section left for too long on water • Poor fixation and/or processing (residual solvent)

Contd...

Faults Detected after Staining and Microscope Viewing: Their Reasons and Remedies

Fault	Appearance	Reasons/ Remedies
Section not flat (poor adherence)		• Poor quality section (wrinkles, bubbles) • Flotation bath too cold • Use of uncoated slide • Section not drained thoroughly after flotation • Insufficient drying time • Drying temperature too low
Dust present		• Dirty slide • Flotation bath not skimmed or contaminated • Slides drained, dried or stored in a dirty environment • Fragments of pencil lead from labeling

An animated diagram shows the handling processes of the histological sections (Fig. 9.21).

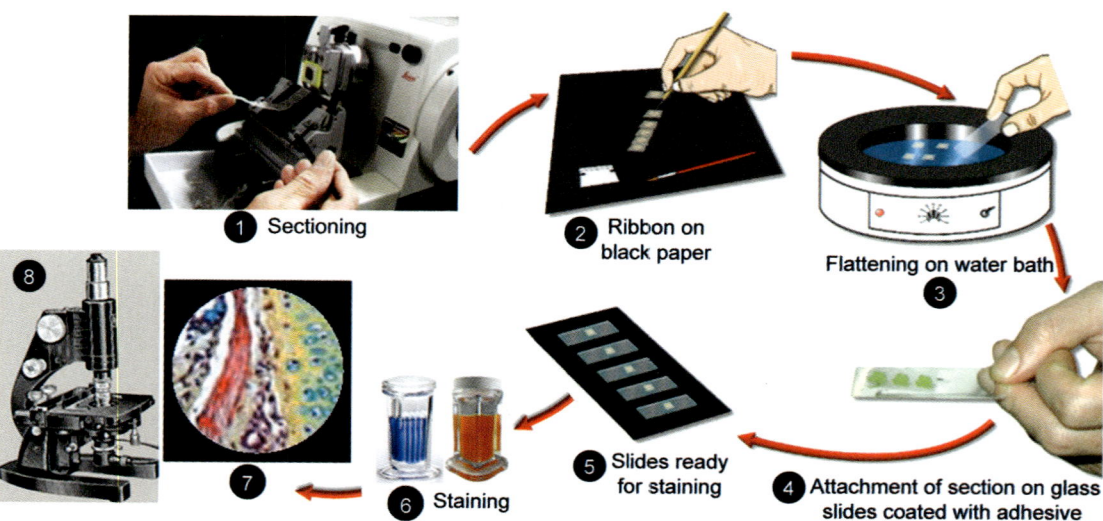

Fig. 9.21: This animated diagram depicts summarized steps in preparing histological sections obtained with a microtome. Sections are stained before being examined in a microscope.

10

Stains and Staining

Staining causes tissue components to change colors when brought into contact with different chemicals. In addition to dyes, antibodies are reacted with tissues to identify specific tumor cell lines with a method called immunohistochemistry. This technique is critical to guiding the patient's physician in selecting the most effective tumor treatment. DNA probes are also applied to tissue sections to identify the presence of bacterial and viral infections and some tumors. When staining is completed, the tissue section is ready for examination under a microscope by a pathologist or other scientific investigator. Without specialized staining techniques, many tissue components would remain invisible.

STAINING AGENTS AND DYES

A *stain* is a substance used to impart color to tissues or cells to facilitate microscopic study and identification. *Staining* is artificial coloration of a substance to facilitate examination of tissue under the microscope. The term staining agents, used in a broad sense, include all substances which are capable to combine with tissue components in such a way so as to improve their visibility by imparting color to them. By staining the parts of a cell or tissue stain differently because of their chemical and/or physical differences, and can thus be more easily distinguished. However, the words *stain* and *staining* no longer have recognizable meanings. It is said that basic fuchsin stains chromatin, silver compounds stain nerve fibers, and sudan black stains lipids. The processes in these three events are fundamentally different from the chemical and physical points of view.

Staining agents include *dyes* and *stains*. A *dye* usually refers to such a staining agent whose chemical formula is known, but it may be used less strictly for a mixture of very closely related compounds whose staining properties are alike. On the other hand, a *stain* generally includes dyes which are metallic salts of animal and vegetable origin. It may also include certain other substances which produce appreciable differences in the color or opacity among various components of organs, tissues, and cells. The stains are usually held by a tissue even when a tissue is processed later by a solvent in which they (stains) were themselves dissolved. *General stains* affect all the components in a tissue sample, but *special (specific) stains* impart color to only some of the components, thus making them prominent and rendering the remaining either colorless or colored with a different intensity. It is for this reason that the simplest and most self-explanatory explanation to cover various processes involved to confer color to tissues (staining) seems to be *coloring agents* or *colorant*; and the corresponding verb is *to color*. Finally, the mounted sections are treated with an appropriate histology stain.

Why is Histology Sample Stained?

Put another way, what is the purpose of histology stains? Biological tissue has very little variation in colors/shades when viewed using either an ordinary light (optical) microscope or an electron microscope. Staining biological tissues is done to both increase the contrast of the tissue and also highlight some specific features of interest—depending on the type of tissue and the stain used.

Classification of Stains

There are many different histology stains. Histology stains are normally selected according to the type of tissue to be observed. Some stains are more widely used than others while some are only used to study very specific types of biological tissue.

The dyes are classified from chemical standpoint into *three* categories:

i. *Basic dyes* when a color 'base' is combined with a non-coloring acid (acetate, chloride, or sulphate radicals). Fuchsin and hematoxylin belong to this category. These dyes may dissolve in water and/or alcohol.

ii. *Acid dyes* when a 'color acid' is combined with a non-coloring metallic base (usually sodium or potassium). Eosin-Y and light green are the examples of this group. The acid dyes may also dissolve in water and/or alcohol.

iii. *Neutral dyes* when compounds of a 'color base' with a 'color acid' are combined. These may dissolve in water, more usually in alcohol, and often give colloidal solutions. The neutral red is an example of neutral dyes.

It is important to keep in mind that in speaking of dyes the term 'acid' or 'basic' refers to the character of the color-acids or color-bases, and not to those of salts. An acid dye may have a neutral or alkaline reaction. A basic dye may similarly have a neutral or acidic reaction.

The dyes are also classified as:

i. *Histological stains* when they serve to define tissues in an organ, and

ii. *Cytological stains* when they serve to define cell components. The cytological dyes may be further subdivided into *cytoplasmic* if they have an affinity to stain the cytoplasm, or *nuclear* when they stain the nucleus of a cell. Sometimes this classification of dyes is referred to as a microtechnique classification.

METHODS AND TYPES OF STAINING

There are different methods of staining. *Routine staining*, done generally with hematoxylin and eosin (H&E), provides only a little differentiation. *Special staining* (e.g. Fuelgen staining) is performed in the selective instances. *Vital staining* is applied to the living tissues; for example, trypan blue and lithium carmine stains are used for the staining of histiocytes within the body of living organism. Alizarin stains the bones red. *Supravital staining* is the term used for the method of staining when the stain is applied to a living tissue which has already been removed before it is stained. The example of supravital staining is the staining of mitochondria by Janus-B green.

In contrast to *nonvital staining* (done for fixed cells), the *vital staining* is preferred for staining the living tissues by perfusing the staining solution in living, anesthetized animals. Either the entire animal is perfused or the blood vascular system of the region to be stained is isolated by tying off communicating arteries and veins and running the staining solution directly into the main artery of supply. The vein draining the organ to be stained is opened at the beginning of the perfusion but clamped near the end. The staining fluid (made by dissolving 0.3 g methylene blue and 9.0 g NaCl in 1000 ml of distilled water) is run into the tissues under pressure raised to 300 to 400 mm of mercury. After the perfusion is completed, the tissue is quickly removed to permit exposure to air. The tissue is then placed in a fluid similar to perfusion fluid (but without the dye). It is covered thinly with fluid and aerated frequently but not allowed to dry. When stain appears to have reached its maximum intensity, the tissue is fixed in a suitable fixative and processed.

The process of staining may be completed by two ways:

i. *Progressive staining*, which depends on the fact that certain dyes, e.g. hematoxylin stain first the nucleus and then the cytoplasm of the cell. In this type of staining, a tissue section is placed in the diluted strength of a particular stain and kept in it deliberately for such length of time that it remains understained. More stain is then added gradually and the staining reaction stopped as soon as the desired intensity of color in the nucleus, or later, in the nucleus and cytoplasm is obtained. The underlying principle in this type of staining is the selective affinity of dyes for different tissue components. The progressive staining is relatively less sharp than the second category of staining.

ii. *Retrogressive* or *regressive staining*, in which a tissue is placed in a stock solution of stain, deliberately over stained and then differentiated or destained partially controlled under a microscope. A *differentiating agent* removes the surplus stain from the section. This type of staining provides a sharper differentiation and is the one routinely used in most of the medical and biological laboratories.

Opinions differ on the comparative merit of these types of staining. While the progressive type of staining is slower, a good rule for the beginner is to use it. A given tissue may be stained either as 'sections', or 'smears', or in 'bulk'. It is always advisable to stain the tissue in sections rather than in bulk.

HEMATOXYLIN AND EOSIN STAINS

Although many stains are available in a histology laboratory, only the two stains, hematoxylin and eosin, used in routine staining of animal tissue sections will be described here.

Hematoxylin

Hematoxylin is a widely used nuclear stain that is extracted from the heartwood of the Central America's logwood tree, *Haematoxylon campechianum*. The freshly cut logwood is colorless, but during atmospheric or chemical oxidation, it becomes dark reddish-brown. This logwood in then processed into hematoxylin, the lifeblood of histology. Hematoxylin has been used since 1520 AD as a textile dye, but it did not become important to histology until three hundred years ago. Robert Hooke reported using this dye as a coloring matter for fluids, but not until 1860 was hematoxylin used in histological staining.

Hematoxylin owes its coloring property not due to itself but to the formation of one of its oxidation product—*hematein* which is a reddish dye, hematoxylin is itself a colorless solid derived from the wood of a tree-*Haematoxylon campechianum*, a leguminous plant of caesalpiniaceae family, indigenous to Mexico but cultivated in the West Indies, having about 2 feet stem circumference. The biological significance of the presence of hematoxylin in the wood is not known. The second 'e' in the word hematein distinguishes this dye from an entirely unrelated colored substance called Haematin, which is the non-protein component of hemoglobin of blood. Still another word haematine is used in the textile industry to designate partly or wholly oxidized hematoxylin. The structural formulas and molecular weights of a hematoxylin and hematein are shown in Fig. 10.1.

Ripening and Mordanting of Hematoxylin

Solutions of hematoxylin must be *ripened*; that is partially oxidized to produce the dye, hematein, before they are ready for use. If solutions of hematoxylin are bottled and left to stand over a period of a few weeks or a few months, there occurs gradual atmospheric oxidation and hematoxylin gets formed into hematein. This process of oxidation is called *natural ripening*. As only hematein can be used for staining purposes, fresh solutions of hematoxylin are not suitable for immediate use. This is due to their being not sufficiently ripened unless they are kept for several months or for several weeks at least. *Artificial ripening* may be done almost instantaneously by adding some *oxidizing agents* (e.g. hydrogen peroxide, chloral hydrate, potassium permanganate, and sodium iodate). Thus the solutions of hematoxylin, containing hematein, are ready for use soon as they are prepared.

However, if large quantities of artificially ripened solutions of hematoxylin are stored for a long period, the negative charge (isoelectric point of about 6.5) possessed by hematein loses its affinity for tissue components and the staining solutions become quite useless. To counteract this, mordanting is an essential step.

A *mordant*, ammonia alum, potash alum and iron alum (alum = aluminium compounds), forms a lake which carries a positive charge and makes the solution a strong basic dye

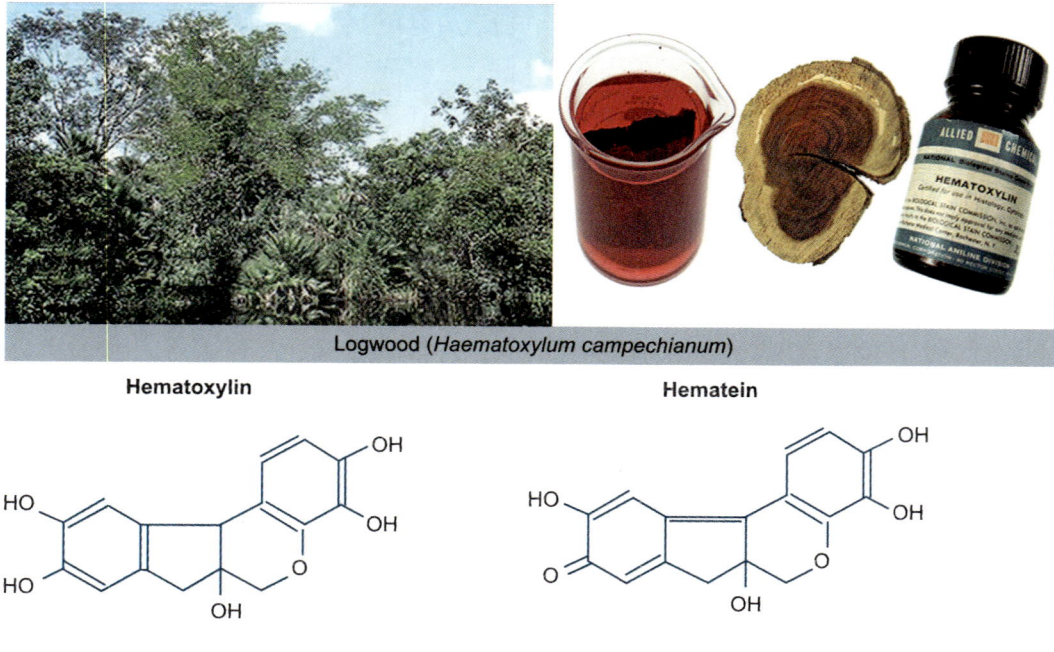

Logwood (*Haematoxylum campechianum*)

Hematoxylin   Hematein

Mol. Wt. 302.272   Mol. Wt. 300.256

Fig. 10.1: Formulas, chemical structures, and molecular weights of hematoxylin and hematein: Note that oxidation (loss of electron) is demonstrated by the loss of hydrogen and its electron from the hematoxylin structure.

suitable for staining. Typically, the water-soluble lakes from aluminium compounds are blue with slight violet color. These are used in progressive staining. On the other hand, the lakes from ferric compounds are blue-black or blue, insoluble in water, and are used in regressive staining.

Different kinds of Hematoxylins

A plain solution of hematoxylin is suitable to stain sections from tissue fixed in fixatives containing bichromate and chromic acid, but it is more useful if a mordant is added to it. This is further improvised by addition of an oxidising agent which converts hematoxylin to hematein. Therefore, depending upon the nature and amount of chemical used as mordant and the use of any oxidizing agent (air or chemical oxidation) gives rise to different kinds of hematoxylin solutions. Broadly speaking there are two main groups of hematoxylin preparations in routine use: (i) *Alum hematoxylins*, and (ii) *Iron hematoxylins*. Whereas the former are better nuclear stains, iron hematoxylins do not stain nuclei well, and are also cumbersome for routine use. The commonly used hematoxylin stains, with mordants and oxidizers are listed in Table 10.1.

The specific formulae of some hematoxylins and directions for their making and use in staining of sections from animal tissue are detailed below.

1. Delafield's Hematoxylin

Solution A

Hematoxylin	1.0 g
Absolute (or 95%) alcohol	6.0 ml

Solution B

Ammonium alum	10.0 g
Distilled water	90.0 ml

Combine solutions A and B and allow the solution to ripen naturally in the atmosphere. After several weeks the mixture appears darkened; indicating completion of ripening. Now add:

Glycerine	25.0 ml
Ethyl (or methyl) alcohol	25.0 ml

Final ripening is checked by testing staining activity every week. Store the stain in a tightly stopper glass bottle after ripening is complete.

2. Ehrlich's acid Hematoxylin

Solution A

Hematoxylin	0.67 g
Absolute (or 96%) alcohol	32.00 ml

When dissolved, add in the following sequence:

Glacial acetic acid	3.00 ml
Glycerine	32.00 ml

Solution B (mordant)

Distilled water	32.00 ml
Potassium alum (crystal)	3.00 g

Table 10.1	Mordants and oxidizers used for different hematoxylin types	
Name of hematoxylin	**Mordant**	**Oxidizer**
Alum: Hematoxylins		
Delafield's	Ammonia alum	Oxygen (of air)
Ehrlich's acid	Ammonia alum	Oxygen (of air)
Harris	Ammonia alum	Mercuric oxide
Mayer	Ammonia alum	Sodium iodate
Iron: Hematoxylins		
Heidenhain's	Ferric sulfate or ferric alum	Oxygen (of air) sometimes oxidizer not required when hematein is used
Weigert	Ferric chloride	Ferric chloride

Alum dissolves faster when water is heated. While the solution B is still warm, slowly add it to the solution A and keep shaking the latter.

Ripening takes place after several weeks until the mixture assumes a deep red color. The solution keeps well for many years and is a progressive stain that has a little tendency to over stain. The stain may be ripened instantaneously if 0.4 gram of sodium iodate is added, as an oxidizer, while preparing the stain.

3. Harris' Hematoxylin

Harris hematoxylin is the most commonly used hematoxylin in regressive methods typically with Eosin as the counter-stain. Harris hematoxylin has an extended shelf life, and exclusively uses BSC certified hematoxylin dye, which produces a vibrant, purple-blue "Nucleus of Excellence." Master Tech's Harris Hematoxylin is available in four convenient sizes: Pint, Liter, Gallon, or a five-gallon cube.

The Harris' hematoxylin is prepared by making two solutions 'A' and 'B' as follows:

Solution A

Hematoxylin	1.0 g
Absolute alcohol	10.0 ml

Solution B

Ammonium alum or	20.0 g
Potassium alum	
Distilled water	200.0 ml

Heat solution B to boiling and add solution A, then add the oxidizer:

Mercuric oxide	0.5 g

Stir the mixture using gentle heat (56°C oven or water-bath) and dissolve the alum in the distilled water using heat (Bunsen) with frequent stirring for 2 or 3 minutes, then cool it quickly by immersing the container in cold water. Wait for overnight, filter and store the stain in a well-stopper bottle. The addition of 8–10 ml of 4 per cent glacial acetic acid improves the keeping quality and improves nuclear staining by inhibiting the staining of cytoplasm.

4. Mayer's acidified Hematoxylin (Lillie's modification)

Distilled water	70.0 ml
Ammonium alum (cryst.)	5.0 g
Hematoxylin	0.5 g
Sodium iodate	0.1 g

Dissolve the above ingredients in the water in the order listed and add the following:

Glycerine	30.0 ml
Glacial acetic acid	2.0 ml

5. Heidenhain's Hematoxylin

Hematoxylin	0.5 g
Alcohol (95%)	10.0 ml
Distilled water	90.0 ml

For use the above solution is mixed with equal parts of distilled water. Staining is done for about 24 hours, and is particularly useful for demonstration of secretory granules.

6. Weignert's Iron Hematoxylin

Solution A

Hematoxylin	1.0 g
Alcohol (95%)	50.0 ml
Glycerine	50.0 ml

Solution B

Ferric ammonium sulfate (violet crystals)	15.0 g
Ferrous sulfate (cryst.)	15.0 g
Distilled water	100.0 ml

Mix equal parts of solutions A and B. Stain for 5 minutes.

Eosin

Eosin was first used in 1875 as "Eosin" for the dyeing of silk and wool.

Eosin is a synthetic anionic dye that combines electrostatically with tissue such as collagen and muscle, the latter in an amphoteric manner belonging to the xanthene group of dyes. Raising the pH of the eosin solution causes more intense staining. The raised pH may be achieved either by the use of a suitable buffer or by dissolving the dye in tap water. It is derived from fluorescein and is available in two main shades.

Eosin B (eosin bluish)

This variety of eosin is more soluble in alcohol. Usually a solution of 0.5 per cent strength is prepared by dissolving the dye in 70 per cent alcohol; and staining time for 2 to 3 minutes is required.

Eosin Y (eosin yellowish)

This is more soluble in water but less so in alcohol. Eosin Y is used most commonly in hematoxylin-eosin (H&E) staining methods. Routinely, a 5 per cent aqueous stock solution (prepared by dissolving the dye in tap-water) is used. A crystal of thymol or a few drops of formalin are added to the solution for preventing the growth of moulds in the staining solution. Eosin Y can also be employed in alcoholic solution. The color acid of eosin Y (precipitated by HCl) is not soluble in water and must be used in 95 per cent alcohol.

Whenever hematoxylin-eosin methods are selected for staining of tissues, as a rule, the hematoxylin is applied first, and after it has been differentiated, the eosin is applied. Hematoxylin usually behaves as the faster of the two dyes, hence the eosin is employed as a counter stain and its intensity can be adjusted without affecting the degree of hematoxylin staining.

Hematoxylin and eosin are used in histology primarily to display structural features

Despite the merits of H&E staining, the procedure does not adequately reveal certain structural components of histological sections, including elastic material, reticular fibers, basement membranes, and lipids. When it is desirable to display these components, other staining procedures, most of them selective, can be used. These procedures include the following:

1. For elastic material—orcein and resorcin-fuchsin staining
2. For reticular fibers and basement membrane—silver impregnation methods

Storage of Stains

Most of the stains keep well at the room temperature, but some require refrigeration for their storage. These stains are Schiff's reagent, aldehyde fuschin, methyl-green, azocarmine, and silver nitrate solutions. During staining, *staining racks* (Fig. 10.2) are available to accommodate a large number of slides with mounted sections. These racks are of particular help when intermittent washing with running tap water is required. When slides are to be passed through the staining solutions only for short periods, the stains are generally kept in

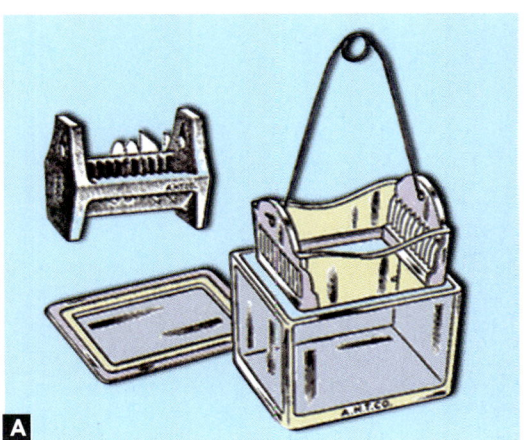
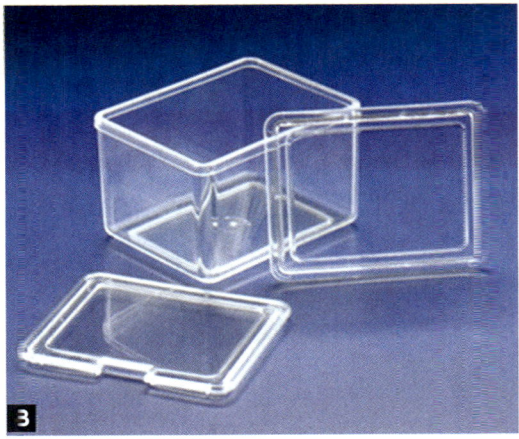

Fig. 10.2: (A) Staining racks, (B) Plastic staining dish for processing up to 20 slides at a time.

Fig. 10.3: Different types of staining jars: (A) Coplin 10 slide staining jar; (B) Columbia staining jar with cover; (C) Tall Coplin jar with plastic screw top; and (D) Vertical staining jar with glass lid.

staining jars or *staining dishes* (Fig. 10.3). Whereas, *Coplin jars* are frequently used and are provided with grooves inside for the glass-slides to be placed vertical, the *Naples jars* are less often used for keeping the staining solutions. For daily use, store the stain in an airtight (prevent moisture entering the stain) amber (semi-opaque) container. Bright light and heat oxidize the stain, especially when in aqueous solution and will cause precipitation of insoluble precipitates, e.g. of methylene violet (Bernthsen). Precipitate formations in a staining solution and/or poor-staining results are the signs which suggest discarding a stain. Evaporation of methanol, absorption of moisture and precipitation of Azure-Eosinate salts are also additional problems during storage that require filtering the stock while aliquotting for daily use.

When not in use, the stock solutions of stains are stored in well-stopper glass bottles. A closable dropper bottle, e.g. TK dropper bottle can be used that should be kept tightly closed when not in use. Keep in a cool place (not refrigerated) and never in direct sunlight. The stock stain should be kept in a tightly stopper light opaque (e.g. amber) container in a cool dark place. Renew every 3 months or earlier if indicated. To obtain optimum color reaction, some suggest that 3–5 days should be allowed before using freshly made stain.

General Procedure for Staining of Sections

Staining

Apart from a few natural pigments such as melanin, the cells and other elements making up most specimens are colorless. In order to reveal structural detail using bright-field microscopy some form of staining is required. The routine stain used universally as a starting point in providing essential structural information, is the hematoxylin and eosin (H&E) stain (Fig. 10.4). With this method cell nuclei are stained blue and cytoplasm and many extracellular components in shades of pink. In histopathology many conditions can be

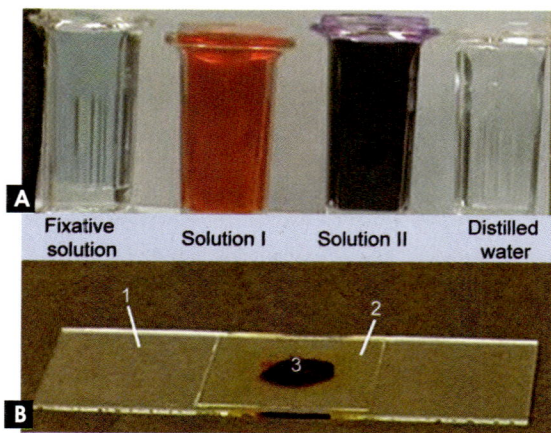

Fig. 10.4: (A) H&E staining jar, (B) Stained section for light microscopy. 1. Microscope slide, 2. Cover slip, 3. Stained tissue section between 1 and 2.

diagnosed by examining an H&E alone. However, sometimes additional information is required to provide a full differential diagnosis and this requires further, more specialized staining techniques. These may be "special stains" using dyes or metallic impregnations to define particular structures or microorganisms, or immuno-histochemical methods (IHC) involving the location of diagnostically useful proteins using labeled antibodies. Molecular methods such as *in situ* hybridization (ISH) may also be required to detect specific DNA or RNA sequences. These methods can all be applied to paraffin sections and in most cases the slides produced are completely stable and can be kept for many years.

After staining, the sections are covered with a glass cover slip and are then sent to a pathologist who will view them under a microscope to make an appropriate diagnosis and prepare a report.

Routine staining in histology laboratories is done almost entirely on paraffin sections. Sections cut with a freezing microtome from either fixed (but unembedded) or fresh tissue samples can also be stained. Whereas, paraffin sections are invariably attached to glass slides before staining, the sections from the fresh material are processed directly. Mounted paraffin sections are deparaffinized, i.e. the wax is removed, before the actual staining is started, since the presence of paraffin is likely to interfere.

When the sections are hydrated (by reaching up to water gradually through the descending strengths of alcohol), the slides are kept in the hematoxylin solution and over staining of tissue is done (regressive staining). After this, the sections are destained until the desired intensity of staining is attained. The degree of over staining with hematoxylin is of a little or no practical consequence because the dye can be removed during the process of destaining called *differentiation*. Before proceeding further, some important points about differentiation must be considered.

Differentiation

The general principal of differentiation is to destain basic dyes by weakly acid medium and acid dyes by a weakly basic one. A section stained with aqueous solution of hematoxylin can be differentiated in acidified alcohol (0.5–1.0% hydrochloric acid in 95% alcohol) to effect sufficiently rapid differentiation. On the other hand, sections stained in eosin (an acid dye) can be differentiated in alcohol containing 0.1–0.5 per cent of concentrated ammonium hydroxide. The rate of fading of a stain during differentiation is dependent upon the concentrations of acid or base in the differentiating fluids.

With hematoxylin stains, the nuclei of cells retain the stain better than the other tissue components. Since the dye is more soluble in water than in alcohol, differentiating in *acid alcohol* has the theoretical advantage that differentiation is not too rapid and thus can be easily controlled under a microscope. A slow differentiating action permits it to be stopped by the next alcohol strength of the dehydration series. However, the hematoxylin stained sections may also be differentiated, not by acid alcohol, but by reagents like ferricyanide, permanganate, oxalic acid or bisulphite. In such case staining is restricted chiefly to myelin sheaths and red blood cells. Thus, unlike results may be obtained with the hematoxylin staining by varying the method of differentiation. In some staining techniques even mordants are used as differentiators.

Hematoxylin and Eosin (H&E) Staining of Paraffin Sections

A conventional sequence of fluids for first deparaffinizing the sections and ending with the preparation wet with water is as follows:
1. Xylene (first)
2. Xylene (second)
3. Absolute alcohol (first)
4. Absolute alcohol (second)
5. 95 per cent alcohol
6. 70 per cent alcohol

7. 50 per cent alcohol
8. 30 per cent alcohol
9. Distilled water

Two changes each of xylene and absolute alcohol, though not absolutely essential, are very desirable because there is likelihood of residual paraffin remaining in the sections. The second change of absolute alcohol ensures complete replacement of xylene with alcohol; thus preventing contamination of fluids in use.

As a rule, the hematoxylin dye is used first, and after differentiation has been attained the eosin (usually alcohol soluble) is used as a *counter-stain*. Although the reverse sequence in the application of these stains is possible, but it is more difficult and hence not conventionally followed. Any of the different types of hematoxylin solutions can be suitably used. For good quality staining (Figs 10.5 and 10.6) it is recommended that the eosin must be used in 95 per cent alcohol.

The following are the steps in routine hematoxylin and eosin (H&E) staining:

Staining Steps

1. Warm gently the slide (with paraffin sections) over burner's flame and place it quickly in xylene for 2–3 minutes.
2. Transfer to next xylene jar until the sections appear absolutely clear.
3. Transfer to absolute alcohol—1 minute (sections will become opaque).
4. Transfer to next absolute alcohol—1 minute.
5. Dip in 95 per cent alcohol.
6. Dip in 70 per cent alcohol.
7. Dip in 50 per cent alcohol.
8. Dip in 30 per cent alcohol.
9. Wash thoroughly with distilled water, keep it for a few minutes.
10. Place in hematoxylin—2 to 20 minutes (varying with the type of hematoxylin

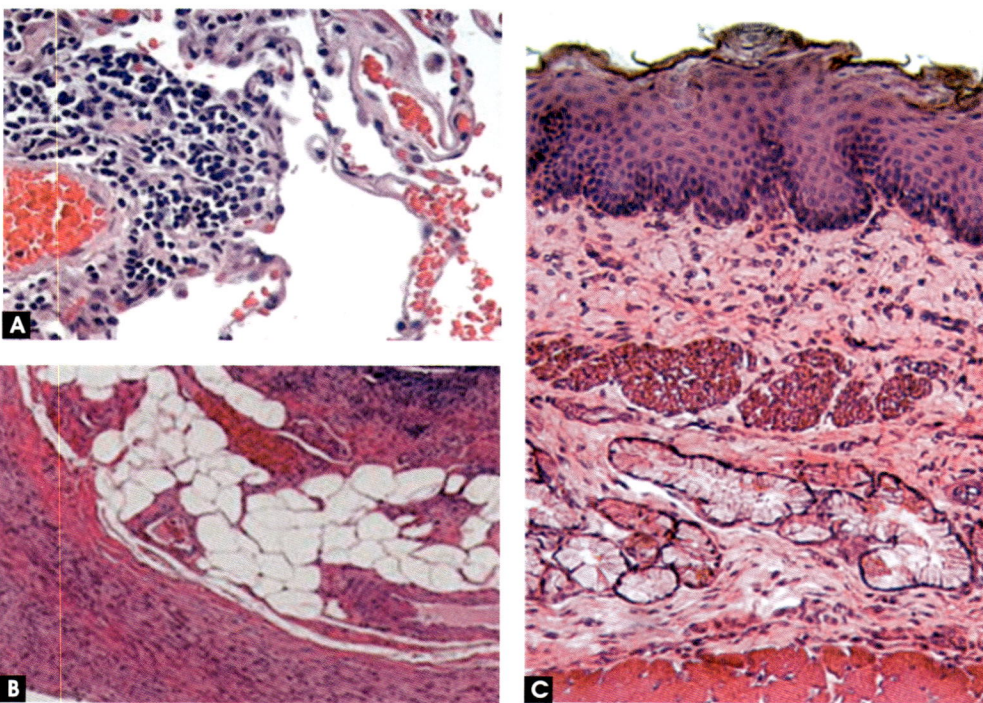

Fig. 10.5: Photomicrographs of human tissues stained with hematoxylin and eosin (H&E): (A) Lung tissue, (B) wall of vermiform appendix, and (C) skin.

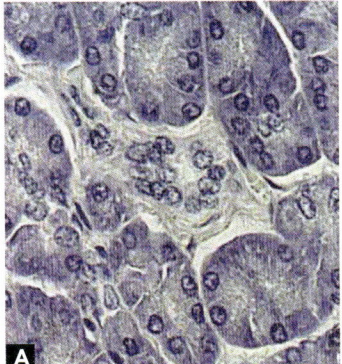 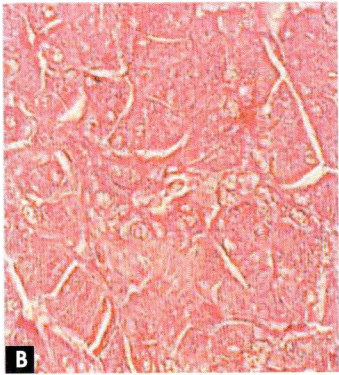 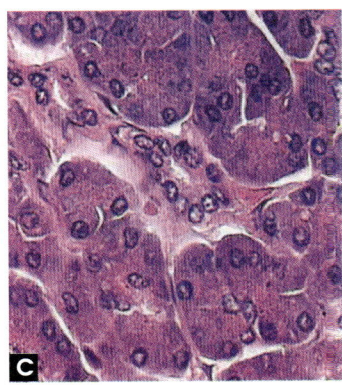

Fig. 10.6: Photomicrographs of sections of pancreas stained with **hematoxylin and eosin (H&E)**: Serial sections of pancreas demonstrate the effect of hematoxylin and eosin used alone and hematoxylin and eosin used in combination. (A) Staining with hematoxylin only—shows general overall staining of the specimen, but those components and structures that have a high affinity for the dye are most heavily stained, e.g. nuclear DNA and areas of the cell containing cytoplasmic RNA. (B) Staining only with counter-stain eosin likewise has an overall staining effect. The nuclei are less conspicuous than in the specimen stained with hematoxylin only (C) Combined staining effect of H&E (after section stained with hematoxylin is then prepared for staining with alcoholic eosin. Note, however, that the hematoxylin that is not tightly bound is lost, and the eosin then stains those components of the section to which it has a great affinity.

solution used; about 3–5 minutes with Harris' and 15–20 minutes with Ehrlich's hematoxylin).

11. Examine the sections, after a rinse is distilled water, under low magnification of microscope to confirm that section is over stained. If not so, transfer the section back to stain, otherwise wash well in distilled water.

12. Differentiate in acidified alcohol confirming the desired destaining with microscope.

13. When the stain is well differentiated (confined mainly to nuclei; the cytoplasmic components almost unstained), 'blue' the sections in tap-water or 1 per cent aqueous lithium carbonate solution. If the tap-water is enough alkaline, 'blueing' of sections is achieved by thorough washing of sections in running tap-water only.

14. Rinse in distilled water.

15. Place the slide in another jar of 30 per cent alcohol for about 3 minutes.

16. Place in 50 per cent alcohol.

17. Place in 70 per cent alcohol.

18. Place in 95 per cent alcohol.

19. Counter stain in 0.5–1.0 per cent eosin in 90 per cent alcohol for 30 seconds to 1 minute; until the cytoplasm and muscle cells if present take a deep pink stain.

20. Dip in 95 per cent alcohol for a few seconds.

21. Keep in absolute alcohol for 3 minutes.

22. Keep in next absolute alcohol for 3 minutes (to ensure full dehydration).

23. Transfer in xylene for 2 minutes.

24. Transfer in next xylene for 2 minutes (until section appears absolutely clear or transparent).

25. Mount in balsam or DPX.

Results

Nuclei—blue to blue-black; *Nucleolus*—dark blue; *Cytoplasm*—pink; other components varying according to tissue (*cartilage*—pink or light blue; *collagen fibres* and *osteoid tissue*—light pink; *muscle cells, thyroid colloids, thick elastic fibers*—deep pink).

Some Common Tissue Stains In Histology

Use	Stain	Comments	Remarks
Routine histological study	Hematoxylin and Eosin (H/E)	Nuclei: Blue; Cytoplasm: Pink	A good routine method
Connective tissues (specially collagen)	Azocarmine	Collagen and reticulin: Dark blue	Valuable for renal globular basement membrane
	Mallory's aniline blue	Collagen: Blue	
	Massons trichrome	Collagen: Blue/green Muscle fibers: Red	
	Van Gieson's	Collagen: Red Muscle fibers: Yellow	
Reticulum	Ammonical silver impregnation	Reticulum fibers: Black	
	Periodic acid Schiff (PAS)	Reticulum fibers: Rose to purple red Collagen: Pale to deep pink	Valuable for most basement membranes
Elastic tissue	Verhoeff's	Elastic fibers: Black	
Muscles	Massons trichrome Van Gieson's Phosphotungstic acid Hematoxylin	Muscle fibers: Red Muscle fibers: Yellow Striations in muscles: Dark blue	
	Bests' Carmine	Glycogen: Red	Fixation in alcohol formalin is recommended
Glycogen	PAS (before and after digestion with saliva or diastase)	Glycogen: Purple red	Digestion removes glycogen resulting in negative PAS reaction
Mucin	Mayer's Mucicarmine	Mucin: Red	
Acid mucopolysaccharides	PAS Toludine blue or thionine Alcian blue	A purple reaction is +ve Acid mucopoly-saccharides: Purple red Acid mucopoly-saccharides: Blue-green	Metachromasia
Fat	Scharlach R (Scarlet red and Sudan III) Osmic acid	Neutral fats and lipids: Orange-red Fat: Black	Frozen sections required; Mineral oil also stained Mineral oil NOT stained
Amyloid	Congo red Iodine	Amyloid: Orange-red Amyloid: Mahogany brown	
	Methyl or cresyl violet	Amyloid: Purple-red	Metachromasia
Free iron	Thioflavine-T	Amyloid-yellow fluorescence in UV light	
	Perl's ferrocyanide reaction (Prussian blue method)	Hemosiderin: Blue	

Contd...

Contd...

Use	Stain	Comments	Remarks
Calcium	Von Koss's Silver method	Calcium: Black	
Argentaffin granules	Masson-Fontana Silver method	Argentaffin granules: Black	
Melanin	Masson-Fontana Silver method	Melanin: Dark brown-black	
	DOPA reaction	Melanoblasts: Black (DOPA+) Melanin: Dark brown-black	Frozen sections needed; for DOPA reaction incubation in dihydroxy-phenylalanine solution to be done
Nervous tissue	Bodian Silver method	Nerve fibers and Neuro-fibrils: Blue-black	Tissue fixation in ammonium bromide; frozen sections
	Weil–Weigert (Lillie's variant)	Myelin: Blue-black	
	Cajal's gold sublimate method	Protoplasmic astrocytes–stain black	

Troubleshooting in H&E Staining

Problem	Cause (s)	Diagram	Prevention
The nuclei are too dark	• Slide exposed to hematoxylin • Sections are too thick	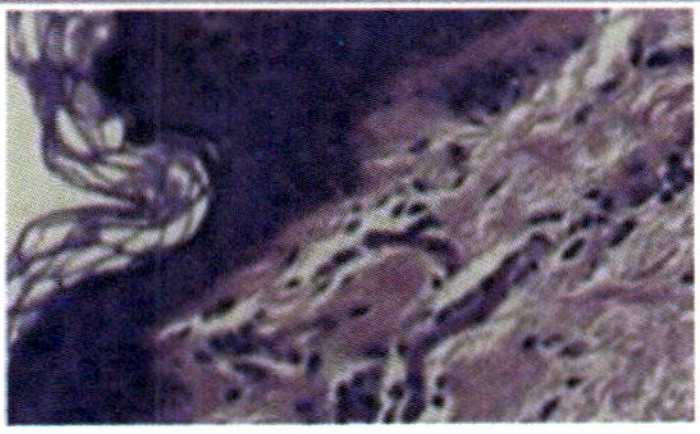 A section of skin with over stained nuclei, so that nuclear detail is lost. The cytoplasm, especially in the epithelium, contains some hematoxylin staining, and the contrast between the nuclei and cytoplasm is poor.	• Decrease hematoxylin exposure • Increase time for differentiation • Cut thinner sections
White patches are seen in the sections after deparaffinization **Incomplete deparaffinization**	• Water left in the tissue • Incomplete drying • Not leaving the slides in xylene long enough	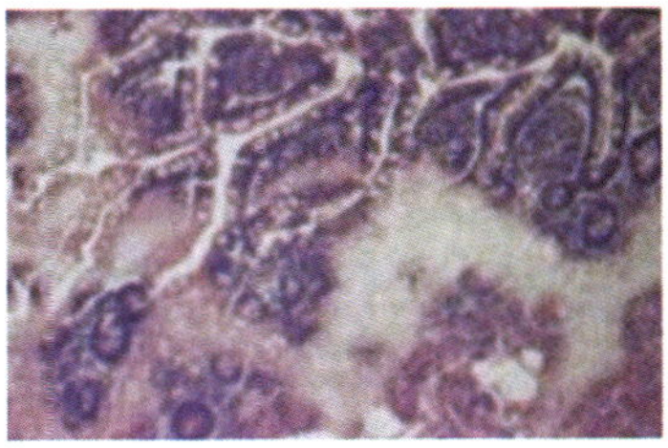 The white (unstained) areas in the section are caused by the incomplete removal of paraffin. Stains will not penetrate the tissue in areas in which paraffin remains.	• Dry section properly before beginning deparaffinization • Allow sufficient time in xylene for deparaffinization • Avoid contaminated xylene, change fluids according to schedule • If the slides have been stained, decolorize and restain

Contd...

Contd...

Problem	Cause (s)	Diagram	Prevention
Pale cytoplasmic staining	• Eosin pH is over 5, which may result from carry-over of the blue-ing agent • The section may be too thin • The sections may be left long in dehydration	 The eosin staining is too much pale	• Check eosin pH • Completely remove blueing agent before transferring slides to eosin • Do not allow stained slides to stand long in the lower concentration of alcohols after the eosin (the more water in the alcohol, the more eosin that will be removed)
Sections detach and float from slides during staining	• The sections are likely to be subjected to strong alkaline or acidic solutions	A process of coating thin paraffin sections for preventing them to float from slides during staining—called **collodionization** is done by coating slides with 0.5% solution of celloidin, and is recommended in cases where glycogen demonstration is desired	

SUMMARY

Stains and Staining

- Staining—use of dyes to provide color to various tissue constituents
- Different tissue constituents react differently to dyes—contrast

- Chromogen
- Chromophore
- Auxochrome—acid/alkali radicals. Responsible for solubility

Stains—classification

Principle	Chemical nature
Chemical nature	*Basic*: Colored organic base + uncolored acetate, chloride or sulfate radical (safranin, methylene blue, crystal violet)
	Acidic: Metallic base (Na. K⁻ + colored organic radical (anilline blue, eosin, orange G)
	Neutral: Combinations of acidic and basic dyes (Giesma stain, Sudan black B)
Affinity to different plant parts	*Nuclear*: Nucleus
	Cytoplasmic: Cytoplasm
Microtechnical purposes	*Histological*: Defines tissues (xylem, phloem, etc.)
	Cytological: Define cell components (nucleus, chromosomes, etc.)

Stains

Natural dyes—dyes obtained from plant/animal (Brazilin, hematoxylin, carmine)

Synthetic dyes—made from coal tar — (orange G, safranine, fast green)

- Brazilin (Timber of *Caesalpinia crista, C. echinota*)
- Hematoxylin *Haematoxylon campechianum*
- Carmine insect *Dactyloplus coccus*

Staining Methods

1. Progressive staining
2. Regressive (retrogressive staining)
3. Counter staining
4. Double, triple and quadruple staining

Methods of Staining

Progressive Staining

- Useful for beginners
- Tissues are understained first
- Gradually more stain is added until the desired intensity attained
- Staining interval required is determined by trial

Regressive (Retrogressive) Staining

- Overstained first
- Then destained until the desired intensity is attained
- Destaining agent—70% alcohol with 1% acetic acid
- Proper washing after differentiation

Methods of Staining

Counterstaining

- Staining certain part of cells/tissues with one stain
- Other parts with a contrasting color

Double/Triple/Quadruple Staining

- Use of 2, 3, 4 colors on same section
- Double staining   : Safranin O and fast green
- Triple staining   : Safranin O, gentian violet green and orange G
- Quadruple staining : Safranin O, methyl violet, fast green and orange G

Hematoxylin and Eosin (H&E) Staining of Paraffin Sections

1. Warm gently the slide (with paraffin sections) over burner's flame and place it quickly in xylene for 2–3 minutes.

2. Transfer to next xylene jar until the sections appear absolutely clear.

3. Transfer to absolute alcohol—1 minute (sections will appear opaque).

4. Transfer to next absolute alcohol—1 minute.

5. Dip in decreasing grades of alcohols (95-70-50-30 per cent)

6. Wash thoroughly with distilled water and keep in it for about 5 minutes.

7. Transfer in hematoxylin stain solution (time varies with the variety of hematoxylin used; usually 3 to 5 minutes with Harris').

8. Rinse in distilled water and quickly examine under low magnification of a light microscope.

9. Differentiate in acidified alcohol until desired destaining achieved.

10. Dehydrate through increasing alcohol strengths.

11. Counterstain with 0.5–1% alcoholic eosin for nearly 30 seconds.

12. Rinse quickly in 95 per cent alcohol for 3 to 4 seconds to get rid off any extra pink color of eosin.

13. Transfer to fresh absolute alcohol.

14. Keep in 2 changes of fresh xylene for proper clearing (the section should appear absolutely clear).

15. Mount in DPX (or balsum) quickly to avoid drying; take care of bubbles coming over the section.

16. Let the slide dry completely before microscopy.

11

Special Staining of Tissue Sections and Preparation of Blood Film

The routine staining of histological sections is with hematoxylin and eosin (H&E), and it displays the structural features by coloring the nucleus blue and the cytoplasm pink. These stains do not adequately reveal certain components of a tissue, namely basement membranes, decalcified bone matrix, lipids, and connective tissue fibers. When it is essential to visualize these components in particular, certain special staining procedures, often being selective, can be undertaken. Although, there are too many staining methods available, some of the conventional methods used to reveal the specific structures are listed in Table 11.1. The more frequently used will be described.

Furthermore, some information about the living cells may be obtained by the following *two* kinds of special staining:

i. *Vital staining*: In the vital staining, a dye is injected into a living animal. Certain cells of a particular organ selectively absorb the dye; and even widely dispersed groups of morphologically different cells can be identified on the basis of their ability to phagocytose foreign particles of the dye used. For example, vital staining—when trypan blue is injected in an experimental animal, the dye is found in macrophage cells of connective tissue, the reticular cells of the spleen, lymph nodes and bone marrow, and in Kupffer cells in the liver.

ii. *Supravital staining:* The *supravital staining* consists in adding a dye(s) to cells that already removed (commonly by surgical manoeuvrings) from an organism, and then placed for a tissue culture. For example, *Janus green* stain, with an affinity for mitochondria, is used to identify later these cellular organelles in the fresh films and spreads of some organs.

STAINING FOR BASEMENT MEMBRANES

The basement membranes, consisting of a dense layer of fine fibrillar material, are usually supported by a layer of reticulin fibers. Formerly, the membranes were thought to be formed as a condensation of the connective tissue ground substance; but recent studies show that these are in part formed by the

Table 11.1	Special stains used for light microscopy sections
Tissue components	**Staining methods used**
1. Basement membranes	PAS technique
2. Connective tissue Fibers	
Collagen Fibers	Van Gieson's staining
Elastic Fibers	Taenzer-Unna orcein method (modified)
Reticulin Fibers	Gomori's staining method
	Golgi silver impregnation*
3. Muscle tissue	Masson's trichrome staining (Lillie's modification)
4. Nervous tissue	
a. *Nerve cell bodies*	
Nissl granules	Thionin staining
	Toluidine blue staining
	Cresyl violet staining
Neurofibrils	Bielschowsky method
b. *Supporting Neuroglia*	
Astrocytes	Cajal's gold-sublimate method
Oligodendrocytes and microglial cells	Del Rio Hortega method
c. *Cell processes (nerve fibers)*	
Axis cylinders/dendrites	Holme's method
Myelin sheaths	Pal-Weigert's method
d. *Nerve endings*	Zinn and Morin's method

overlying epithelial cells. They stain strongly with *Periodic Acid Schiff* (PAS) technique and with silver methnamine. Only the former method will be described here.

Periodic Acid Schiff (PAS) Technique

Specific for: Many normal and pathological tissue constituents like polysaccharides, muco-proteins, glycoproteins, glycolipids of cells; basement membranes.

Fixation recommended: Most fixatives can be used.

Sections: Paraffin, cryostat, and frozen-sections.

Preparation of solutions:

i. **0.5% Periodic Acid (HIO_4) in distilled water**

ii. **Schiff's reagent**

Basic fuchsin	1 g
Distilled water	200 ml

Boil to dissolve with constant stirring.
Cool to 50°C and filter. Then add

1 N Hydrochloric acid	20 ml

Cool to 25°C, and add

Anhydrous sodium bisulfite	1 g

The reagent should be stored in dark. After about 2 days its colour turns orange or straw-colored, and then it is ready to be used.

iii. **Sulfurous acid rinse**

10% Sodium metabisulfite	6 ml
1 N Hydrochloric acid	5 ml
Distilled water	100 ml

Staining Procedure

1. Bring sections through xylene, and graded alcohols up to water.
2. Place for 5 minutes in 0.5% aqueous periodic acid.
3. Rinse in tap water.
4. Rinse in distilled water.
5. Place in Schiff's reagent for 15 minutes.
6. Place in running tap water briefly.
7. Dehydrate in ascending strengths of alcohols.
8. Clear in xylene.
9. Mount in Canada balsam or DPX.

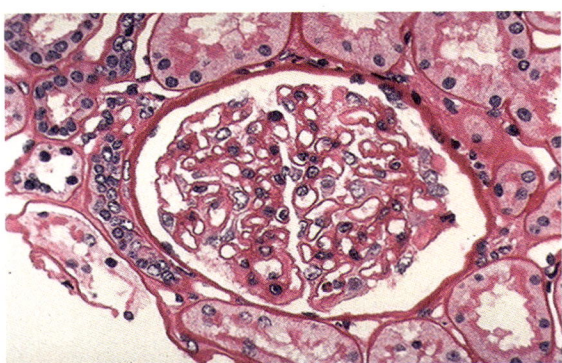

Fig. 11.1: Photomicrograph of a normal renal glomerulus is stained with PAS to highlight basement membranes of glomerular capillary loops and tubular epithelium. The capillary loops of this normal glomerulus are well-defined and thin. The endothelial cells are seen in capillary loops. The mesangial regions are of normal size. Podocytes are present and forming the visceral epithelial surface. Bowman's space is seen along with parietal epithelial cells.

Result

Basement membrane takes stain on the red side of purple (Fig. 11.1).

Remarks

i. Periodic acid keeps fairly well in a Coplin jar even at room temperature.
ii. When Schiff's reagent becomes stained with pink or red, it remains no longer usable; hence should be discarded.
iii. Sulfurous acid rinses tend to become pink to purple as they accumulate Schiff's reagent. Using a system of number 1, 2 and 3 for this, with no. 1 the first used after the Schiff's reagent, it is good practice to discard no. 1 when it shows any color. Moving no. 2 and no. 3 up to 1 and 2 and putting up a new no. 3.

STAINING FOR CONNECTIVE TISSUE FIBERS

Van Gieson's Staining for Collagen Fibers

Specific for: Collagen fibers, nuclei (Fig. 11.2)

Fixation recommended: Formalin (10%), better results with fixatives containing mercuric salts.

Sections: Paraffin, nitrocellulose, or frozen sections.

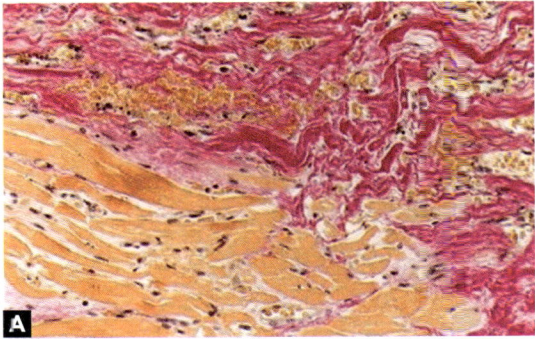

A

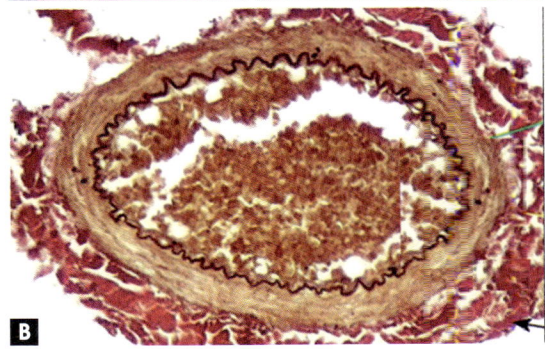

B

Fig. 11.2: Photomicrographs from Van Gieson stained method reveals: (A) Showing collagen (red), cytoplasm (yellow), and nuclei black/ blue black; (B) Showing three tunics in the wall of the anterior spinal artery.

Preparation of staining solution

1% Acid Fuchsin in distilled water	10 ml
Saturated aqueous picric acid	100 ml

Staining Procedure

1. Stain sections with any of the alum hematoxylins (generally, Harris hematoxylin is used) in the usual manner.
2. Wash thoroughly in tap-water.
3. Stain in the above staining solution for 3–5 minutes.
4. Rinse in tap-water and differentiate in 95 per cent alcohol.
5. Dehydrate in absolute alcohol.
6. Clear in xylene through 2 or 3 charges and apply a cover slip with balsam or DPX.

Results

1. White fibrous (*collagen*)—red or deep red
2. *Nuclei*—blue to black

3. Other tissue components—bright lemon yellow (in epidermis) or dull olive color (in case of muscle, and nerve cells)

Taenzer-Unna Orcein Method for Elastic Fibers

Specific for: Elastin fibers.

Fixation recommended: Any fixative may be used. Choice is not critical.

Embedding: In paraffin.

Sections: Thin (8–10 micron) paraffin sections.

Preparation of staining solution:

Orcein	1 g
Alcohol (80%)	100 ml
Hydrochloric acid (conc.)	1 ml

Staining Procedure

1. Dewax the sections.
2. Bring sections to 70 per cent alcohol.
3. Stain in the above solution for 30 minutes to 2 hours.
4. Wash thoroughly in 70 per cent alcohol.
5. Differentiate in 1% acid alcohol for destaining of collagen Fibers.
6. Wash thoroughly in tap water.
7. Counterstain nuclei with methylene blue.
8. Dehydrate in alcohol.
9. Clear in xylene.
10. Mount in DPX or Canada balsam.

Results

1. *Elastin fibers*—dark brown
2. *Nuclei*—blue (Fig. 11.3).

Gomori's Staining Method for Reticulin Fibers

Specific for: Reticulin fibers.

Fixation recommended: 10% formalin (any other also).

Embedding: In paraffin

Sections: Thin sections affixed to slides with adhesive.

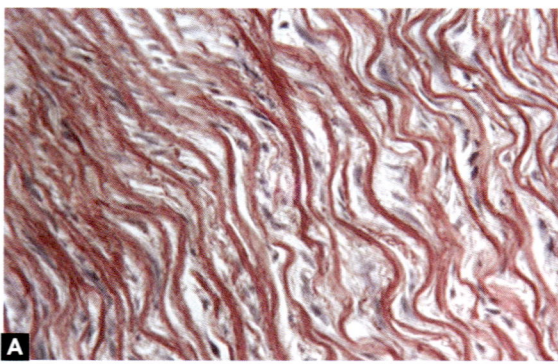

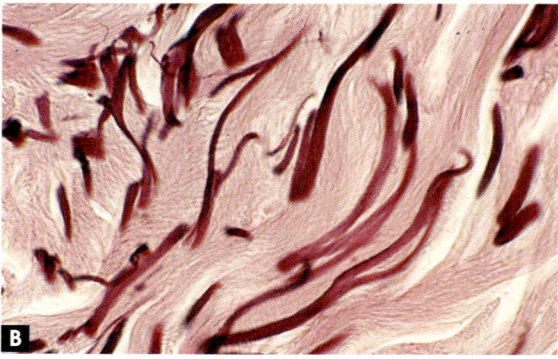

Fig. 11.3: Photomicrographs showing collagen fibers from: (A) abdominal aorta, showing elastic fibers—stain: Orcein; (B) Skin dermis, selectively stained for elastic fibers. Dark elastic fibers are interspersed with pale red collagen fibers.

Preparation of staining solution: (**should be freshly made**)

Silver nitrate (10% sol.)	20 ml
Potassium hydroxide (10% sol.)	4 ml

Note

The precipitate formed is dissolved by adding strong ammonia drop by drop. Add more 10 per cent silver nitrate solution drop by drop shaking the mixture after each drop until the resulting precipitate dissolves on shaking. Then add equal amount of distilled water.

Staining procedure

1. Deparaffinize the sections and hydrate to water.
2. Place the slides in 1 per cent potassium permanganate solution for 1 to 2 minutes.
3. Rinse in tap water.

4. Transfer the slides to 3 per cent potassium metabisulphite until sections appear bleached.
5. Wash in tap water for about 15 minutes.
6. Place in 2 per cent iron alum solution.
7. Wash thoroughly in tap water; and then in two changes of distilled water.
8. Keep in ammoniacal silver solution for 1 minute.
9. Rinse quickly with distilled water.
10. Transfer in 10 per cent neutral formalin solution. Keep for 3 minutes (reduction will take place).
11. Wash thoroughly in tap water.
12. Tone the sections by placing them in 0.2 per cent gold chloride for up to 10 minutes.
13. Rinse through several changes of distilled water.
14. Transfer for 1 minute in 3 per cent potassium bisulfite solution.
15. Rinse with distilled water.
16. Fix for 1 to 2 minutes in 2.5 per cent hypo solution.
17. Wash thoroughly in tap water.
18. Dehydrate through changes of alcohol.
19. Clear in xylene and mount in DPX.

Results

Nuclei—gray; *Reticulin fibers*—black (Fig. 11.4); *Collagen fibers*—greyish purple.

Warning

It is important to remember that ammoniacal silver solution has *explosive* nature. Hence it should be made always *immediately before use*, prevented from *being exposed to sunlight*; and the *unused reagent immediately made inactive* by adding excess of dilute hydrochloric acid.

STAINING FOR CARTILAGES

Specific for: General histological staining

Fixation recommended: Bouin's fixative or 10% formalin in 95% alcohol for 3 days.

Embedding: Paraffin

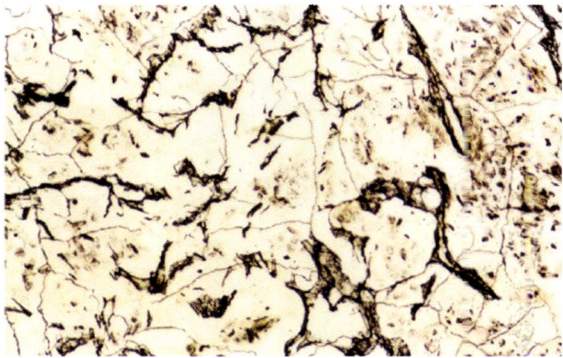

Fig. 11.4: Photomicrograph showing reticulin fibers appearing black in a section stained with Gomori's silver-impregnation-stain for reticulin fibers.

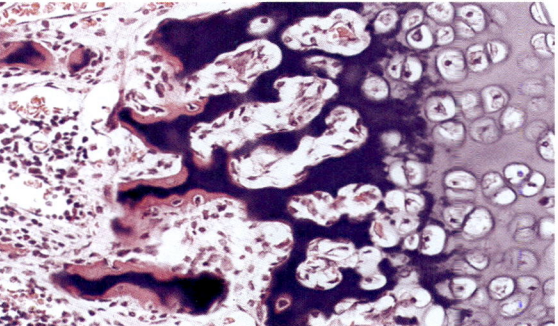

Fig. 11.5: Photomicrograph of H&E stained epiphyseal growth plate.

Sections: 8–10 μm paraffin sections (Figs 11.5 and 11.6).

STAINING FOR MUSCLE TISSUE

Masson's Trichrome Staining (Lillie's Modification)

Specific for: Cells; cytoplasm; muscle; collagen; hypophysis cerebri; and thyroid gland.

Fixation recommended: Bouin's fixative or 10% formalin in 95% alcohol for 3 days.

Embedding: Paraffin

Sections: 8–10 μm paraffin sections (Fig. 11.7).

Preparation of solutions:

i. **Mordant**

Phosphomolybdic acid	5 g
Phosphotungstic acid	5 g
Distilled water	200 ml

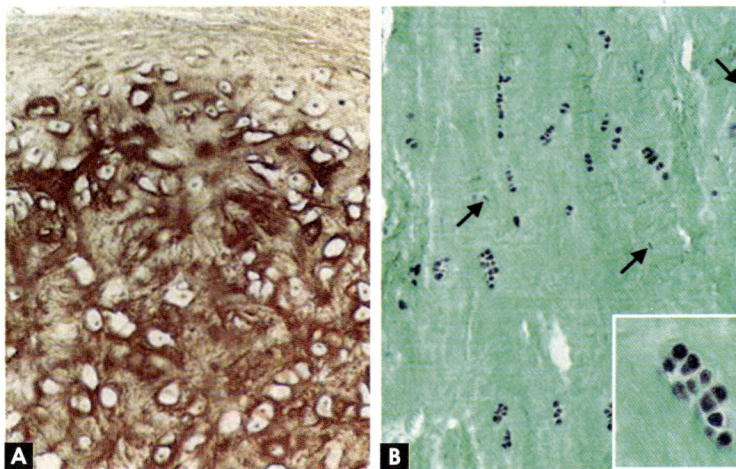

Fig. 11.6: (A) Photomicrograph of elastic cartilage specimen stained with orcein reveals perichondrium on the top of the photomicrograph; chondrocytes of various sizes, stained brown, within the cartilage matrix; (B) A low-power (60x) photomicrograph of a section of the fibrocartilage from an intervertebral disc stained with gomori trichrome preparation is seen. Notice that the collagen fibers are stained green. A few small-sized fibroblasts with dark round nuclei are visible. The chondrocytes are arranged either in rows among the collagen fibers or in isogenous groups. The inset at the lower right corner of the photomicrograph is an higher magnification (700x), and shows chondrocytes contained within lacunae.

ii. ***Staining solutions***: The following three solutions are to be prepared separately.
   a. Weigert's iron haematoxylin (*see* p. 150)
   b. 1 per cent Biebrich scarlet in 1 per cent acetic acid
   c. 2.5 per cent fast green in 2.5 per cent acetic acid

Staining Procedure
1. Dewax sections and hydrate to water.
2. Place for 2 minutes in saturated alcoholic picric acid solution.
3. Wash with running tap water for several minutes.
4. Transfer to stain in Weigert's iron-hematoxylin for 6 minutes.
5. Wash in running tap water.
6. Stain for 4 minutes in scarlet solution.
7. Rinse in tap water.
8. Keep in mordant for 1 minute.
9. Stain for 4 minutes in fast green solution.
10. Differentiate for 1 minute in 1% acetic acid.
11. Dehydrate through alcohol strengths.
12. Clear and mount in DPX.

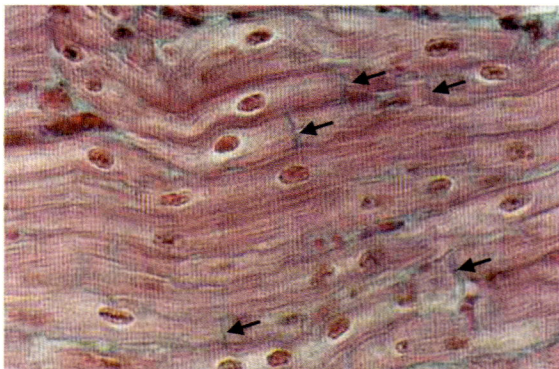

Fig. 11.7: A high-power (420x) photomicrograph of a Mallory-stained section showing cardiac muscle fibers that also exhibit striations. The cardiac muscle fibers are composed of individual cells that run in a longitudinal array. The arrows indicate intercalated discs at the sites of the junction of adjoining cells.

Results

Nuclei—black; *Cytoplasm*—pink to brown; *Muscle*—red; *Collagen*—green; *Erythrocytes*—brilliant scarlet; and *Myelinated nerve fibers*—red.

STAINING FOR NERVE CELLS

Thionin Staining (Schedule I)

Specific for: Nissl substance in nerve cells. Thionin is also used for staining of decalcified bone, mucin, mucopolysaccharides, mast cells, and sex-chromatin.

Fixation recommended: Bouin's fixative, others satisfactory.

Embedding: In paraffin

Sections: 10 μm thick sections (thinner has no advantage).

Preparation of staining solutions

Solution A

Lithium carbonate	5.5 g
Distilled water	1000.0 ml

Solution B

Thionin (CI 52000)	0.25 g
Lithium carbonate (0.55%)	10.00 ml

Staining Procedure

1. Deparaffinize the sections with xylene.
2. Hydrate to distilled water.
3. Place for 5 minutes in lithium carbonate solution (solution A).
4. Overstain in thionin stain (solution B) for 5–10 minutes.
5. Rinse in distilled water.
6. Dip the slides for a few seconds in 70 per cent alcohol.
7. Dehydrate (preferably in butyl alcohol) with 2 changes of 2–3 minutes each.
8. Clear in xylene.
9. Mount in Canada balsam or DPX.

Results

Nissl substance—bright blue (Fig. 11.8)

Remarks

If differentiation is necessary, briefly rinse the slides in 95 per cent ethyl alcohol following step 6 and place in aniline, then in lithium carbonate saturated in 95 per cent alcohol. Then proceed to the step 7.

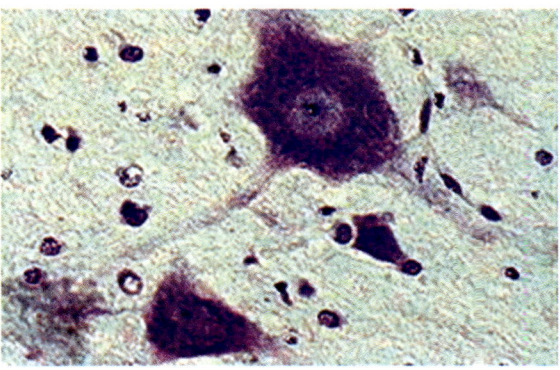
Fig. 11.8: A high-power (420x) photomicrograph of a nerve cell.

Thionin Staining (Schedule-II)

Specific for: Nissl substance.

Fixation recommended: 10 per cent formalin solution for 1–2 days.

Embedding: Paraffin (also nitrocellulose).

Sections: 8–10 μ thick sections.

Preparation of staining solutions
Two types of thionin solutions (varying pH) are commonly used; formulae are listed below:

	for pH 3.7	*for pH 4.5*
Acetic acid 0.6% aqueous (0.1 M)	90.0 ml	60.0 ml
Sodium acetate 0.8% aqueous (0.1 M)	10.0 ml	40.0 ml
Thionin 1.0% aqueous solution	2.5 ml	2.5 ml

Staining Procedure

1. Deparaffinize the sections with xylene.
2. Hydrate the sections.
3. Stain for 20–40 minutes in any of the two thionin solutions of pH 3.7 or 4.5.
4. Rinse off excess dye in 50 per cent alcohol.
5. Transfer the sections for **differentiating fluid**:

Ethyl alcohol (95%)	100.00 ml
Glacial acetic acid	0.25 ml

6. Dehydrate in ascending series of alcohol.
7. Clear in xylene.
8. Mount in Canada balsam or DPX.

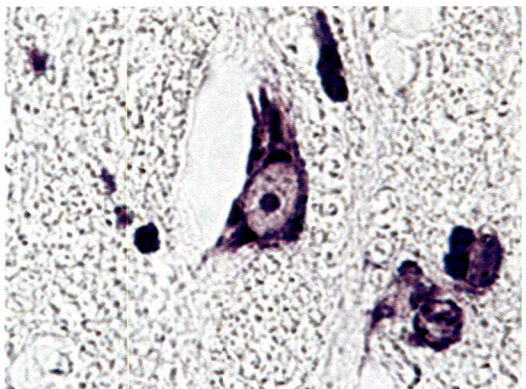

Fig. 11.9: Photomicrograph of a multi-polar nerve cell (thionin stain). The smaller nuclei are neuroglia (oligodendrocytes and microglia).

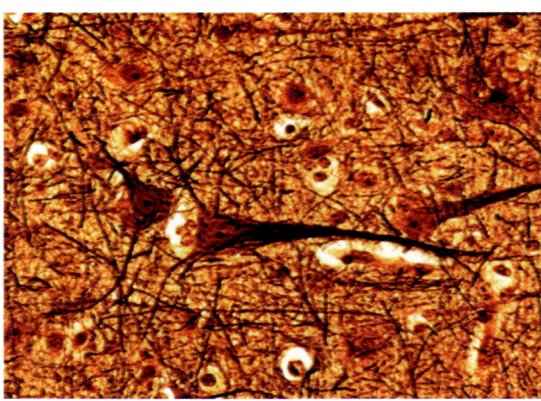

Fig. 11.10: Photomicrograph of nerve cells and processes stained with Bielschowski stain.

Results

Nissl substance— bright blue (Fig. 11.9).

- With thionin solution of pH 3.7—only Nissl granules, nucleoli, and chromatin are stained against a practically clear background.
- With thionin solution of pH 4.5—there is darker staining of Nissl substance, but with a pale-bluish background.

Remarks

If differentiation is too rapid, dilute the differentiating solution with equal volume of alcohol; if it is too slow, double the amount of glacial acetic acid.

STAINING FOR NEUROFIBRILS

Bielschowsky Method

Specific for: Nerve cells and processes; neurofibrils (Fig. 11.10)

Fixation recommended: 10% formalin (sometimes 0.25 to 0.5% trichloracetic acid added).

Embedding: In paraffin (after bulk staining of specimen)

Preparation of staining solution

Solution A

Dissolve 1.5 g silver nitrate (Ag NO₃) in 100 ml of distilled water.

Solution B

Strong ammonia water (sp. gr. 0.9)	5 ml
Sodium hydroxide (2% aqueous)	40 ml

Shake well and add:

Silver nitrate (8.5% aqueous)—till solution becomes permanent opalescent.

Staining Procedure

1. Wash tissues in distilled water for about an hour after fixation.
2. Place in pure pyridine for 1 or 2 days.
3. Wash 2 to 6 hours in distilled water.
4. Keep the block in solution A for about 3 days at 37° C.
5. Wash in distilled water for 30 minutes.
6. Impregnate for 6 to 24 hours in solution B.
7. Reduce for 1 to 12 hours in 1: 100 formalin
8. Wash in tap water.
9. Dehydrate the block.
10. Embed in paraffin.
11. Section and mount on slides.
12. Deparaffinize with xylene and cover.

Results

Nerve fibers and *neurofibrils*—brown to black; *Background*—yellow to brown

STAINING FOR NEUROGLIA

Cajal's Gold Sublimate Method

Specific for: Protoplasmic and fibrous astrocytes (selective for mammalian tissues) (Fig. 11.11)

Fixation recommended: Formalin-ammonium-bromide (FAB) for 3 days to several weeks.

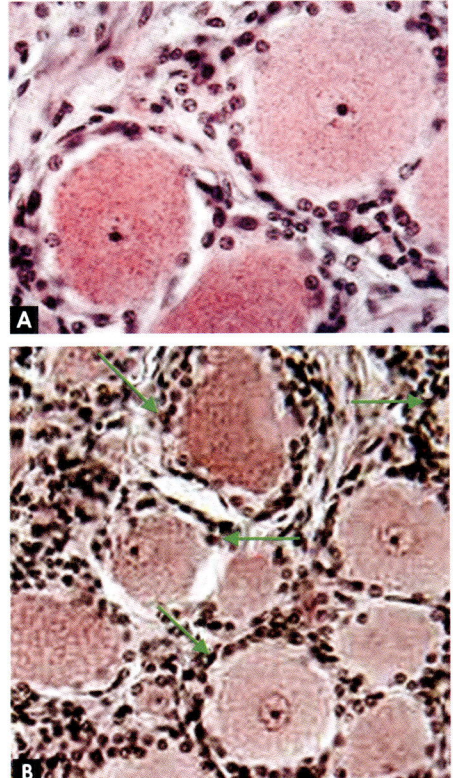

Fig. 11.11: Neuroglia from a canine spinal ganglion (H&E stain). Unipolar neuron cell bodies are surrounded by satellite glial cells (arrows).

Sections: 30–40 µm frozen sections (thinner may also be used). Store sections in FAB until ready for staining

Preparation of staining solution:

Distilled water	100 ml
1% aqueous gold chloride	20 ml
Mercuric chloride	1 g

The ingredients are heated slowly and shaken until the mercuric salt has dissolved, cool and filter. The staining solution should be taken in a flat covered dish, adjusting the temperature to 18–20°C.

Staining Pocedure

1. Remove sections from FAB solution carefully, using glass instruments.
2. Rinse well in distilled water (2–3 changes).
3. Transfer the sections to the staining solution and allow the sections to remain therein for 4–8 hours.
4. Wash rapidly in distilled water.
5. Fix for 5–10 minutes in 5–10 per cent sodium thiosulfate (common name **Hypo**) solution to which is added about 1 ml of 1 N sodium bisulfite per 100 ml.
6. Wash in distilled water (or 30% alcohol).
7. Float sections on to clean slides (generally one or two sections per slide), and blot the sections with soft, fine-grained filter paper.
8. Dehydrate sections on the slide itself by dropping increasing grades of alcohol through absolute.
9. Clear in carbol-xylene.
10. Wash with xylene, apply a mountant and cover.

Results

Astrocytes—reddish purple to nearly black.

Nerve cells—pale rose to violet.

Nerve fibers—unstained or stained pale.

Remarks

i. The time of fixation and the temperature of the staining solution play a very critical role in staining of astrocytes. The photoplasmic astrocytes lose their stainability after 2–3 weeks fixation; but fibrous types are stained suitably only after 2 months or longer of fixation. Optimum fixation time is 5–10 days.

ii The temperature of the staining solution should not be below 10°C, and should never exceed above 30°C.

Del Rio Hortega's Method

Specific for: Oligodendrocytes and microglia

Fixation recommended: In 10 per cent formalin for 12–48 hours; better results with fixation in formalin-ammonium-bromide (FAB). In FAB, the brain tissue should be fixed for 12–48 hours for oligodendrocytes, and for 2–3 days for microglia.

Sections: Frozen sections 15–20 µm thick.

Preparation of ammoniacal silver carbonate

Silver nitrate (10% aqueous)	5 ml
Sodium carbonate (5% aqueous)	15 ml

Ammonium hydroxide (28% conc.) is added drop by drop in just sufficient quantity to dissolve the precipitate. Add distilled water to make up 75 ml.

Staining Procedure

1. Wash the frozen sections in 1 per cent ammonia water, then in plain distilled water.
2. Impregnate for 1–5 minutes (exact time to be determined by trial and error) in the above ammoniacal silver carbonate solution.
3. Wash for a few seconds in distilled water.
4. Submerge quickly in 1 per cent formalin and allow the sections to remain undisturbed. This *reduces* the sections.
5. Wash thoroughly in tap water.
6. Tone in 0.2 per cent aqueous gold chloride until sections appear grey all over.
7. Fix in 5–10% aqueous hypo solution.
8. Wash in tap water.
9. Dehydrate.
10. Mount in balsam.

Results

Oligodendrocytes (cytoplasm and processes)—black

Microglia and *processes*—black (distinguished by their elongated shape and branching processes).

Astrocytes, nuclei—practically unstained.

Background—grey (Fig. 11.12) Del Rio-Hortega (1922)

STAINING FOR NEURITES

Holme's Staining Method

Specific for: Axons of central and peripheral system (Fig. 11.13); Also good for staining of cross striations of skeletal muscle cells.

Fixation recommended: 10 per cent formalin, fixatives with mercuric chloride give best results.

Sections: Paraffin sections (10–20 μm thick) mounted on glass slide with an adhesive.

Staining solutions:

i. **Buffer solutions**:

1. Boric acid	12.4 g
Distilled water	1000.0 ml
2. Borax ($Na_2B_4O_7.10H_2O$)	19.0 g
Distilled water	1000.0 ml

ii. **Reducer**:

Hydroquinone	1.0 g
Sodium sulfite (crystals)	10.0 g
Distilled water	100.0 ml

(Keeps well only for a few days during which it can be used repeatedly).

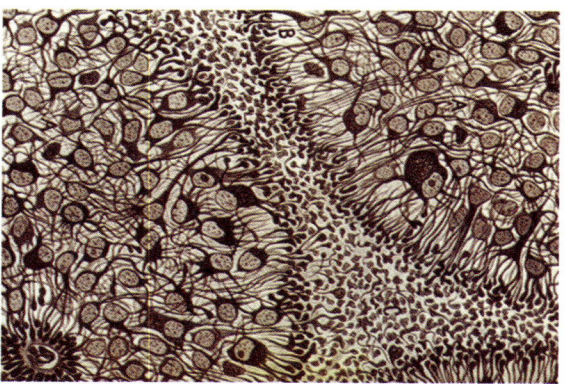

Fig. 11.12: Structure of the pineal gland-Del Rio-Hortega method (1922).

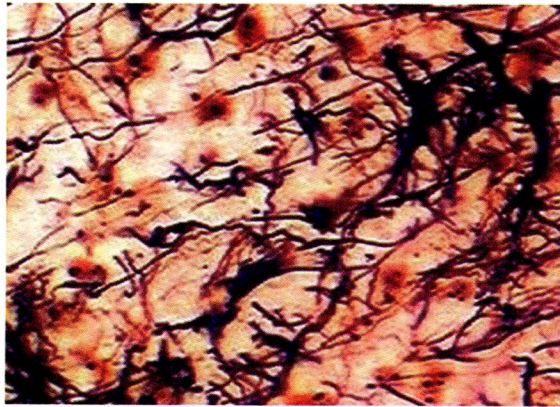

Fig. 11.13: Holme's staining of neurites.

Staining Procedure

1. Dewax the sections with xylene, and hydrate to water. If mercuric salt fixative was used, iodine treatment must be followed.
2. Wash for 10 minutes with running tap-water first, then followed by wash in distilled water.
3. Place in dark in 20 per cent silver nitrate solution at room temperature for nearly 2 hours. The sections may be kept in this solution overnight.
4. Wash sections in 3 changes of distilled water for 10 minutes each.
5. Transfer the slides to the *impregnating medium* by mixing the boric buffer solution as below:

Boric buffer solution	45 ml
Distilled water up to	494 ml
1% silver nitrate	1 ml
10% pyridine sol (aqueous)	5 ml

Note

Silver nitrate and pyridine solutions should be taken with *separate* pipettes. Whole solution is to be mixed thoroughly. Keep the slides at 37°C overnight.

6. Take out the slides from the incubator and keep them in the *reducer* for at least 2 minutes.
7. Wash thoroughly in running tap-water for a few minutes, and rinse then in distilled water.
8. Tone the sections with 0.2 per cent gold chloride (yellow) until the brown color of sections disappears.
9. Rinse in distilled water.
10. Transfer the slides in 2.0 per cent oxalic acid solution for 5–10 minutes and watch intermittently for axons assuming a blue black color against light.
11. When axons appear blue or black, take out the slides from oxalic acid solution, and wash thoroughly in distilled water.
12. Fix the sections with 5 per cent sodium thiosulfate (hypo) solution for about 5 minutes.
13. Wash thoroughly in tap-water.
14. Counter stain with neutral red (optional).
15. Dehydrate through ascending strengths of alcohol.
16. Clear in xylene, and mount in balsam or DPX.

Results

Axons—blue black or black
Background—pale-stained or greyish

STAINING FOR MYELIN SHEATH

Luxol fast blue is a copper phthalocyanine dye that is soluble in alcohol and is attracted to bases found in the lipoproteins of the myelin sheath. Under the stain, myelin fibers appear blue, neuropil appears pink, and nerve cells appear purple. Luxol fast blue stain, abbreviated LFB stain or simply LFB, is a commonly used stain to observe myelin under light microscopy, created by Heinrich Klüver and Elizabeth Barrera in 1953. LFB is commonly used to detect demyelination in the central nervous system (CNS), but cannot discern myelination in the peripheral nervous system.

Pal-Weigert Method

Specific for: Myelin sheaths of brain and spinal cord.

Fixation recommended: 10% formalin for 7 days or longer, with or without 2% ammonium bromide (NH_4Br).

Embedding: In either celloidin or paraffin. Frozen sections may also be cut (fixing fluid with 2% NH_4Br recommended).

Preparation of staining solution

Solution A

Lithium carbonate (saturated aqueous)	7 ml
Distilled water	93 ml

Solution B

Hematoxylin	1 g
Ethyl alcohol (absolute)	10 ml

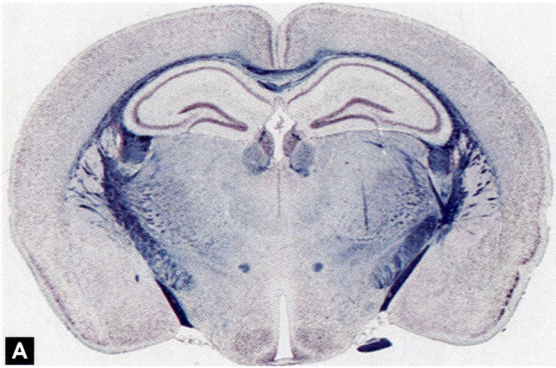

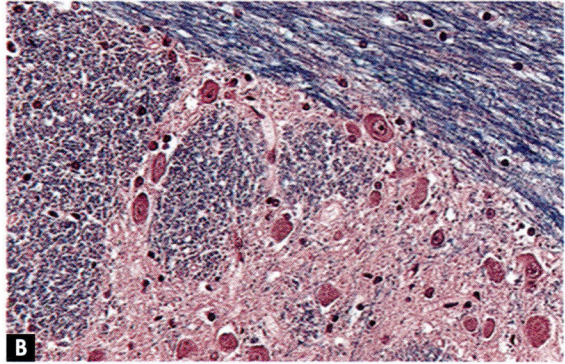

Fig. 11.14: (A) Coronal section of a mouse brain stained with hematoxylin and LFB; (B) Micrograph of the pons using a hematoxylin and eosin-LFB stain.

Mix one part of solution A with nine parts of solution B. The mixture is ready for use; no ripening is required.

Staining Procedure

1. Transfer sections from water to a mordanting solution, 4 per cent ferric ammonium sulfate [iron alum, $FeNH_4(SO_4)_2 \cdot 12H_2O$].
2. Wash in tap water.
3. Place in the staining solution for 1–2 hours.
4. Wash for 2–3 minutes in water.
5. Differentiate in 0.4 per cent aqueous $KMnO_4$ until grey and white matter are clearly distinguishable when the sections are viewed against a white table top or window. The overall color of the sections becomes brown.
6. Rinse quickly in tap water.

7. Complete the de-colorization by immersing in the following freshly prepared solution:

Oxalic acid (1% aqueous solution)	1 part
Sodium sulfite (1% aqueous solution)	1 part

The sections should be kept in this solution until the brown color produced by the permanganate solution disappears and the gray matter becomes completely clear and colorless.

8. Wash for 2–3 minutes in tap water.
9. Soak for 5 minutes in the solution A till the blue color is restored.
10. Wash thoroughly in tap water.
11. Dehydrate, clear and mount.

Results

Myelin sheaths—dark blue

Other structures—unstained unless some counter stain is used.

STAINING FOR NERVE ENDINGS

Zin and Morin's Method

Specific for: Nerve endings.

Fixation recommended: Small pieces of muscle be fixed for 10 minutes in 0.001 M citric acid, unless the piece becomes translucent.

Sections: Not required.

Staining procedure

1. Wash thoroughly the translucent muscle piece in several changes of distilled water.
2. Pour off the water and blot the muscle piece.
3. Cover the piece with 1 per cent aqueous gold chloride solution until a golden yellow color appears. Generally it takes about 30 minutes.
4. Rinse in several changes of distilled water.
5. Transfer the piece gently to the following mixture:

Formic acid	1 part
Distilled water	2 parts

6. Wash thoroughly in many changes of distilled water.
7. Blot the piece with a fine filter paper.
8. Transfer in a few drops of glycerine placed over a slide. Tease the tissue under a binocular dissecting microscope to isolate muscle fibers bearing motor end-plates.
9. Mount in glycerine-jelly, prepared as follows:

Gelatin	10.00 g
Distilled water	60.00 ml
Phenol (crystal)	0.25 g
Pure glycerine	70.00 ml

Dissolve gelatin in distilled water over a water-bath till gelatin melts completely. Mix glycerine and phenol while still fluid. Decant in a small jar. It will solidify after some time. For use, melt in a water-bath.

10. Ring the cover slip.

Results

Nerve endings and *nerve fibers*—stain black or deep purple (Fig. 11.15).

Muscle cells—reddish.

STAINING FOR BLOOD FILMS

A *blood smear* or peripheral blood film (PBF) is a thin layer of blood smeared on a glass microscope slide and then stained in such a way as to allow the various blood cells to be examined microscopically. The PBF is of two types: (i) Thin blood film, and (ii) thick blood film (Fig. 11.16).

Three basic steps to make blood film are: (i) Preparation of blood smear, (ii) fixation of blood smear, and (iii) staining of blood smear (Fig. 11.17).

Preparation of Leishman Stain

Materials

| Leishman stain powder | 0.6 g |
| Methanol (water free methyl alcohol) | 400 ml |

Procedure

- Weigh the Leishman powder and transfer it to a completely dry brown bottle
- Add a few glass beads (to assist dissolving the dye). A magnetic stirrer or carefully warming to 37°C may help in dissolving.
- Using a dry cylinder, measure the methanol and add this to the stain.
- Mix well at intervals until the dye is completely dissolved.
- Label the bottle and mark it **Flammable and Toxic.**
- Store it at room temperature in the dark

For use: Filter 50–100 ml of the stain into a stain dispensing container which can be closed

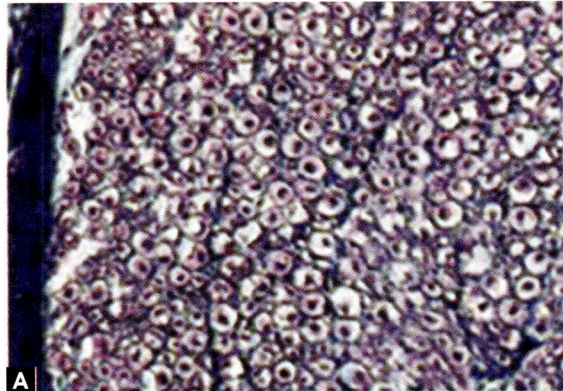
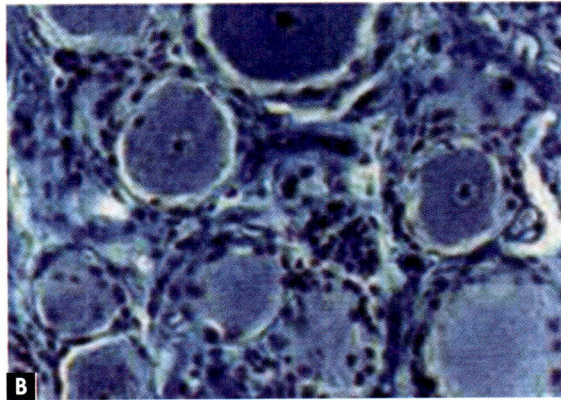

Fig. 11.15: Photomicrographs of: (A) Mallory-stained section of a peripheral nerve. Notice a vast number of thread-like myelinated axons (appearing as small, red, dot-like structures) held together by connective tissue (stained blue); (B) Azan-stained section of a nerve ganglion, showing spherical nerve cell bodies and the nuclei of the small stellate cells surrounding the nerve cell bodies.

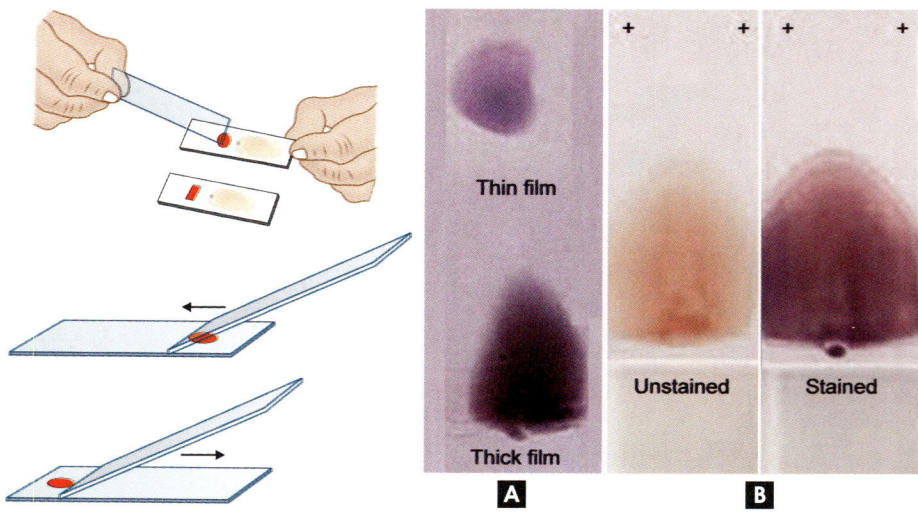

Fig. 11.16: Making of blood film on a glass slide: (A) Properly prepared peripheral smear called a blood film is a thin layer of blood smeared on a glass microscope slide and then stained in such a way as to allow the various blood cells to be examined microscopically; (B) Two push-type peripheral blood smears suitable for characterization of cellular blood elements. Left smear is unstained, right smear is stained with Leishman stain (*refer to* Fig. 11.17).

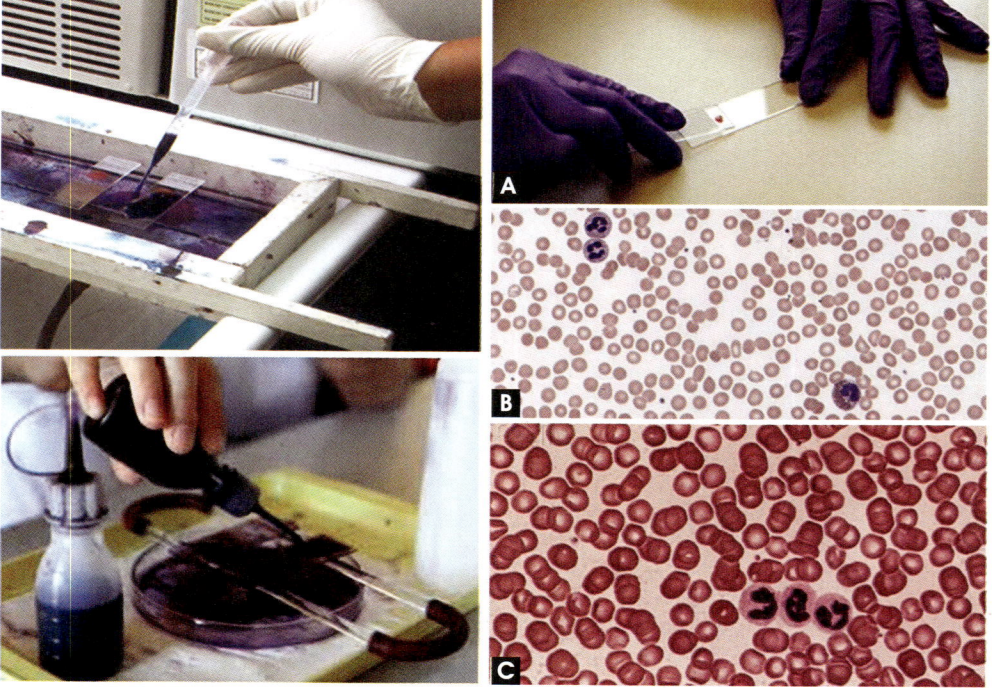

Fig. 11.17: Techniques for blood film staining for differential white blood cell count is shown on the left side. On the right is shown: (A) Preparing a blood smear on a microscope slide, (B) human blood smear, microscope view, and (C) RBC smear sample and three white blood cells (WBCs).

when not in use. *An aliquot of the stain (e.g. usually 50–100 ml) should be filtered into a dispensing unit for daily use, and the following storage conditions should be followed.*

Storage

Bright light and heat oxidize the stain, especially when in aqueous solution and will cause precipitation of insoluble precipitates, e.g. of *methylene violet* (Bernthsen). Evaporation of methanol, absorption of moisture and precipitation of Azure-Eosinate salts are also additional problems during storage that require filtering the stock while aliquotting for daily use.

For daily use, store the stain in an airtight (prevent moisture entering the stain) amber (semi-opaque) container. A closeable dropper bottle, e.g. TK dropper bottle can be used that should be kept tightly closed when not in use. Keep in a cool place (not refrigerated) and never in direct sunlight.

The stock stain should be kept in a tightly stoppered light opaque (e.g. amber) container in a cool dark place. Renew every 3 months or earlier if indicated. To obtain optimum color reaction, some suggest that 3–5 days should be allowed before using freshly made stain.

Notes

i. When kept tightly stoppered, the stain is suitable for several weeks
ii. Moisture must not be allowed to enter the stain.

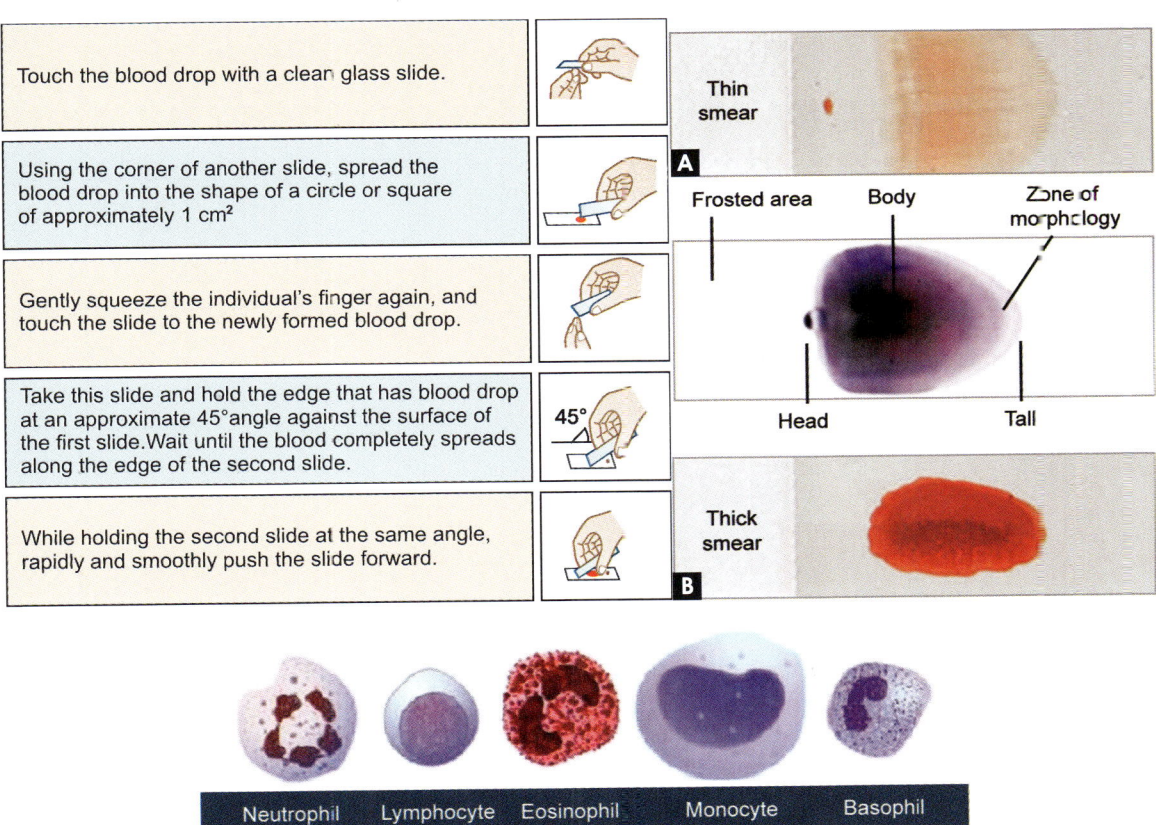

Fig. 11.18: Steps showing preparation of a thin and a thick blood film on the same slide are listed on the left. Different zones of a blood film are depicted in the middle side. These can be easily identified in a thin smear (A) but not so in a thick smear. (B) In the zone of morphology the five normal types of WBCs can be identified.

How to Stain Blood Film by Leishman Stain?

1. Stain with Leishman stain for 3 minutes (Fig. 11.17)

2. Pour with buffer solution for 10–15 minutes, make sure the blood film is flooded with buffer solution.

3. Rinse with distilled water to clean the remaining stain.

4. Air dry.

5. Apply a uniformly thin layer of immersion oil on the dried film (no cover glass)

6. Examine under microscope using oil-immersion objective

7. Identify the blood cells (Fig. 11.18).

Preparation of a Thin and a Thick Blood Film on the Same Slide

Sometimes a blood film consisting of a thin and thick smear of blood is needed. It is relevant to know the regions in a typical blood film (Fig. 11.18). Before making a thin and thick blood film on the same slide, ascertain to follow the following instructions.

1. Write the identification number on the slide.

2. Wait until the thick blood film is completely dry.

3. Stain the slide and examine under the oil-immersion objective of a light microscope.

The steps blood films consisting of both thin and thick smears are summarized in Fig. 11.18.

SUMMARY

List of staining methods used for the tissue components indicated against each is given in Table 11.1. Details of any specific method may be searched in the text.

12

Mounting, Covering and Labeling of Sections

After a histological section of tissue has been stained and cleared, it needs to be mounted in a suitable mounting medium that should be *stable* with *preservative* qualities. Its consistency should be such that it *solidifies* within reasonable time, and should preferably be *colorless* and with *raised refractive index* (for enhancing the visibility of the section). It should not act upon the stains which have been used for the staining of that tissue.

MOUNTING AND MOUNTANTS

Mounting

The term *mounting* has been used loosely among histologists. According to some, it means *attachment* and is used to designate the technique of affixing histological sections to glass slides either directly or after the cleaned slides have been coated with some adhesive. However, more often *mounting* is meant for the placement of some material over a section(s) before covering them with cover slips. In the present chapter, the term mounting will be used in the latter sense.

Mountants

A *mountant* or *mounting medium* is a substance used to permeate the finished stained section and fill the space between it and the cover glass.

Classification of Mounting Media

The commonly used mounting media for routine histological studies can be broadly divided into *two* groups:

1. Water-soluble media

A *water-soluble mountant* is always used for sections that have not been dehydrated. Such sections are usually frozen sections floated on albumenized slides and allowed to dry like paraffin sections. The mounting medium is applied immediately before the edges of the section have time to dry and curl, and after putting a drop or two of the medium a cover slip is placed on them (Figs 12.1 and 12.2).

The common kinds of water-soluble mounting media are:

  i. *Glycerol*: It is generally used to mount sections that are to be retained only for a

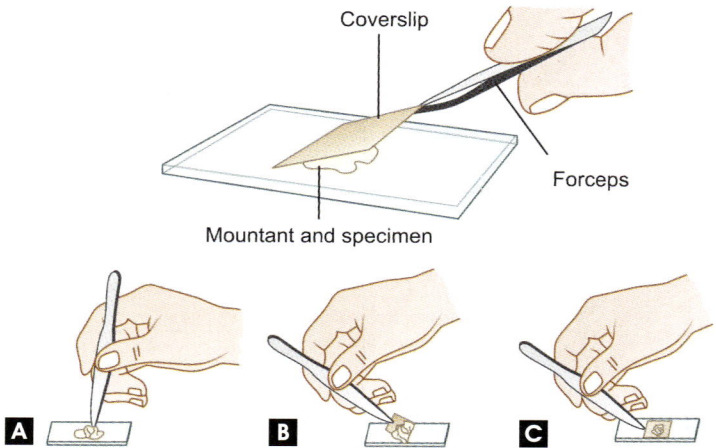

Fig. 12.1: Steps in mounting a stained tissue section on the glass slide: (A) Add a drop of mountant to a stained section on the slide; (B) Place the edge of a cover slip so that it touches the edge of the mountant; and (C) slowly lower the cover slip to prevent forming and trapping air bubble.

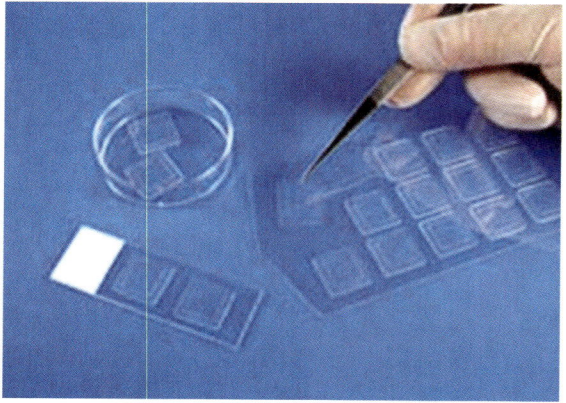

Fig. 12.2: Mounting of sections in a water-soluble mountant shown on a large-sized glass sheet. On the left the sections are mounted on a normal slide used in a histology laboratory.

day or two. If the edges of the cover slip are sealed, the durability of sections mounted with glycerol may be enhanced. However, if left unsealed, glycerol creeps out over the slide and is likely to absorb moisture from the air.

ii. *Kaiser's glycerine jelly*:

Gelatin	10.00 g
Distilled water	60.00 ml
Glycerol	70.00 ml
Phenol	0.25 g

Gelatin should be dissolved in the distilled water using just sufficient heat. After adding glycerol (pure glycerine) and phenol, complete mixing is done and while still fluid, the mixture is decanted into small jars. Before use the mixture must be melted since it gels at room temperature. As phenol causes fading of certain stains, the use of Merthiolate and Zephiran is recommended.

iii. *Apathy's gum syrup*:

Gum acasia (crystals)	50 g
Sucrose (cane sugar)	50 g
Distilled water	100 ml

Mix and dissolve the ingredients with gentle heat. These set quite hard, hence sealing of cover slips is not required.

2. Water-insoluble (or resinous) mounting media

Unlike the aqueous mountants, a water-insoluble medium is applied to sections which must be well dehydrated and wet with a fluid (xylene) that is either the same as that used to dissolve the mounting medium or one that is freely miscible in it.

Water-insoluble or resinous media, like Canada balsam and DPX, which are generally produced by coniferous trees or are synthetic products. Mounting media or *mountants* are usually kept in jars provided with loose-fitting glass-lids. They are usually applied to sections in drops from the end of a glass rod which remains in the jar. Contamination with dust and dirt and evaporation may be avoided by keeping a small quantity of mountant in small containers.

The resinous mounting media may be either *natural* or *artificial*. Canada balsam and DPX are the main examples of commonly used resins.

Canada balsam or Xylene balsam: It has inherent acidity and yellows with age. It is a mixture of turpenes, carboxylic acids and their esters obtained from blisters on the bark of a tree *Abies balsamea*. To memorise the source of Canada balsam is to remember three Bs (B-B-B) for *Balsam-Blisters-Bark*. About 60 per cent by weight of dried natural Canada balsam is dissolved in xylene in a paraffin wax oven. The solution should be filtered and the desired consistency of the mountant obtained by varying the amount of xylene. The refractive index of balsam ranges between 1.52 and 1.54.

DPX: It is a commonly used synthetic resin that is a mixture of *three* substances **d**istrene, a *plasticizer* (**d**ibutylphthalate or tricresyl phosphate), and **x**ylene mixed in the following proportions:

Distrene 10 g
Dibutylphthalate 5 ml
Xylene 35 ml

DPX is nowadays routinely used as a resin mountant. It is colorless and preserves the stains well. Unlike Canada balsam, it dries quickly; and the extra quantity may be peeled off the slide after cutting round the cover slip with an ordinary knife.

The point of practical importance in selecting a mounting medium is the solubility of resin and viscosity of resulting solution. It is convenient to use a resin which is viscous at 30 per cent concentration than a resin which still flows freely at 70 per cent concentration.

Refractive Indices of Mounting Media

One of the important functions of mounting media is to enhance the transparency of the section so that after the cover slip is cemented to the slide (over the section), it can be examined under a microscope. Most of the mounting media, therefore, possess their refractive indices close to that of the glass. The refractive indices of some of the mounting media and other materials in common use are shown below.

Name of the media	Refractive index
1. Air	1.000
2. Absolute alcohol	1.367
3. Alcohol (methyl)	1.323
4. Aniline oil	1.580
5. Canada balsam	
Solid	1.535
In xylene	1.524
6. Castor oil	1.481
7. Cedarwood oil	
Thickened	1.520
Not thickened	1.510
8. Clove oil	1.533
9. Oil of Bergmot	1.464
10. Egg albumen solution	1.350
11. Glycerine jelly	1.470
12. Glycerine	
Pure (100%)	1.470
Aqueous (50%)	1.397
13. Glass (crown)	1.518
14. Liquid paraffin	1.471
15. Xylene	1.497
16. Water	
Distilled	1.336
Sea	1.343

The greatest transparency is obtained when the refractive index of a mounting medium is same as that of the tissue elements. The media having a lower index than that of tissue give a diminished transparency, but greater boldness of details. The media with a higher index than that of tissue elements give greater transparency but diminished visibility of section's details.

Assessing the Refractive Index of Mountants

Majority of the histological tissue sections are mounted in Canada balsam (dissolved in xylene), the refractive index of which is 1.524 and is closer to the most of the tissue components. However, there is always some minor difference in the refractive index of a tissue and that of the mountant around and within it. A procedure, often termed as *Becke Line Test*, helps one to get an approximate idea of the qualitative difference in the refractive index of the surrounding media and that of the specimen mounted.

Becke Lines

The Becke lines are defined as the broad, dark or bright lines (due to refraction and/or diffraction) formed in the image at the boundary between media of different optical path lengths. They move in the direction of the longer optical path when the distance between the objective and the object is increased. The Becke line disappears in the region of the object that lies in exact focus.

Formation of Becke lines is illustrated in the figure below. In (A) the transparent specimen has a higher refractive index than the surrounding medium. When the objective is raised above focus a bright Becke line appears inside the specimen, but the Becke line appears to enlarge and surrounds the specimen when the objective is moved below the focus point. If the specimen has a lower refractive index than that of the medium (B), the situation is reversed and raising the objective above focus produces a bright Becke line surrounding the specimen.

Grains and fibers are often convex in shape and either diverge or converge light, depending upon whether their refractive index is lower or higher than the surrounding medium (usually oil). In either case, light is converged into the medium having a **higher** refractive index. A specimen having parallel faces will produce Becke lines at the boundaries having the higher refractive index by total internal reflection.

Reliable Becke lines may be obtained with medium-power objectives (10x through 50x)

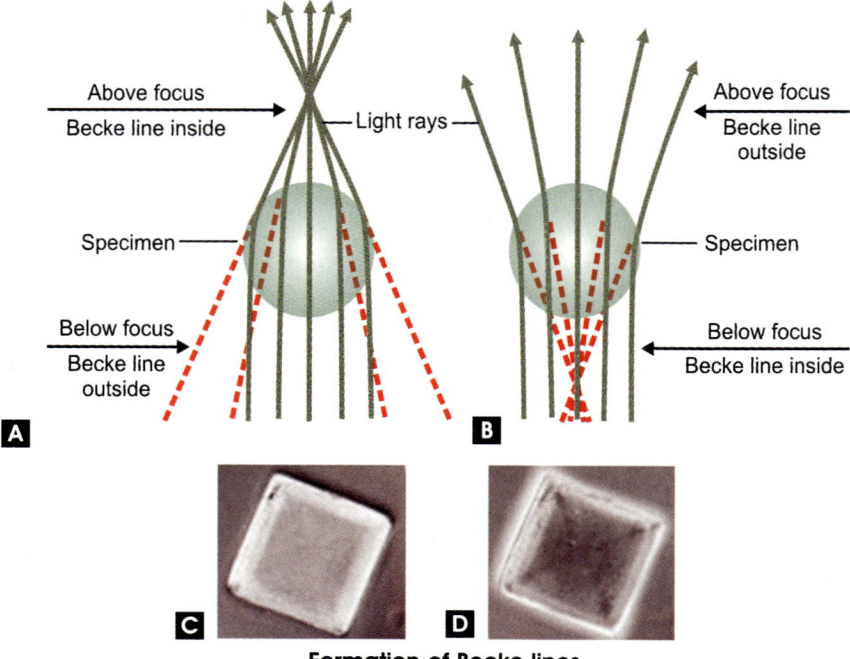

Formation of Becke lines

with numerical apertures ranging from 0.5 to around 0.7. The condenser aperture should be reduced until strong Becke lines are observed through the microscope eyepieces, with increasingly smaller aperture sizes being employed as the refractive index difference between the specimen and medium is narrowed. Adjustment of the condenser aperture should be undertaken while viewing the specimen and progressively closing the diaphragm until good Becke lines are observed.

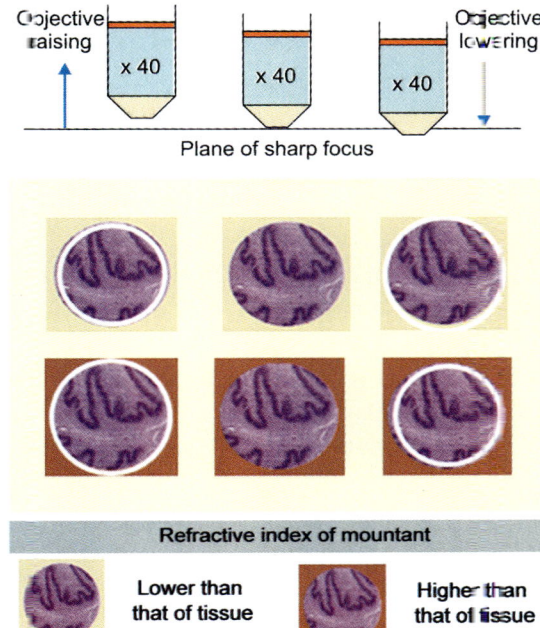

Note

This phenomenon is used to recognize relative differences in refractive index of two adjacent media, for example, a particle and the surrounding mounting medium. When the refractive indices are matched, the Becke line disappears.

Becke Line Test

For the test a given stained mounted section is sharply focused with a medium power microscope objective. If the focus is now raised a bright (but ill-defined) line may be seen with a little care. This line is termed the *Becke line* that was described by Friedrich Johann Karl Becke (1855–1931). As a rule, this line on raising the focus moves to the medium of higher refractive index (Fig. 12.3). Similarly, when the focus is lowered than the correct

Fig. 12.3: Becke line shifting test for assessing the refractive index of the mountant. The line shifts in the medium of higher refractive index (darker shade) on raising the focus just above the correct plane of focus. Or lowering the focus the line shifts into the medium of lower index.

focus plane, the Becke line appears to move in the area of lower refractive index. Thus, a rough idea about the difference in the refractive indices of tissue and that of mounting medium can be obtained.

Friedrich Becke (1855–1931)

The **Becke line test** is a technique in optical microscopy developed by **Friedrich Johann Karl Becke** that helps determine the relative refractive index of two materials. Becke developed a method for determining the relationship between light refraction and refractive index differences observed in microscopic specimens. The phenomenon, which is now referred to as the formation of **Becke lines** (named for Karl Becke). It is done by lowering the stage (increasing the focal distance) of the microscope and observing which direction the light-line appears to "move" toward. The movement will always go into the material of higher refractive index.

MOUNTING TISSUE SECTIONS

To preserve and support a stained section for light microscopy, it is mounted on a clear glass slide, and covered with a thin glass cover slip (Fig. 12.4). The slide and cover slip must be free of optical distortions, to avoid viewing artefacts. A mounting medium is used to adhere the cover slip to the slide. Aqueous based mounting media are available, which allow the mounting of tissues directly from the staining procedure. However, the water solubility of some stains allows them to bleed and/or fade in such mountants, necessitating the use of resinous mounting media. To use a nonaqueous mountant, the section must first be dehydrated (again!) and cleared. Any water carried over to the mounting stage will show up as bubbles or vacuole-like structures, as the water droplets aggregate and distort the tissue. It is important to note also that the clearing

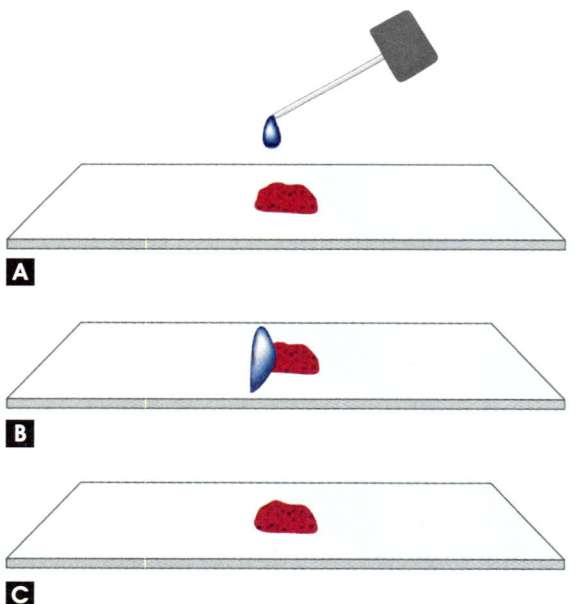

Fig. 12.4: To mount a slide, (A) Apply a single drop of mounting medium upon tissue section. (B) Hold cover slip at 45° allowing the drop to spread along the edge of the slip. (C) Let go of slip and allow medium to spread slowly. On the right is shown the 'mounted' slide using some aqueous mountant.

agent used must be compatible with the mounting medium, or the sections must be thoroughly dried prior to mounting.

COVERING AND LABELING OF SECTIONS

Covering

After a suitable mounting medium is placed over a section, it is to be *covered* by placement of a cover slip over it. The cover slip protects the mounted section and provides a plain and transparent surface through which it can be visualized under a microscope. Care should be taken to prevent air bubbles being trapped between the slide and cover slip during covering of sections.

If the mounting is done with such mounting media which fail to set completely hard quickly, air bubbles are unavoidable. This happens due to evaporation of the solvent of mounting medium. In case of using such mounting media, the edges of cover slip should be coated with some non-porous *ringing medium*. A viscous solution of ringing medium must be applied through a fine brush (or bent wire). Paraffin wax is used as a temporary ringing medium in many laboratories. Nail polish, and some commonly marketed adhesives like Durofix and Quickfix may also be used for the purpose of ringing. These last indefinitely.

Labeling and Sealing

The necessity of keeping the glass slides (with histological sections) correctly *labeled*, from the beginning to the end of their staining processes and afterward cannot be over-emphasized. It is a very embarrassing moment when the slides with sections of different tissues get mixed.

The easiest method of labeling the slides is with the help of a *wax pencil*. But the wax of the marking pencil is soluble in some of the fluids used and is likely to be washed off, hence a *paper-label* is preferable (Fig. 12.5). Such labels may be written with either a waterproof ink

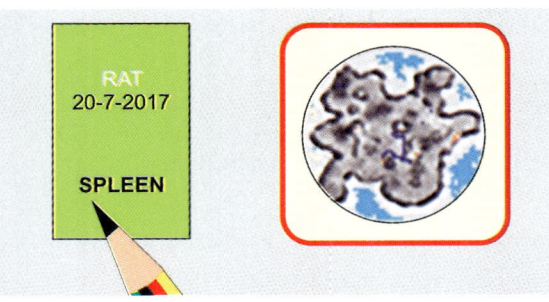

Fig. 12.5: Label pasted on stained, mounted histology slide. Correct labeling of the slide cannot be overemphasized. Ringing of the cover slip may be preferred for keeping the slide indefinitely.

or with India ink and immersed in melted paraffin when the ink is dry. Printed labels are available commercially. The finished slides may be stored in *racks* (Fig. 12.6).

Storage of slides for long periods of time is not generally recommended for enzymes; even at −40°C they can be kept only for about 6 hours. If, however, storage is unavoidable, quenching at −160°C and storage at −85°C will preserve many enzymes in the tissue. Some enzymes on slides can be kept for 3 months at −25°C, but −85°C is even better and serves for a year.

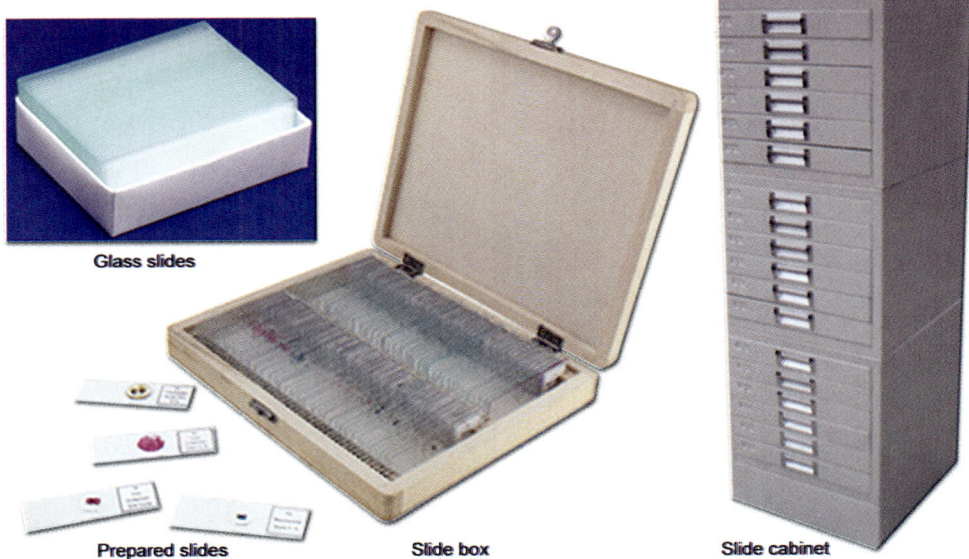

Glass slides

Prepared slides

Slide box

Slide cabinet

Fig. 12.6: Slide cabinet and slide boxes for storing prepared slides.

SUMMARY

Mounting of Slides

Mounting is a term used in histology to designate the technique of affixing histological sections to glass slides (direct or coated with adhesive). More often the term is meant for the placement of some material over a section(s) before applying cover slips.

Types of Mounting Media

A mounting medium or mountant is a substance to permeate a finished stained tissue section and to fill the space between it and the cover slip. The two main groups of mountants are:

i. Water-soluble media
   • Glycerol
   • Kaiser's glycerine jelly
ii. Water-insoluble (Resinous) media
   • Canada balsam—a mixture of turpenes, carboxylic acid, and its ester. It is obtained from blisters on the bark of a tree—*Abies balsamea* (3 **B**s: **B**alsam-**B**listers-**B**ark)
   • **DPX**—a plasticizer (**D**ibutyl **P**hthalate **X**ylene) is a colorless, stain preserving, quickly drying used commonly in routine practice.

Refractive Indices (approximate) of Common Mountants

• Canada balsam (in xylene)   1.524
• Xylene   1.497
• Glycerine (100% pure)   1.470
• Distilled water   1.336

The greatest transparency is achieved when the refractive index of a mounting medium is same as that the tissue components.

Assessment of Mountant's Refractive Index

• A Becke line (named after Karl Becke) is defined as a sharp bright line caused due to refraction and/or diffraction formed in the image at the boundary between media of different refractive indices. The line disappears when a specimen is in correct focus; however, the line moves in the medium of higher index when the distance between the objective lens and section (under examination) is increased slightly by raising the microscope objective (changing the correct focus). A reliable Becke line may be obtained with medium-power (10X) objective of a light microscope.

• **Becke line test**—is a technique for assessing the difference between refractive indices of section and mountant over the section under a cover slip. The test is given to a given mounted section— sharply focusing with a medium power objective. Slight raising the objective (ill-defined focus) brings the Becke line to the medium having higher refractive index. The reverse is true when the objective is slightly lowered bring the Becke line in the medium of lower refractive index.

13

Specialized Techniques in Anatomy

Besides the conventional method of *paraffin-wax* embedding and subsequent sectioning of any given biological sample, another way of preparing sections is *freezing* or *cryostat* method, which will be described in brief only.

USE OF FREEZING MICROTOME AND CRYOSTAT

Freezing Microtome

The pieces of tissue do not require prior fixation and are placed directly on the stage of a freezing microtome (or cryostat) after moistened by dipping briefly in saline solution. The tissue block is then held firmly against the stage with the finger. Bursts of carbon dioxide, under pressure, are allowed to pass close to and under the stage of microtome. This produces an icy seal between the moistened tissue and microtome. Finger is then removed and carbon dioxide released again until whole tissue is *frozen* and appears white. Next step is *thawing*, which is done by dipping the finger

into water and transferring a water drop to the surface of frozen tissue. Usually several drops and momentary pressure over the tissue is required. Sections are now cut. The sections initially obtained will be too thick but as additional cuts are made suitably thin sections appear.

Cryostat

It is easy to consistently create high-quality sections with the Leica CM1860 high precision microtome and stepper motor control. The microtome is completely encapsulated, saving your time with maintenance-free reliability (Fig. 13.1).

Advantages

1. The sections can be cut very quickly. Hence the method is useful where immediate results are needed.
2. Sections may be stained with specific fat stains because tissue is not passed through clearing agents which dissolve fats.

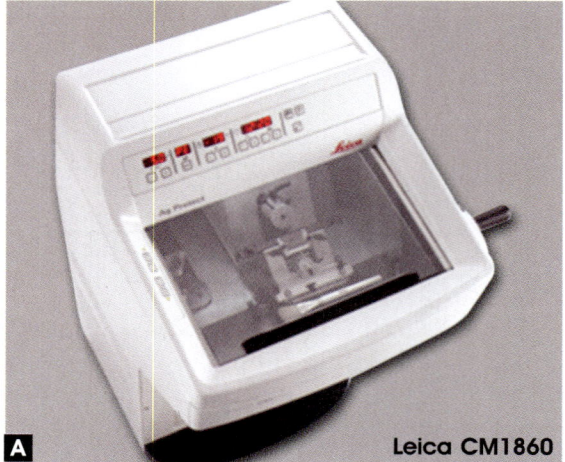

A   **Leica CM1860**

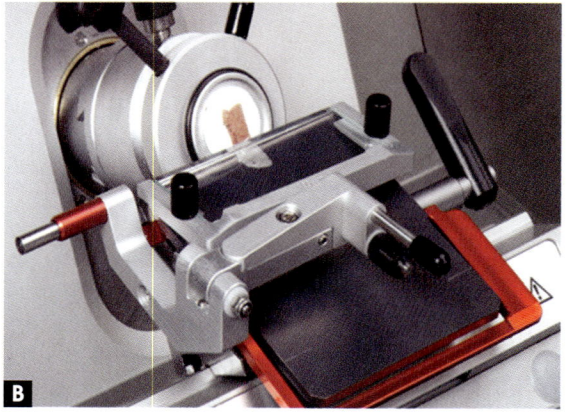

B

Fig. 13.1: (A) A Leica CM 1860 model of cryostat. (B) The interior of the instrument with a section cut from a fresh (unfixed) sample taken during a surgical operation is shown.

Disadvantages

1. The sections are thicker than those obtained by paraffin wax block.
2. Too many details in the specimen cannot be seen.

The flexible and efficient Leica CM1860 workspace includes a new object plate holder, a movable shelf and designated slide slots and tool storage areas on top of the instrument. Control buttons labeled with intuitive icons shorten the time to sectioning. No need to scroll through multiple long menu options.

Save time by directly setting the function you need in a well-equipped Clinical Histopathology Laboratory. The unique premium blade holder safety features combined with the ergonomics of a designated finger guard, blade ejector and palm rest for brush technique users improve safety and comfort while sectioning.

ELECTRON MICROSCOPY OF ANIMAL SPECIMENS

Working Principle of Electron Microscopy

An electron microscope uses an 'electron beam' to produce the image of the object and magnification is obtained by 'electromagnetic fields'; unlike light or optical microscopes, in which 'light waves' are used to produce the image and magnification is obtained by a system of 'optical lenses'.

It has already been discussed that the smaller is the wavelength of light, the greater is its resolving power. The wavelength of green light (= 0.55 μm) is 1,10,000 times longer than that of electron beam (= 0.000005 μm or 0.05 Å; 1 μm = 10,000 Å).

That is why, despite its smaller numerical aperture, an electron microscope can resolve objects as small as 0.001 μm (= 10 Å), as compared to 0.2 μm by a light microscope. Thus, the resolving power of an electron microscope is 200 times greater than that of a light microscope. It produces useful magnification up to 400000x, as compared to 2000x in a light microscope. Thus, the useful magnification is 200 times greater in an electron microscope than that in a light microscope. The scanning by a scanning electron microscope looks at surface only (Figs 13.2 and 13.3).

Fixatives, Staining, and Procedure

Fixation

Good tissue preservation is of paramount importance in electron microscopy, because bad preservation will waste all subsequent painstaking and time consuming efforts in

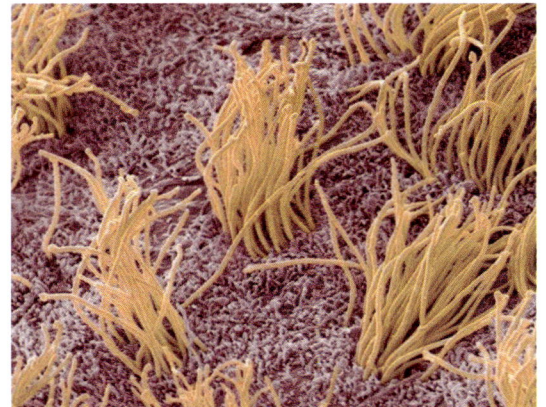

Fig. 13.2: Colored scanning electron micrograph (SEM) of the lining of the brain, showing the ciliary hairs (orange) of ependymal cells.

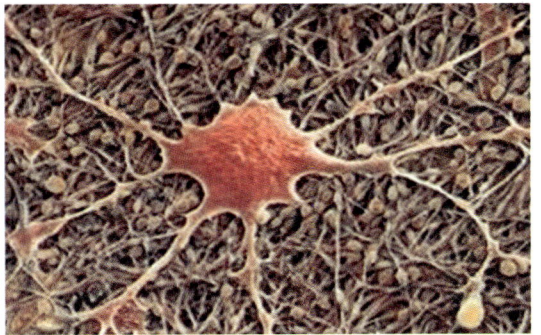

Fig. 13.3: Scanning electron micrograph of an astrocyte—neuroglial cell.

section cutting and microscopy. For initial fixation, most laboratories use Karnovsky's fluid which penetrates much better than buffered osmium tetroxide. The latter, besides being very expensive, the vascular perfusion causes extreme vasoconstriction so that actually only a little or no fixative goes beyond the point of blood vessel puncture site. For electron microscopic work, fixation with formaldehyde is not desirable since its solution contains an unwanted contaminant—methanol, that reacts unfavourably with the buffer used.

The two routinely used paraformaldehyde and Karnovsky's fixatives are prepared as follows.

Paraformaldehyde fixative

The paraformaldehyde powder is dissolved in distilled water by heating the glassware up to 60–65°C and adding a few drops of concentrated sodium hydroxide (NaOH) until the turbidity of the solution clears. The pH of the solution thus obtained generally ranges between 7.1 and 7.3. Usually a 4% paraformaldehyde solution in phosphate buffer (pH 7.3) is used.

Karnovsky's fixative

Paraformaldehyde	2 g
Distilled water	25 ml
Glutaraldehyde (25%)	10 ml
Cacodylate buffer (0.2 M, pH 7.4)	15 ml
Calcium chloride (anhydrous)	25 mg
or	
Calcium chloride (1% aqueous)	5 drops

Paraformaldehyde and water mixture is heated to 60–70°C with constant stirring. Drop by drop 1 N NaOH (sodium hydroxide) is added till the solution becomes clear. It is then cooled under running tap-water and glutaraldehyde added. In the end, cacodylate buffer solution and calcium chloride are added.

After an anesthetized animal is perfused with the one of the above fixatives, careful dissection is done to obtain the sample, taking all precautions mentioned for procurement of tissues (Chapter 3). The block of tissue is cut into tiny pieces of pinhead size (less than 1 mm as a rule) in order to allow good penetration of the fixative solution. The pieces are washed with the same fixative (used for perfusion), with 2–3 quick changes, in order to wash out any adherent particles sticking on the cut surfaces. These pieces are kept overnight immersed in the fixative. These are then washed in 0.2 M cacodylate buffer for 1 hour and *post-fixed* in buffered osmium tetroxide for 1 to 3 hours.

Preparation of Buffered Osmium Tetroxide

Osmium tetroxide (4%)	5 ml
Cacodylate buffer (2%)	5 ml

One gram capsule of osmium tetroxide (OsO_4) is carefully broken in a hood over a jar containing 50 ml of distilled water. It is left overnight to dissolve, and stored in a refrigerator.

Dehydration

The osmicated post-fixed pieces of the given sample are washed several times in deionized water and dehydrated with graded strengths of ethanol. Changes of 10 minutes each in 30, 50, 70 and 80% ethanol are followed by changes of 15 minutes duration each in 90 to 95% ethanol. Then the pieces are kept first in two changes of absolute ethanol; and then in two changes of absolute acetone.

Embedding

Following dehydration, the pieces are embedded in a plastic embedding medium. The embedding is done at room temperature ($23° \pm 1°$ C) in a working plastic mixture for 60 minutes each in 2:1 absolute acetone and plastic mixture, 1:2 absolute acetone and plastic mixture, and pure plastic mixture. Thereafter, the pieces are transferred to fresh plastic mixture in a polythene trough, and kept overnight in an incubator at 60–70° C for curing and polymerization of plastic. The blocks become suitable for microtomy after about two days.

Plastic Embedding Medium

a. Stock mixture

Epon 812	25 ml
Araldite	15 ml
Di butyl phthalate (DBP)	3 ml

Note

The components should be thoroughly shaken for proper mixing.

b. Working mixture (to be prepared fresh)

Stock mixture	9 ml
Do Decenyl Succinic Anhydride (DDSA)	11 ml
2, 4, 6-tridimethyl-aurinomethyl phenol (DMP-30)	30 drops

Note

The DBP is called *plasticizer* and it controls the hardness of the block. DDSA is the *hardner*, and DMP-30 the *accelerator*.

Ultramicrotomy

It is the technique of obtaining ribbons of ultrathin sections from the given sample which has been fixed and dehydrated, and a plastic-block is prepared. This is done by an ultratome device consisting of a glass-knife (or diamond knife) and a block moving system with automatic thermal feed.

The procedure starts by cutting semi-thin sections of about 1 to 2 µm thicknesses. These are picked up on clean glass slides and dried over flame. Staining, generally done with toluidine blue, helps in identification of the desired specific areas in the section which need to be seen with a transmission electron microscope. Nowadays, computerized facility for cutting the sections of desired thickness is provided in sophisticated ultratomes. Routinely, sections of grey or silver grey color are cut; the thickness ranging between 600 and 900Å. Some idea about the section thickness can be made by the following chart.

Color of Section(s)	Section thickness (Å)
Grey	600
Silver	600–900
Gold	900–1500
Purple	1500–1900
Blue	1900–2400
Green	2400–2800
Yellow	2800–3200

Picking the Ribbon of Ultrathin Sections

Unlike ribbons of paraffin embedded sections (which are lifted and attached on adhesive coated clean glass slides), the ultrathin sections cut for transmission electron microscopy are picked on uncoated copper grids. The new copper grids (300 mesh) are cleansed by immersing them in 1 N hydrochloric acid for 3 to 4 minutes. Thereafter, the grids are thoroughly washed with deionized water, with a few rinses in absolute ethanol. For picking up a ribbon, clean copper grid is held with a fine forceps, and bent in such a way so as to make an obtuse angle between it and the forceps. Kept in the same position, grid is gently immersed in the water in the 'boat' and brought carefully under the ribbon. One end of the grid is touched to an end of ribbon; and with the help of a mounted 'eye-lash', the whole ribbon is manipulated over the grid guided by viewing through an optical microscope. The grid is then taken out of the 'boat' with gradual careful tilting. Any excess water on the grid surface is drained off by touching the grid on blotting-paper. After drying, the grids are immediately kept stored in a grid box.

Staining the Ultrathin Sections on a Grid

Staining is done with the use of following two solutions one after the other:

Uranyl acetate

Uranyl acetate	0.5 g
Deionized water up to	100.0 ml

The above solution is prepared in a 250 ml conical flask. It is shaken several times and kept overnight for complete dissolution. The solution is filtered and stored in a glass-stopper bottle in a refrigerator.

Lead citrate

1.33 g lead nitrate [$Pb(NO_3)_2$], 1.76 g sodium citrate ($Na_3C_6H_5O_7.2H_2O$) and 30 ml deionized water is placed in a 50 ml volumetric flask, and shaken intermittently for 1 minute. After about 30 minutes, 8.0 ml 1 N NaOH is added. The suspension is then diluted to 50 ml with deionized water and mixed by inversion. The solution, thus prepared, has an average pH about 12.

Viewing Sections in a Transmission Electron Microscope

The grid with stained ribbons is placed in the specimen holder, and inserted into stage via the specimen airlock. In the TEM, generally an accelerating voltage of 75 kV and suitable condenser and objective apertures are used for examination of the section. The regions of interest in the specimen are located and moved to the center of the fluorescent screen of the microscope. After proper focusing and checking that image is 'stationary', and the illumination 'uniform', exposure of film/plate attached with the camera is made. The exposed film is then removed from the TEM and chemically processed to get the electron micrograph.

Strengths and Limitations of Scanning Electron Microscopy

Strengths

There is arguably no other instrument with the breadth of applications in the study of solid materials that compare with the SEM. The SEM is critical in all fields that require characterization of solid materials. Scanning electron microscopy is most concerned with geological applications. Most SEMs are comparatively easy to operate, with user-friendly "intuitive" interfaces. Many applications require minimal sample preparation. For many applications, data acquisition is rapid. Modern SEMs generate data in digital formats, which are highly portable.

Limitations

Samples must be solid and they must fit into the microscope chamber. Maximum size in horizontal dimensions is usually on the order of 10 cm, vertical dimensions are generally much more limited and rarely exceed 40 mm. For most instruments samples must be stable in a vacuum on the order of 10^{-5}–10^{-6} torr. However, "low vacuum" and "environmental" SEMs also exist, and many of these types of samples can be successfully examined in these specialized instruments. Most SEMs use a solid state X-ray detector (EDS), and while these detectors are very fast and easy to utilize, they have relatively poor energy resolution and sensitivity to elements present in low abundances when compared to wavelength dispersive X-ray detectors (WDS) on most electron probe microanalyzers (EPMA). An electrically conductive coating must be applied to electrically insulating samples for study in conventional SEMs, unless the instrument is capable of operation in a low vacuum mode.

METHODS USED TO STUDY CHROMOSOMAL MORPHOLOGY

Fundamental Knowledge About Chromosomes

The chromosomes (*chrom* = color; *soma* = body) may be identified by their lengths, the position of centromere (kinetochore or primary constriction), and the absence or presence of secondary constriction. A method called *karyotyping* is applied to a systematized array of the chromosomes of a single cell prepared either by photomicrography or drawings made either with free hand or with the help of camera lucida. The resulting preparations are termed *karyogram* or *idiogram* respectively. The latter term is often used only for the diagrammatic representation of a karyotype, usually based on measurements of the length of chromosomes.

Chromosomal spreads for microscopic examination may be commonly obtained from either direct venous blood or cultures of leucocytes from nasal or buccal swabs, or bone marrow cells (Fig. 13.4); and squash preparations from tissue (mainly skin) are other kinds of materials used for the chromosome studies in man.

Procedure for Karyotyping

For the preparation of a karyotype, about 5 ml venous blood is taken with a syringe (Fig. 13.5). The RBCs are then separated and the remaining fluid with suspended WBCs or leucocytes is added to *phytohemagglutinin* (M or P). The culture medium is incubated a 37°C for 3 days under sterile conditions. The mitosis is interrupted at the metaphase stage with spindle inhibitors (like colchicine or viniblastine). Thereafter, the leucocytes are separated after 1 hour and the hypotonic saline solution added.

This results in swelling of individual cells. Cell spreads are made by dropping from some height—a drop of suspension over a clean glass slide (Fig. 13.5), which causes dispersal of chromosomes. The slides are air-dried and stained with either aceto-orcein or aceto-carmine. A good photomicrograph or accurate drawing of a good stained preparation of chromosomes is obtained. The images of individual chromosomes are cut out and trimmed. The chromosomes are arranged in their descending lengths with the centromeres placed along a horizontal line drawn on a sheet of paper. A karyogram or idiogram is thus obtained for the chromosomal analysis.

Other Methods for Chromosome Identification

The procedure of *aminocentesis ultrasonography*, and *foetoscopy* are some of the other commonly used prenatal diagnostic tools to detect chromosomal and/or genetic defects. These procedures are generally done during the early mid-trimester of pregnancy. Detailed descriptions of these procedures are beyond the scope of the present textbook. Also, certain inborn errors of metabolism may be detected if the cells are processed for biochemical analysis or enzyme activities.

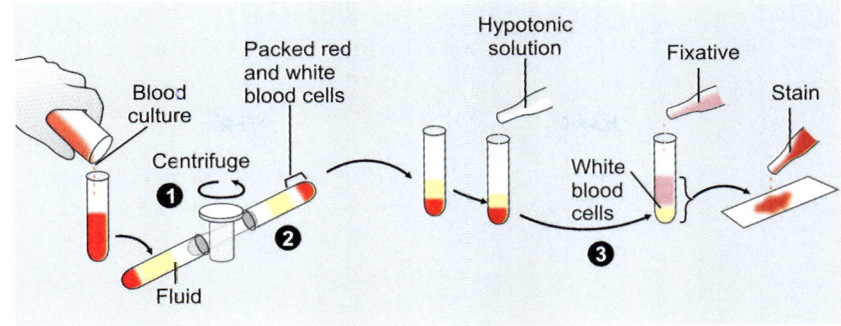

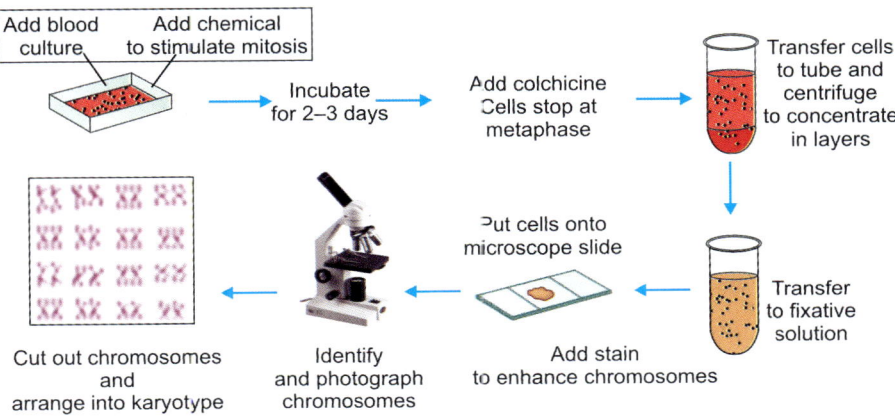

Fig. 13.4: Procedure of karyotyping from blood /tissue culture for the study of chromosomes.

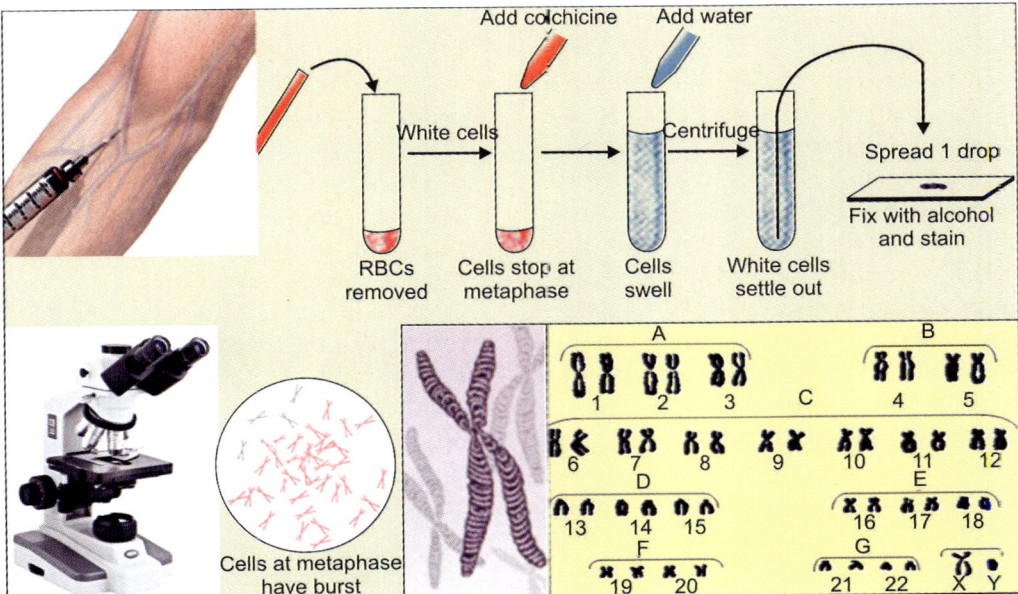

Fig. 13.5: Procedure of karyotyping for the study of chromosomes.

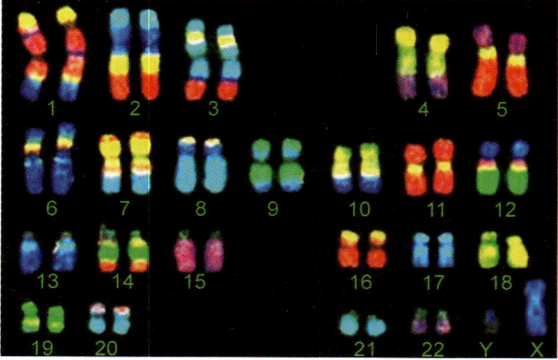

Fig. 13.6: A color karyotype (of a normal male) is prepared through a computerized set-up. Notice that except the sex chromosomes, all other chromosome pairs have homologous chromosomes.

AMNIOCENTESIS

Amniocentesis (also referred to as amniotic fluid test or AFT) is a medical procedure used in the beginning of pregnancy to detect prenatal diagnosis of chromosomal abnormalities in the fetus and fetal infections, and also for sex determination, in which a small amount of *amniotic fluid*, which contains fetal tissues, is sampled from the *amniotic sac* surrounding a developing *fetus*, and then the fetal DNA is examined for genetic abnormalities.

This procedure was first introduced by American obstetrician **Fritz Friedrich Fuchs** and Danish gastroenterologist **Polv Riis** in 1956 for fetal sex determination and up to mid 1970s amniocentesis were done 'blind'. Doctors Jens Bang and Allen Northeved from Denmark were the first to report amniocentesis done with the guide of an ultrasound in 1972. Chorionic villus sampling (CVS) was first performed by Italian biologist Giuseppe Simoni in 1983. Now real-time ultrasound is used during all invasive procedures because it provides for the safety of the fetus and accuracy of results. The most common reason to have an "amnio" is to determine whether a baby has certain genetic disorders or a chromosomal abnormality, such as *Down syndrome.*

Amniocentesis (or another procedure, called *chorionic villus sampling* can diagnose these problems in the womb. Amniocentesis is performed when a woman is between 14 and 16 weeks gestation. Women who choose to have this test are primarily those at increased risk for genetic and chromosomal problems, in part because the test is invasive and carries a small risk of miscarriage. This process can be used for *prenatal sex discernment* and hence this procedure has legal restrictions in some countries.

Amniocentesis is done when a woman is between 16 and 22 weeks pregnant; a sample of amniotic fluid being taken from the amniotic sac surrounding the unborn baby and its DNA is examined for genetic abnormalities. The amniotic fluid has cells that the skin of the developing baby has shed, as well as his/her waste products. Each cell from the baby in the fluid contains their complete set of DNA (genetic information). When these cells are analyzed, there is great help for the doctor to assess the condition of fetus to detect any potential problems. The entire procedure for amniocentesis lasts approximately 45 minutes, most of which involves a detailed ultrasound examination.

Steps for Amniocentesis Procedure

1. Before the start of the procedure, a local anesthetic can be given to the mother in order to relieve the pain felt during the insertion of the needle used to withdraw the fluid.

2. After the local anesthetic is in effect, a physician punctures the sac (by a sterile syringe needle) in an area away from the fetus with the aid and guidance of ultrasound, through the mother's lower anterior abdominal wall, then through the wall of the uterus, and finally into the amniotic sac (Fig. 13.7).

3. A small amount (approximately 20 ml) of amniotic fluid is withdrawn through the needle.

4. The amniotic fluid contains fetal cells, which are separated from extracted sample of the amniotic fluid if used for prenatal genetic

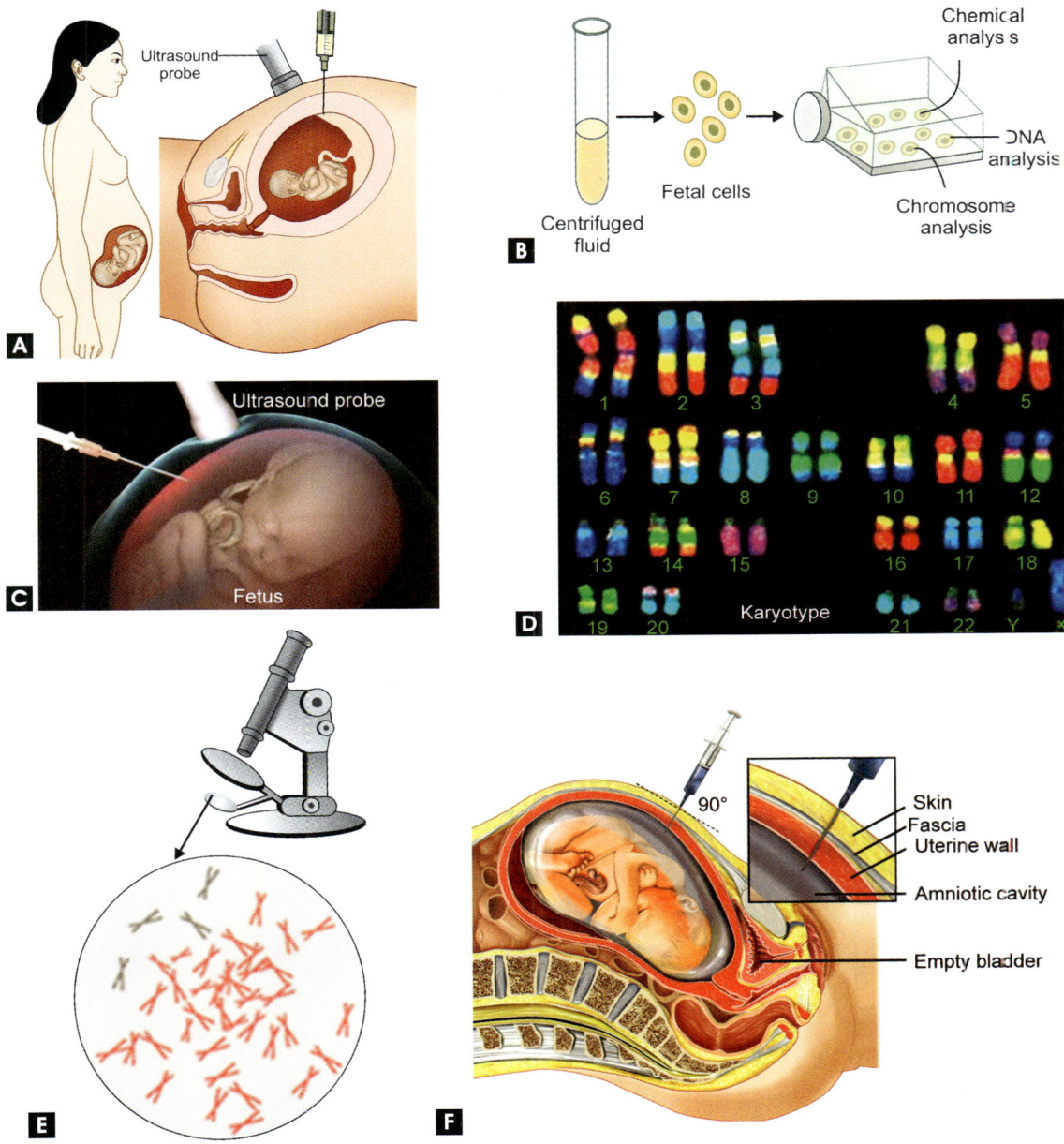

Fig. 13.7: Showing the main steps in the procedure for amniocentesis (see text for detail).

diagnosis. Fetal cells are separated from the extracted sample. The cells are grown in a culture medium.

5. These separated fetal cells are cultured in a suitable medium, then fixed and stained.

6. Tests are then performed on the cultured cells for analysis of: Chemicals, DNA, and chromosomes.

Under a microscope the chromosomes are examined for abnormalities. The most common abnormalities detected are Down syndrome (trisomy 21), Edwards syndrome (trisomy 18), and Turner syndrome (monosomy X). In regard to the fetus, the puncture seals and the amniotic sac replenishes the liquid over the next 24–48 hours.

Results of Amniocentesis

The results of amniocentesis may either be 'positive' (when the developing fetus has the disorder for which the test was done) or 'negative' (when the developing baby will not have the disorders that were tested for). However, it is possible to have a negative result from amniocentesis but the baby may still be born with the condition tested for or another chromosomal condition. This is because a normal test does not exclude every chromosomal disorder. On the contrary, positive test results do not any cure for the majority of chromosomal conditions. Hence, the couple needs to aware about the 'options'

for the further continuation of the pregnancy carefully. The options may include:

- Continuing with her pregnancy while getting information and advice about the condition so she is prepared for caring for her baby.
- Ending her pregnancy—depending on how many weeks pregnant she is when making the decision.

Complications of Amniocentesis

It is important to be aware about the possible complications during or after amniocentesis (Table 13.1).

EMBRYOSCOPY AND FETOSCOPY

These procedures are used for direct visualization of embryo and fetus to detect any abnormalities. These new techniques use fiber optics to look inside the uterus (hysteroscopy) and has been called embryoscopy or

Table 13.1	Complications of amniocentesis
Complication	Remark/feature
Injury from needle	• The placenta (that links a pregnant woman's blood supply to her unborn baby's) may be punctured. The puncture wound usually heals without any more problems developing. • As a safety measure ultrasound (where high frequency sound waves are used to create images) is now commonly used to guide amniocentesis needle. This significantly reduces the risk of injury to the placenta.
Miscarriage (loss of pregnancy)	• There is small risk (around 1 in 100) of causing miscarriage
Club foot	• Baby is born with deformed ankle and foot (higher risk if procedure of amniocentesis is done before week 15 of your pregnancy)
Rhesus disease	• This is caused where proteins in a pregnant woman's blood attack her baby's blood cells. The condition is only possible if the mother's blood is Rh^- and the baby's blood is Rh^+. If this is the case, amniocentesis could trigger rhesus disease if the mother's blood is exposed to the baby's blood during the procedure.
Infection	• In very rare cases (estimated to be 1 in 1,000), an infection may develop if the procedure introduces bacteria in the amniotic sac surrounding the fetus. This can cause one or more of the following symptoms: ⬩ High fever of 38°C (100.4°F) or above ⬩ Tenderness of abdomen ⬩ Contractions of abdomen (tightens then relaxes) Medical advice may be urgently sought if she has any of these symptoms

fetoscopy. Like a dilatation and curettage procedure (commonly abbreviated as D&C), a hysteroscopy is usually performed in a hospital or surgical center under anesthesia. The cervix is dilated large enough to allow passage of a fiber optic telescope. Salt water (saline) is used under high pressure to hold the uterine cavity open to allow adequate visualization.

Embryoscopy

The direct examination of an embryo is performed in the first trimester of gestation by insertion of a lighted instrument called *embryoscope* through the mother's abdominal wall and uterus. The technique may be used to obtain tissue specimens for analysis for evaluation of the cause of miscarriage or to perform needed surgery.

Procedure

- A rigid endoscope is inserted via the uterine cervix in the space between the amnion and the chorion, under sterile conditions and ultrasound guidance, to visualize the embryo (Fig. 13.8) for the diagnosis of structural malformations.

- With direct visualization, the implantation site of the pregnancy can be located in the uterus. The fetus grows inside of a spherical structure called the gestational sac. The sac can be opened and the placental tissue and sometimes the fetus can be identified inside.

- The hysteroscope has multiple channels. These channels allow for the insertion and removal of fluid to keep the uterine cavity clear to the observer. Another channel transmits the optics and yet another provides the light. A final channel allows for the passage of instruments into the uterus. Using grasping instruments, the placental tissue or fetus can be removed.

- There are several benefits of the hystero-scopy approach for the evaluation of the causes for miscarriage. In some cases, when a fetus is present, abnormalities in the development of the fetus can be seen. Collecting the fetal tissue in this way, the specimen is kept sterile and so the chances for the cells to be grown by the genetics lab are enhanced. As stated above, the mixing of the mother's cells with the fetal cells does not occur, so there is much better reliability.

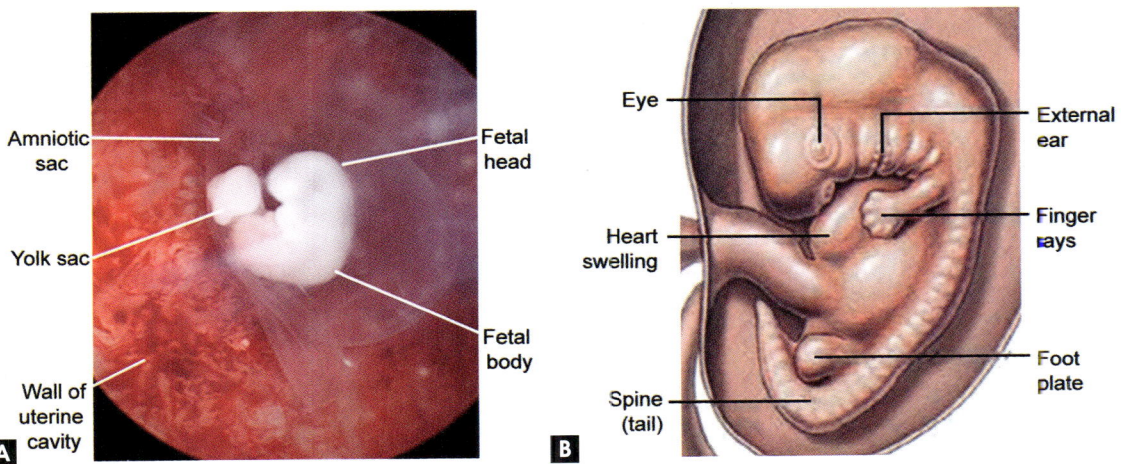

Fig. 13.8: (A) Actual photograph of an embryo of approximately 7th week of pregnancy taken with the help of an embryoscope; (B) Diagram shows main external features of an embryo of 7th week gestation.

Fetoscopy

Fetoscopy is an endoscopic procedure done during the second trimester pregnancy (after 16 weeks gestation) to allow access to the fetus, the amniotic cavity, the umbilical cord, and the fetal side of the placenta. A small (3–4 mm) incision is made in the abdomen, and an endoscope is inserted through the abdominal wall and uterus into the amniotic cavity. Fetoscopy allows medical interventions such as a biopsy or a laser occlusion of abnormal blood vessels or the treatment of spina bifida. The field of surgical fetoscopy was developed by Dr Ruben Quintero.

Procedure of Fetoscopy

- In this technique, a fine caiber endoscope is inserted into the amniotic cavity through a small maternal incision, under sterile conditions and ultrasound guidance, for the visualization of the embryo/fetus (Fig. 13.9) to detect the presence of subtle structural abnormalities
- Fetoscopy is also used for fetal blood and tissue sampling
- The procedure is associated with 3–5% risk of miscarriage

FREEZE-FRACTURE AND FREEZE-ETCHING

Freeze-fracture and *Freeze-etching* are the preparatory techniques whereby specimens, particularly water-containing, can be studied in their natural states.

Freeze-fracture

Freeze-fracture of bulk specimens gives sufficient data concerning the distribution and concentration of electrolytes in a wide variety of animal tissues. The technique involves fracturing or cleaving of a frozen specimen (at low temperatures of 173–177 K) under liquid nitrogen with subsequent transfer to a chamber under high vacuum $<2 \times 10^{-6}$ Torr (Fig. 13.10).

The main unit for freeze-fracture is installed in place of the standard specimen stage of a scanning electron microscope. It consists of a processing chamber for fracturing and coating of frozen specimen which is first fixed by quick freezing in liquid nitrogen and then placed on a cold stage in the vacuum system. Thereafter, the frozen specimen is cut by a cold knife to expose various parts of its cell structure. Fracturing is accomplished by a manually controlled rotary microtome whose depth of

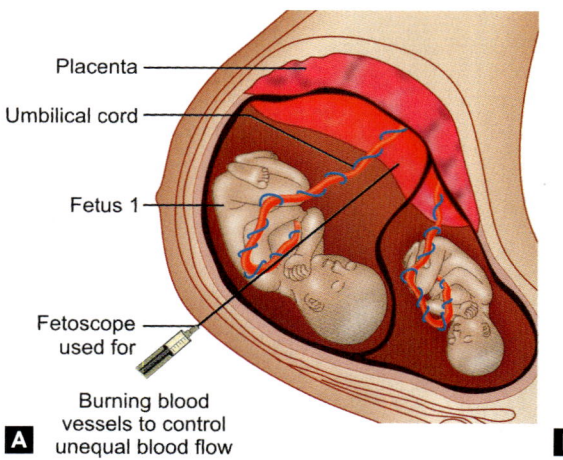

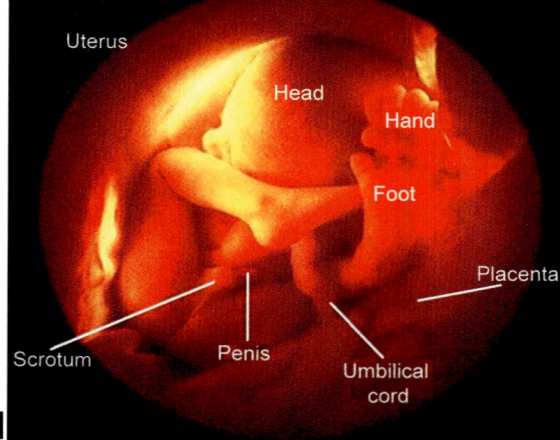

Fig. 13.9: (A) A camera mounted on the fetoscope guides doctors to blood vessels that cause the unequal flow. The scope then burns those vessels, closing them; (B) View of the fetus inside the amniotic sac as seen through a fetoscope.

Fig. 13.10: Freeze-fracture image of protoplasmic-face of plasma membrane of an animal cell. Notice elevated regions which lack the intramembranous particles. Secretory granules are located just beneath the plasma membrane in these regions.

cut is selected by means of a micrometer. The freeze-fracture is through the hydrophobic plane of two lipid layers. Cellular membranes separate into two parts each containing the outer and the inner membrane leaflets as well as unfractured membranes. Freeze-fracture replicas of biological membranes exhibit characteristic *intramembrane particles* (IMPs), at least some of which contain integral proteins. If the fracture plane passes within the plasma membrane of a secretory cell, many elevated areas (about 100–200 nm in diameter) are seen on the photoplasmic-face of the membrane (Fig. 13.11).

The specimens are shadowed in order to obtain a heavy metal-carbon replica of the fractured-faces.

Disadvantages of Freeze-fracture Technique

The classical freeze-fracture technique suffers from one major drawback; the unpredictability of the path of the fracture plane. In addition, other disadvantages in this technique are:

i **Thermal stress cracking** on rapid freezing necessary to minimize ice crystal growth; and

ii **Holes caused by the growth of ice crystal.** However, sample preparation methodology for scanning electron microscopy by freeze-fracture has much less linear shrinkage (20% on a volume; i.e. 7% on a linear) than about 50% shrinkage when critical-point-drying is done for visualizing a specimen with a SEM. Shrinkage in freeze-fracture occurs in two steps: (a) A 6% linear shrinkage corresponding to the loss of ice and (b) a 1% further shrinkage due to the loss of firmly-bound structural water in the given specimen.

Freeze-etching

A method of specimen preparation for electron microscopy in which a replica is made from a sample that has been rapidly frozen and then fractured along natural planes of weakness to

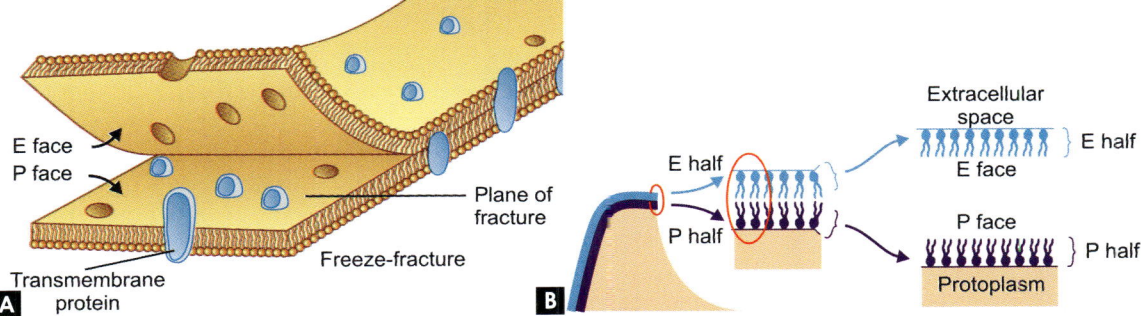

Fig. 13.11: (A) In certain areas of the cell, one also sees protrusions or bumps. These are colored red in the cartoon. Sometimes one can see structure within the bumps themselves. These are the transmembrane proteins. (B) Nomenclature for describing the aspects of membranes revealed by freeze-fracture.

reveal its internal structure. A method used to study unfixed cells by electron microscopy, in which the object to be studied is placed in 20% glycerol, frozen at –100°C, and then mounted on a chilled holder.

Freeze etching is the sublimation of surface ice under vacuum to reveal details of the fractured face that were originally hidden. A metal/carbon mix enables the sample to be imaged in a SEM (block-face) or TEM (replica). It is used to investigate, for instance, cell organelles, membranes, layers and emulsions (Fig. 13.12).

EMBALMING AND PLASTINATION

Embalming

Embalming is the art and science of preserving human remains by treating them with chemicals to forestall decomposition:

• Suitable for public display at a funeral, for religious reasons, or for medical and scientific purposes (anatomical specimens)

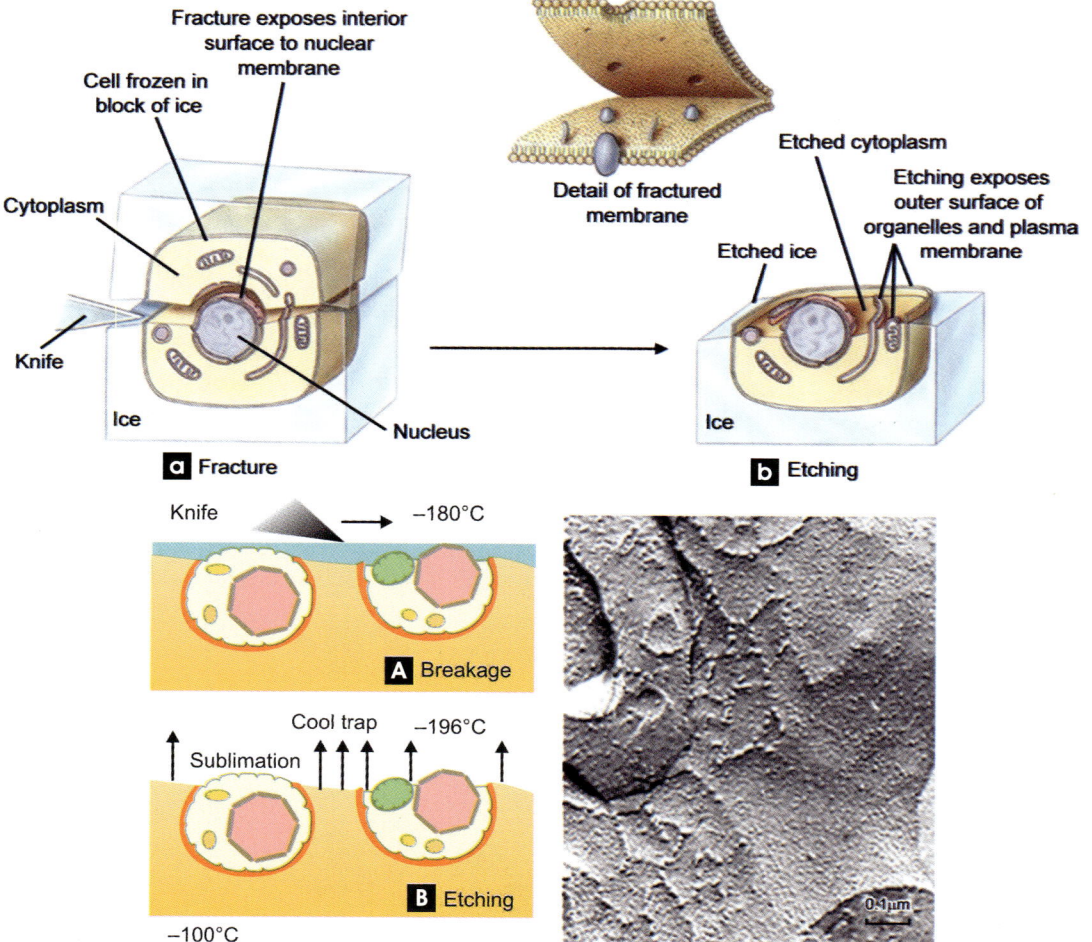

Fig. 13.12: Upper diagram shows (a) freeze-fracture with detail of fractured cell membrane of a typical eukaryotic animal cell, and (b) freeze-etching, which exposes outer surface of organelles and plasma membrane. The lower picture depicts breakage (A) and etching (B). On the right is shown an etched picture of a blood capillary.

- Replacement of blood with preservative chemicals, similar to transfusion.

Embalming is the art and science of preserving human remains by treating them (in its modern form with chemicals) to forestall decomposition. The intention is to keep them suitable for public display at a funeral, for religious reasons, or for medical and scientific purposes such as their use as anatomical specimens. The three goals of embalming are sanitization, presentation and preservation (or restoration). Embalming has a very long and cross-cultural history, with many cultures giving the embalming processes a greater religious meaning.

Types of Embalming

The embalming is of the following *six* types: (i) arterial embalming, (ii) cavity embalming, (iii) hypodermic embalming, (iv) surface embalming, (v) embalming of autopsied body, and (vi) embalming of AIDS body.

Embalming Chemicals

Embalming chemicals are a variety of preservatives, sanitising and disinfectant agents and additives used in modern embalming to temporarily prevent decomposition and restore a natural appearance for viewing a body after death. A mixture of these chemicals is known as **embalming fluid** and is used to preserve cadavers, sometimes only until the funeral, other times indefinitely. Typically embalming fluid contains a mixture of **formaldehyde** (ranging 5 to 29%), **methanol** (ranging from 9 to 56%), and other solvents.

Actions of embalming fluid

- To fix (denature) cellular proteins—so that they cannot act as a nutrient source for bacteria
- Kills the bacteria themselves
- *Formaldehyde* fixes tissue or cells by irreversibly connecting a primary amine group in a protein molecule with nearby nitrogen in a protein or DNA molecule through a

$-CH_2-$ linkage called a *Schiff base.* The end result also creates the simulation, via color changes, of the appearance of blood flowing under the skin.

Modern embalming is not done with a single fixative. Instead, various chemicals are used to create a mixture, called an arterial solution, which is generated specifically for the needs of each case. For example, a body needing to be repatriated overseas needs a higher index (percentage of diluted preservative chemical) than one simply for viewing

Embalming Procedure

Embalming fluid is injected into the arterial system of the deceased. Many other bodily fluids may be drained or aspirated from the body and replaced with the fluid as well. The process of embalming is designed to slow decomposition.

1. **Make sure the body is face up.** If the body is front-side down, gravity will pull the blood down to the lowest parts of the body, particularly the face. This can discolor and bloat the facial features, making it more difficult to create a life-like appearance for the viewing.

2. **Remove any clothing that the person is wearing.** You will need to see the skin for signs that embalming is working, so the body will remain uncovered throughout the procedure. Also remove any IV needles or catheters that are in place.

   a. Typically, you will need to catalog any property found on the person, as well as any cuts, bruises, or other discolorations at this time on your embalming report. This will also be used to document the procedure and chemicals used in the process. The report acts as insurance if the family chooses to sue the funeral home for any reason.

   b. Respect the body at all times. Use a sheet or towel to cover the genitals, and do not

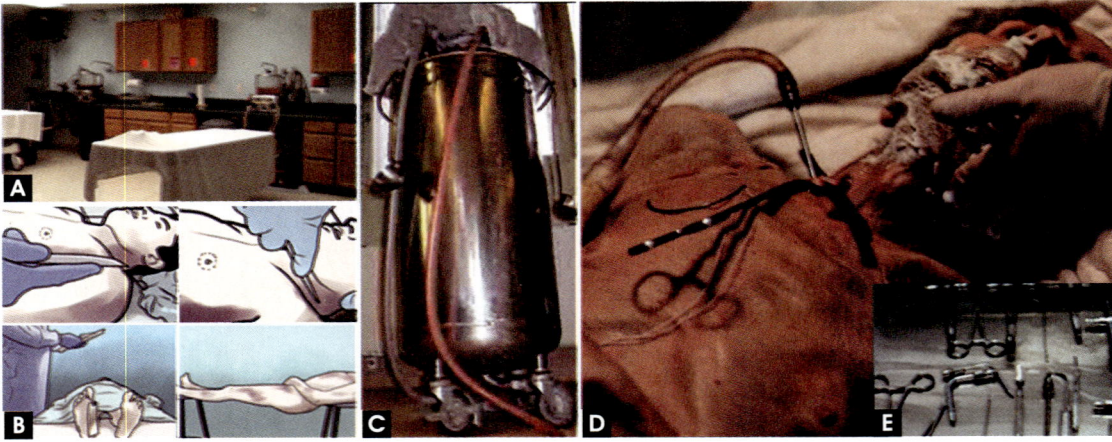

Fig. 13.13: Set-up for embalming of cadavers in an anatomy department: (A) Embalming room equipped with table, embalming chemical, and associated instruments, (B) diagrams showing dead body prepared for embalming, (C) a portable container for embalming solution fitted with tubing and perfusion arrangement, (D) a cadaver being embalmed through canulation in the neck region, (E) dissection instruments set.

leave tools laying around on it while you are working. Assume the family may pop in at any moment.

3. **Disinfect the mouth, eyes, nose, and other orifices.** Powerful disinfectant is used to clean the features, inside and out.

Inspect the deceased in regards to the type of fluid you will need. Some embalmers will use this opportunity to mix all the fluid they will need for the procedure, to get it ready. Usually 16 ounces of fluid with 2 gallons (7.6 liters) of water is a good dilution.

4. **Shave the body.** Typically the face is shaved at this point, as you would shave yourself. Men are usually always shaved, though women and children are also often shaved to remove any stray hairs or "peach fuzz" from the face.

5. **Break the rigor mortis by massaging the body.** Massage the major muscles groups to relieve the tension and move stiff joints to loosen them up. If the muscles are tight, they will increase extra-vascular pressure, diverting embalming fluid away from where it needs to go.

Plastination

Plastination of gross anatomical prosections offers a means of keeping anatomical without the usual deterioration problems associated with wet specimens (desiccation and mould)

Principle for Plastination

Plastination procedure utilizes a silicon base material to produce dry, odorless anatomical specimens that are cost effective in terms of storage and technical preparation time.

Procedure of Plastination

The following method is of general applicability. Plastination is carried out in *four* stages: (i) preparation, (ii) dehydration, (iii) forced impregnation, and (iv) gas cure (Table 13.2).

Preparation

- Specimens are derived from embalmed cadavers.
- The use of low formaldehyde concentration (2.1%) embalming fluid which meets the requirements of the control of substances.
- Before starting dehydration the specimen must be fully dissected and arranged in its final presentation form.

Stage	Process	Procedure	Time taken
\multicolumn{4}{l}{**Table 13.2** **Stages involved in the plastination process**}			
1.	Preparation	Specimen dissected, weighed and soaked in normal saline	24 hours
2.	Dehydration	i. Dehydrate in ascending concentrations of industrial methylated spirit, starting at 50% with 10% increments to 100%	2 weeks at each concentration
		ii. Dehydrate in absolute (100%) acetone using 3 successive baths (Fig. 13.14)	2 weeks in bath 1, 7 to 10 days in bath 2, 7 days in bath 3
3.	Forced impregnation	Dehydrated specimen is placed in plastination polymer, Biodor S10 and S3 and acetone is removed under vacuum	Varies depending on size and density of specimen. For example, heart takes 4 to 6 days; a pelvis takes 12 to 18 days
4.	Gas curing	After thorough draining the specimen is dried in gas cure	Varies depending on size and density of specimen. For example, heart takes 2 to 4 days; a pelvis takes 4 to 7 days

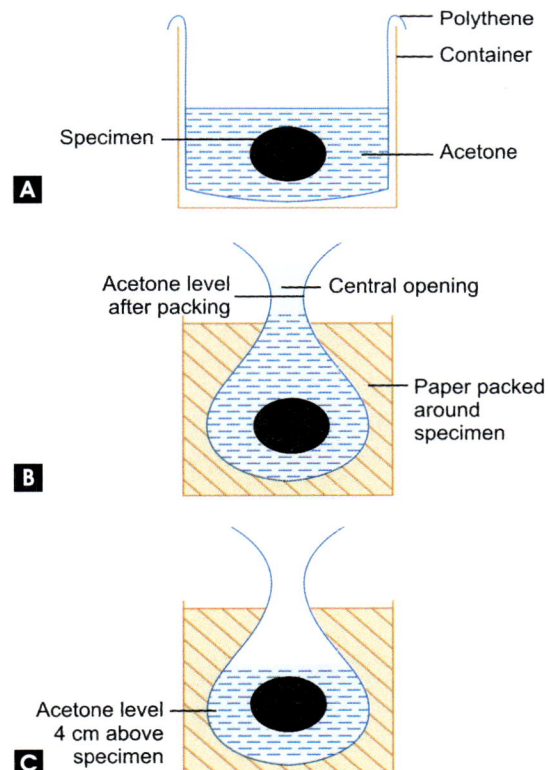

Fig. 13.14: Diagrammatic representation of steps for plastination of a small specimen: (A) Construction of a 'nest' for specimen, (B) shows how the volume of the 'nest' is reduced, (C) shows the amount of acetone in the 'nest' to provide the appropriate volume of polymer.

- The weight of the specimen must be recorded so that the correct volume of dehydrating fluid can be calculated during the dehydration stages.
- The specimen is then soaked for 24 hours in 0.1 M saline.

Dehydration

The ratio of dehydration solvent to specimen weight is 10:1[w/v]. Dehydration is first carried out in industrial methylated spirit starting at 50% and increasing by 10 to 100%. The specimen remains in each concentration for 2 weeks (Table 13.2), stirring daily. The container containing dehydrating fluid should have air-tight lids. The methylated spirit which is not greatly discolored can be reused for baths of 50 to 70% concentrations for as long as it remains clear. Further dehydration is in absolute 100% acetone which involves three successive baths. The specimen remains in the first bath for 14 days, 7 to 10 days in the second, and 7 days in the third, stirring daily. The final acetone bath, if it is not greatly discolored, can be reused for first acetone bath as long as it remains clear. Acetone is decanted outside the building using gloves and safety goggles, and it is stored in sealed containers in a fume cupboard, consistent with safety measures in force at the place of work.

Display of Plastinated Organs

Some of the plastinated soft body parts from a human dead body are shown in Fig. 13.15. They are odorless and almost in their natural color. However, the plastination is not commonly affordable because of its too high cost in preparation; the specimens must be very carefully handled.

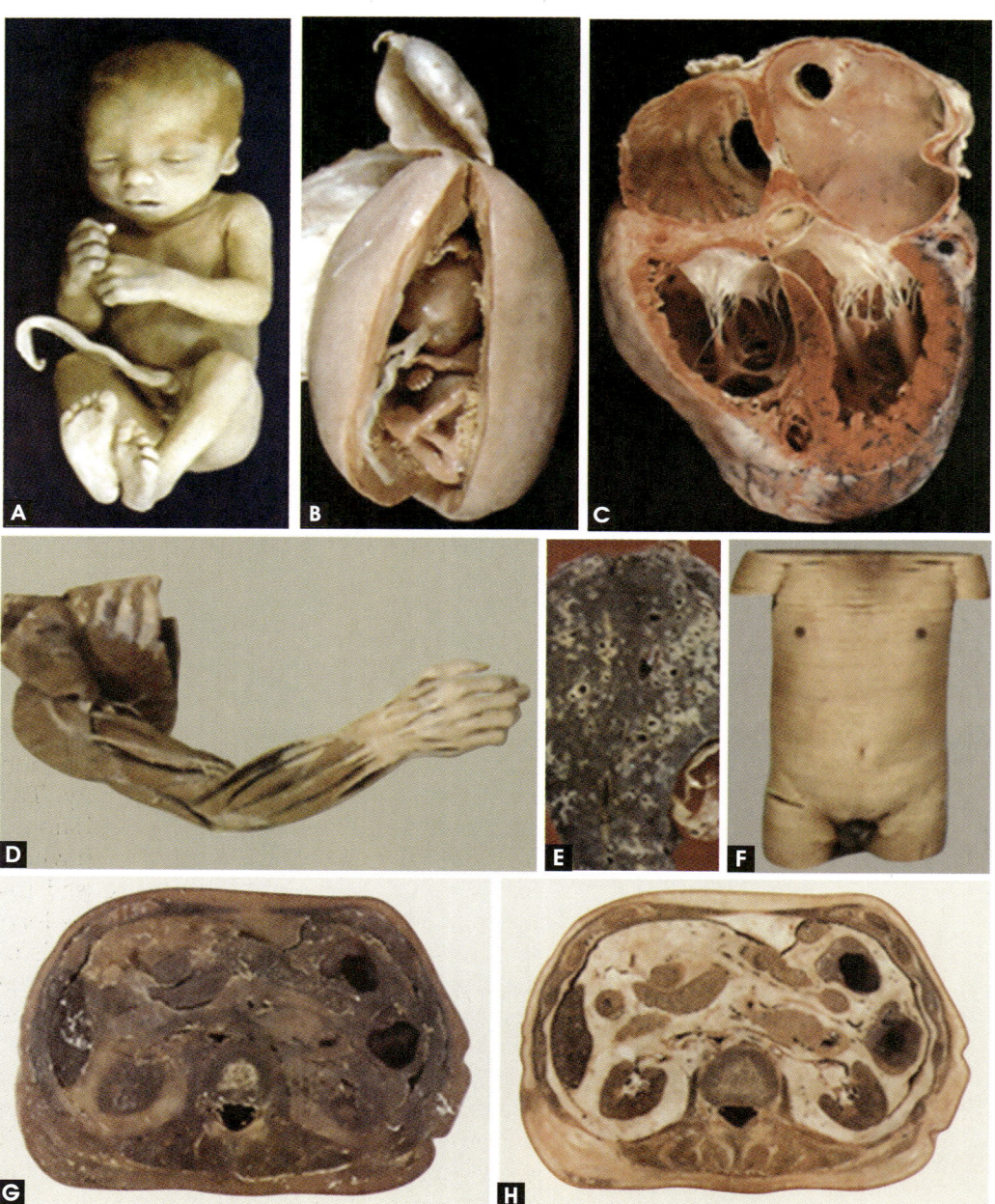

Fig. 13.15: Some of the plastinated soft part from dissection of human body: (A) Near term fetus, (B) unborn fetus inside the uterine cavity, (C) heart chambers, (D) muscles of the right upper limb, (E) vertical section through a lung, (F) anterior abdominal wall and groin region, (G and H) transverse sections through abdomen.

SUMMARY

Freezing Microtomy

- Tissues do not require prior fixation and are placed directly on the stage.
- Block is then held firmly against the stage with the finger.
- Bursts of CO_2, under pressure, are allowed to pass under the stage of microtome allowing icy seal between the moistened tissue and microtome.
- Finger is then removed and CO_2 released again until whole tissue is *frozen* and appears white.
- *Thawing* is done by dipping the finger into water and transferring a water drop to the surface of frozen tissue.
- Sections cut initially, are too thick but subsequently suitably thin sections follow.

Cryostat-sectioning

- Sections can be cut very quickly (useful where immediate results are needed).
- Sections suitable for specific fat stains because specimen is not processed through clearing agents which dissolve fats.
- But, sections are thicker than those obtained by paraffin block cutting.

Electron Microscopy of Animal Specimens

- An 'electron beam' produces the image of the object
- Magnification is obtained by 'electromagnetic fields' unlike light microscopes
- Electron microscope can resolve objects as small as $0.001\ \mu m$ ($= 10\ \text{Å}$), as compared to $0.2\ \mu m$ by a light microscope; the resolving power of an electron microscope is 200 times greater than that of a light microscope.
- Usually a 4% paraformaldehyde solution in phosphate buffer (pH 7.3) is used as fixative
- After proper dehydration, embedding is done at room temperature ($23° + 1°C$) in a working plastic mixture for 60 minutes each in 2:1 absolute acetone and plastic mixture, 1:2 absolute acetone and plastic mixture, and pure plastic mixture.
- Ribbons of ultrathin sections (ultramicrotomy) are cut by an ultratome with a glass-knife (or diamond knife); the block moves with automatic thermal feed.
- Staining is done with the use of following two solutions one after the other: Uranyl acetate and lead citrate
- A grid with stained ribbons is placed in the specimen holder is inserted into stage via the specimen airlock.

- In the TEM, accelerating voltage of 75 kV and suitable condenser and objective apertures are used for examination of the section.

Methods Used to Study Chromosomal Morphology

- *Karyotyping* is systematized array of the chromosomes of a single cell prepared either by photomicrography or drawings.
- The resulting preparation, termed *karyogram, shows* chromosomal spreads for microscopic examination commonly obtained from either direct venous blood or cultures of leucocytes from nasal or buccal swabs, or bone marrow cells.

Procedure for Karyotyping

About 5 ml venous blood is taken with a syringe; RBCs are then separated and the remaining fluid with suspended WBCs is added to *phytohemagglutinin;* the culture medium is incubated a 37°C for 3 days under sterile conditions. Mitosis is interrupted at metaphase with spindle inhibitors (like colchicine or viniblastine). Thereafter, the leucocytes are separated after 1 hour and the hypotonic saline solution added. This results in swelling of individual cells. Cell spreads are made by dropping from some height—a drop of suspension over a clean glass slide, which causes dispersal of chromosomes. The slides are air-dried and stained with acetocarmine. A good photomicrograph of stained preparation of chromosomes is obtained.

Amniocentesis

Amniocentesis is a medical procedure used in the beginning of pregnancy to detect prenatal diagnosis of chromosomal abnormalities in the fetus and fetal infections, and also for sex determination. (*see* Figure on the next page).

Embryoscopy and Fetoscopy

These procedures are used for direct visualization of embryo and fetus to detect any abnormalities. These new techniques use fiber optics to look inside the uterus. Fetoscopy is an endoscopic procedure done during the second trimester pregnancy to allow access to the fetus, through a 3–4 mm incision is made in the abdomen.

Freeze-fracture and Freeze-etching

Freeze-fracture of bulk specimens gives sufficient data concerning the distribution and concentration of electrolytes in a wide variety of animal tissues. The

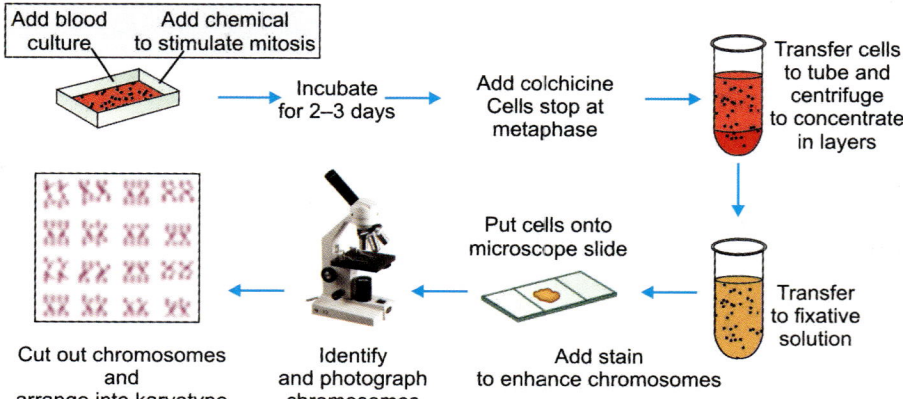

technique involves fracturing or cleaving of a frozen specimen (at low temperatures of 173–177 K) under liquid nitrogen with subsequent transfer to a chamber under high vacuum $<2 \times 10^{-6}$ Torr. *Freeze-etching* is a method of specimen preparation for electron microscopy in which a replica is made from a sample that has been rapidly frozen and then fractured along natural planes of weakness to reveal its internal structure.

Embalming and Plastination

Embalming is the art and science of preserving human remains by treating them with chemicals to forestall decomposition. The embalming is of the following *six* types: (i) arterial embalming, (ii) cavity embalming, (iii) hypodermic embalming, (iv) surface embalming, (v) embalming of autopsied body, and (vi) embalming of AIDS body. **Embalming chemicals** are a variety of preservatives, sanitising and disinfectant agents and additives used in modern embalming to temporarily prevent decomposition and restore a natural appearance for viewing a body after death.

Plastination procedure utilizes a silicon base material to produce dry, odorless anatomical specimens that are cost effective in terms of storage and technical preparation time.

Annexure
Basic Histological Techniques

Common Questions with Answers during viva voce of Practical Examination	
Questions	**Answers**
What is histology?	• The branch of anatomy that studies the microscopic anatomy of animals and plants • In its broader aspect, the word histology is used as if it were a synonym for...? —*microscopic anatomy*
What is extracellular fluid derived from?	• The plasma of the blood
What steps are required to prepare tissues for light microscopy?	1. Tissue collection 2. Fixation 3. Dehydration and clearing 4. Embedding 5. Sectioning 6. Mounting and staining
What does fixation refer to?	• The treatment of tissue with chemical agents that prevent degradation and maintain normal tissue architecture
According to the book, what are the two most common chemical fixative agents used in light microscopy?	• Formaldehyde • Bouin's fluid
What effect do both of these substances have on a tissue?	• They cross-link proteins
What does this help to maintain?	• A life-like image of the tissue
What are used to remove all of the water from a tissue sample?	• A graded series of alcohol baths
What is the step following dehydration called?	• Clearing
What does dehydration consist of?	• Replacing the dehydrant (alcohol) with a substance that is miscible in the embedding medium
What is the dehydrant usually replaced with?	• Xylene
What is the appearance of the tissue in xylene?	• Transparent
What is the usually embedding medium?	• Paraffin Wax
Describe what occurs during embedding.	1. Tissue is placed in a suitable contain of melted paraffin until it is completely infiltrated 2. Tissue is placed into a small receptacle and covered with melted paraffin 3. Tissue is allowed to harden, forming a paraffin block containing the tissue
What is the step following embedding called?	• Sectioning or microtomy

Contd...

Questions	Answers
What is the task performed with?	• A microtome
What is a microtome?	• A machine equipped with a blade and an arm that advances a block of tissue in specific equal increments
For light microscopy, what is the thickness of each section?	• About 5 to 10 micrometers (µm)
Sectioning can also be performed on specimens frozen in either...?	1. Liquid nitrogen 2. The rapid-freeze bar of a cryostat
What is a cryostat?	• An apparatus for microtomy maintaining a very low temperature
How are cryostat sections mounted?	• By the use of a quick-freezing mounting medium
How are cryostat sections sectioned?	• At subzero temperatures using a pre-cooled steel blade
What happens to the samples after they are sectioned?	• They are placed on pre-cooled glass slides and permitted to come to room temperature
On what type of slides are paraffin sections mounted?	• Adhesive-coated glass slides
What are the three main classes of histological stains?	1. Stains that differentiate between acidic and basic components of the cell 2. Specialized stains that differentiate the fibrous components of the ECM 3. Metallic salts that precipitate on tissues, forming metal deposits on them
What are the most commonly used stains in histology?	• Hematoxylin and eosin (H&E)
Is hematoxylin acidic or basic?	• Basic
Therefore, what components of a cell does it preferentially color?	• Acidic components
What color does it stain tissues?	• Blue
What regions stain blue?	• The nucleus and regions of the cytoplasm rich in ribosomes
What are these components referred to as?	• Basophilic
Is eosin acidic or basic?	• Acidic
Therefore, what components of a cell does it preferentially color?	• Basic components
What color does it stain tissues?	• A pinkish color
What region(s) of a cell typically stain a pinkish color?	• The cytoplasm
Why?	• Because many cytoplasmic constituents have a basic pH
What are these elements said to be?	• Acidophilic
What color does treatment with hematoxylin result in?	• Blue

Contd...

Questions	Answers
What parts of the cell does hematoxylin stain?	1. Nucleus 2. Acidic regions of the cytoplasm 3. Cartilage matrix
What color does treatment with eosin result in?	• Pink
What parts of the cell does eosin stain?	1. Basic regions of the cytoplasm 2. Collagen fibers
What color does treatment with Masson's trichrome result in?	1. Dark blue 2. Red 3. Light blue
What components of the cell does Masson's trichome stain dark blue?	• Nuclei
What components of the cell does Masson's trichome stain red?	1. Muscle 2. Keratin 3. Cytoplasm
What components of the cell does Masson's trichrome stain light blue?	1. Mucinogen 2. Collagen
What color does treatment with orcein's elastic stain result in?	• Brown
What components of the cell does it stain?	• Elastic fibers
What color does treatment with Weigert's elastic stain result in?	• Blue
What components of the cell does it stain?	• Elastic fibers
What color does treatment with silver stain result in?	• Black
What components of the cell does it stain?	• Reticular fibers
What color does treatment with iron hematoxylin result in?	• Black
What components of the cell does it stain?	1. Striations of muscle 2. Nuclei 3. Erythrocytes
What color does treatment with periodic acid-Schiff result in?	• Magenta
What components of a cell does it stain?	1. Glycogen 2. Carbohydrate-rich molecules
What color does treatment with Wright's and Giesma stains result in?	1. Pink 2. Blue
What is this particular type of stain used for?	• The differential staining of blood cells
What type of blood cells and cell components stain pink?	• Erythrocytes • Eosinophil granules
What type of blood cells and cell components stain blue?	• Cytoplasm of monocytes • Cytoplasm of lymphocytes
What happens to molecules of some stains when exposed to high concentrations of polyanions?	• They polymerize with each other
Compare the color of these aggregates to the color of individual molecules.	• They differ

Contd...

Questions	Answers
Toluidine blue stains tissues blue except for those that are rich in polyanions. What color do these stain?	• Purple
What is a tissue or cell component that stains purple with this stain said to be?	• Metachromatic
Toluidine blue is said to exhibit…?	• Metachromasia
What are modern day light microscopes known as?	• Compound microscopes
Why are they referred to as such?	• Because they use more than just a single lens
What is the light source within a compound microscope?	• An electric bulb with a tungsten filament
How is light gathered into a focused beam within a compound microscope?	• By a condenser lens
Where does the light beam originate?	• At the bottom of the microscope
What does light passing through the specimen enter?	• One of the objective lenses
Where do the objective lenses sit?	• On a movable turret located just above the specimen
How many objective lenses are usually available on a single turret?	• Four
Generally, by what factor do the lenses magnify a specimen by?	1. ×4 times 2. ×10 times 3. ×40 times 4. ×100 times
The image from the objective lens is gathered and further magnified by?	• The ocular lens of the eyepiece
What does this lens usually magnify by a factor of?	• 10
For total magnifications of?	• 40, 100, 400, and 1000
What is focusing of the image performed by?	• The use of knurled knobs that move the objective lenses up or down above the specimen
The quality of an image depends not only on the capability of a lens to magnify but also on it…?	• Resolution
What is the resolution of a lens?	• Its ability to show that two distinct objects are separated by a distance
What is the theoretical limit of resolution of a light microscope?	• 0.25 nano meters
What is this restriction determined by?	• The wavelength of visible light
What two techniques can be used to localize the specific chemical constituents of tissues and cells?	1. Histochemistry 2. Cytochemistry
How are reactions of interest monitored?	• By the formation of an insoluble precipitate that takes on a certain color
A common histochemical reaction makes use of what?	• The periodic acid-Schiff (PAS) reagent
What color precipitate does PAS reagent form?	• Magenta
With what type of molecules does it form precipitate?	1. Glycogen-rich molecules 2. Carbohydrate-rich molecules
How can one ensure that a reaction is specific for glycogen (not carbohydrate)?	• Consecutive sections are treated with amylase
What color or sections not treated with amylase?	• Magenta

Contd…

Questions	Answers
What color are sections that ARE treated with amylase?	• No stain
When histochemical procedures are produced, what exactly is visualized?	• The product of an enzymatic reaction
How is the reagent designed?	• So that the product precipitates at the site of the reaction
What two major steps comprise immunocytochemistry?	Steps of immunocytochemistry: (i) Cell seeding, and immunostaining, (ii) imaging and image analysis (Fig. 1.11)

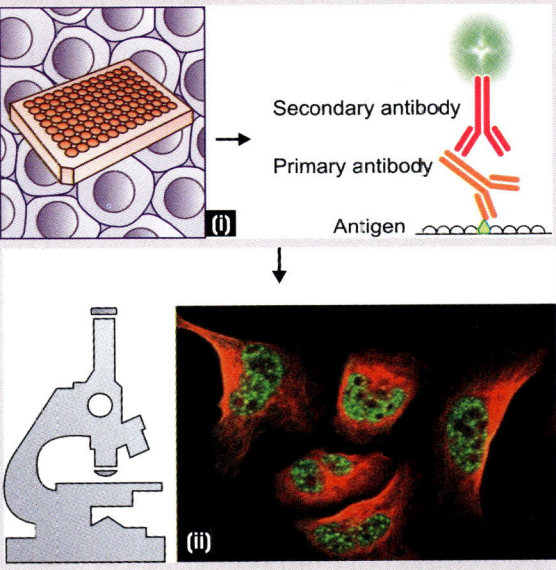

(i) The cells are seeded on a solid support, e.g. into a 96 well plate with glass bottom or on a glass slide. Depending on the type of cell and seeding technique, an incubation time might be necessary before proceeding with immunostaining; e.g. in the case of seeding adherent cells, the cells will attach to the solid support surface during the incubation, which varies from half an hour to 24 h for the different cell types. The cells are immunostained: Cells are fixed, permeabilized, and stained with antibodies. Fixation retains the proteins in the cell and preserves their chemical and structural state at the time of fixation. It can be done for example by cross-linking or by precipitating the proteins. During permeabilization, lipids are often removed from membranes allowing the antibodies to cross the membranes.

Contd...

Questions	Answers

Without this step the antibodies are restricted to the outside of the cell due to their size.

(ii) The cells are visualized using a microscope and images acquired. Images are analyzed and cellular structures are annotated.

Marie François Xavier Bichat (1771–1802), a French anatomist and physiologist, is best remembered as the **Father of modern histology**. Despite working without a microscope, he was the first to introduce the notion of tissues as distinct entities.

Microscopy and Micrometry

14

Microscope and Microscopy: Overview and Historical Account

HISTORY AND BACKGROUND

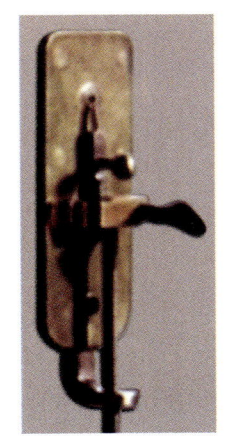

Robert Hooke (1635–1703), upon examining a piece of cork with a rudimentary microscope, saw an abundance of empty small compartments—the cells (Latin, *cellula*—"cell". In the year 1665—there was seen transition from "empty cell" to the actual cell. **Hooke** and **Marcello Malpighi** were the first to observe the true units that form the tissues of animals. In 1678, after **Leuwenhoek** reported of discovering *little animals* (animalcules)—bacteria and protozoa to the Royal Society. Hooke confirmed Leewenhoek's findings; and noted that Leewenhoek's simple microscopes gave clearer images than his compound microscope, but found simple microscopes difficult to use.

Discovery of Nucleus

The first description of the nucleus was carried out by Leewenhoek, in 1700, when examining the red blood cells (RBCs) of the salmon. The first description of the nuclear envelope was accomplished by **Jan Evangelista Purkinje** (1787–1869), a Czech biologist, in 1830. **Robert Brown** (1773–1858), a Scotish botanist introduced the term nucleus in microscopy after the examination of epidermal cells of some orchids in 1831.

Cell Theory

The "cell theory" was put forward by stating cell as the fundamental unit that constitutes all the generality of animals and plants. **Matthias Schlelden** (1804–1881), a German botanist, in 1838, saw under the microscope, thousands of plant specimens, and inferred that all the vegetables are made of cell. Theodor Schwann (1810–1882), a German zoologist and physiologist, in 1839 came to the conclusion that all the then known living beings were composed of cells.

"Tissue"—texture

A French pathologist **Marie François Bichat** (1771–1802) verified that certain textures ("21 Textures") presented a thin thickness and were very flat, to the extent of being compared to pieces of cloth. The first notion of "tissue" (French, *tissue*—means to weave), a texture was a "tissue"—a body component as perceived by its macroscopic physical properties.

Tissue is each of the elementary multicellular components, microscopically and functionally distinct, that constitute either animals or plants where they associate to form organs and systems. Tissues are composed of cells and extracellular matrix. There are four fundamental tissues:

• Epithelia
• Connective tissue
• Muscle tissue
• Nervous tissue

DEVELOPMENT OF MICROSCOPE

A **microscope** (from the Ancient Greek: *mikrós*, "small" and *"skopeîn"*, "to look" or "see") is an instrument used to see objects that are too small for the naked eye. The science of investigating small objects using such an instrument is called microscopy. Microscopic means invisible to the eye unless aided by a microscope. The word *microscope* is derived from the Greek *micros* (small) and *skopeo* (look at). Ever since the dawn of science, man has been interested in being able to look at smaller and smaller details of any given specimen.

Microscopes are instruments designed to produce magnified visual or photographic images of small objects. The microscope must accomplish *three* tasks:

1. Produce a magnified image of the specimen
2. Separate the details in the image, and
3. Render the details visible to the human eye or camera.

The **optical microscope**, often referred to as the "**light microscope**", is a type of microscope which uses visible light and a system of lenses to magnify images of small samples.

It is actually unknown who invented the first microscope. **Euclid (**400 BC) investigated the properties of curved reflecting surfaces. Seneca (4 BC–65 AD) reported that water-filled globules of glass would assist in seeing minute objects. **Claudius Ptomely** (127–151 AD) studied some problems of magnification and made some *burning glasses*. Greeks and Romans in 1000 AD made the first use of lenses to manipulate images. The magnifying glasses (less thinner at the edge than at the center) were, however, not extensively used until the invention of spectacles in the thirteenth century.

Between 1590 and 1610 **Zaccharias Janssen** and his son **Hans** invented a compound microscope (Fig. 14.1) by placing two lenses in a tube. They combined lenses for seeing object and discovered a second lens would magnify the enlarged image from a magnified glass. Their magnification was about 9 times.

About 75 years later, in 1665, **Robert Hooke** constructed a microscope (Fig. 14.2) consisting of an objective lens, a field glass, and an eyeglass. His microscope could magnify objects about 14 to 42 times. Robert Hooke made use of a microscope power of 30X to describe and coin the phrase "cell"—hence he is regarded as the "**father of microscopy**".

A decade later, in 1675, **Antony van Leeuwenhoek** (Fig. 14.3) developed a remarkably simple microscope by mounting a lens between two flat pieces of metal (Fig. 14.4) with only one lens and was the first to describe bacteria. The instrument made by

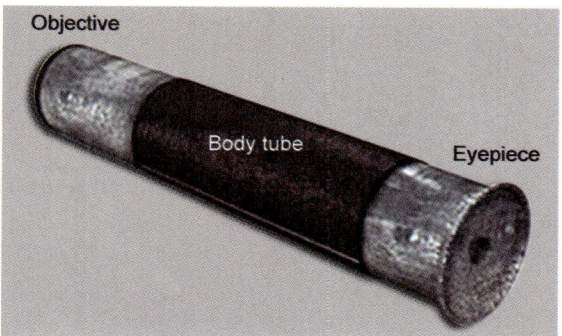

Fig. 14.1: The first compound microscope (circa 1595). Also known as Janssen microscope.

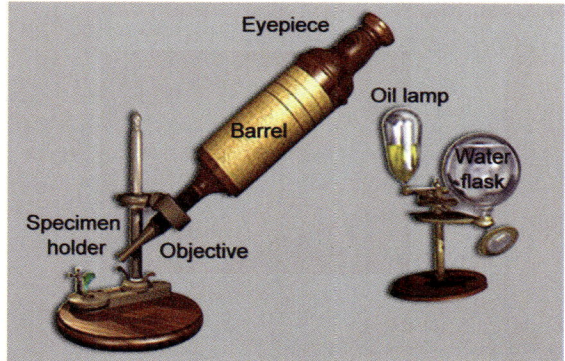

Fig. 14.2: Microscope constructed by Robert Hooke (circa 1670).

Fig. 14.3: Antonie van Leeuwenhoek (1632–1723).

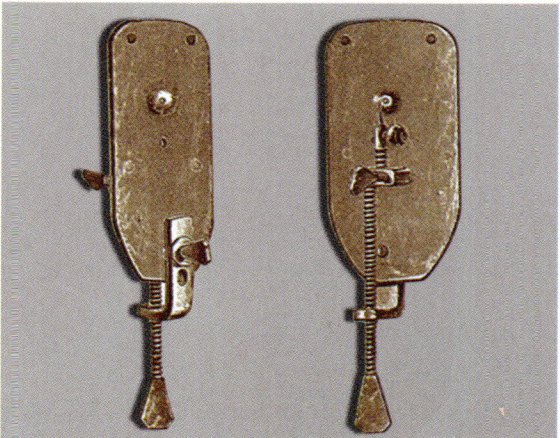

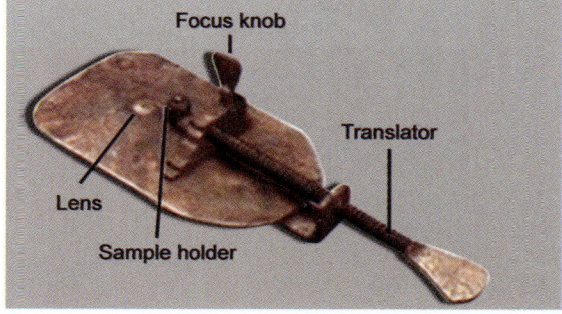

Fig. 14.4: von Leeuwenhoek microscope (circa late 1500s).

this Dutch-man consisted of a powerful convex lens and an adjustable holder for the specimen being examined. He added a pivoted joint for holding the specimen. He was able to achieve a magnification of about 300 times. With this simple microscope Leeuwenhoek discovered protozoa, spermatozoa, and bacteria and was able to classify red blood cells by shape.

In 1645 Rhetia designed the first binocular microscope. It permitted the use of both eyes in looking objects. In 1691 Bonannus improved the microscope. It now consisted of a source of light, a condenser to concentrate light, and a rack of pinion attachment for more efficient focusing. About 1710, Wilson developed a screw-barrel type of microscope. Spencer (1847) and Tolles (1858) made improvements of objectives and invented the homogenous immersion objectives. A proper type of liquid was placed on the cover slip on the slide and the immersion objective was made to touch the liquid, which acted as a type of lens to assist in higher magnification. Until the end of nineteenth century the making of complete microscopes was mainly done. They made one microscope at a time. The metal parts were made by hand. The lenses were ground and polished.

In twentieth century many improvements in the manufacture and usefulness of various types of microscopes have been made. The idea was conceived to add another lens to magnify the image produced by the first lens. The compound microscope—consisting of an objective lens and an eyepiece together with a means of focusing, a mirror or a source of light and a specimen table for holding and positioning the specimen—is the basis of light microscopes today. Janssen (1959) is reported to have constructed the first compound microscope. The more recent improvements include ultramicroscope, ultraviolet microscope, dark-field microscope, phase-contrast microscope, and electron microscope.

SOME TERMS RELATED TO MICROSCOPE

The following terms related to a compound optical microscope are to be understood clearly for understanding the basic principle about microscopy.

Definition

The definition of a microscope is a term to express the capacity of an objective to render outline of the image of an object *clear* (transparent, not clouded, not dim) and *distinct* (easily discernible separate or differing in identity). It depends upon the elimination of the optical aberrations (*see* Chapter 15) inherent in the glass of microscope lenses.

Microscopy

There are many different forms of microscopy but the one most commonly employed is "brightfield" microscopy where the specimen is illuminated with a beam of light that passes through it (as opposed to a beam of electrons as in electron microscopy). The general requirements for a specimen to be successfully examined using bright field microscopy are:

- That the cells and other elements in the specimen are preserved in a "life-like" state (this process is called "fixation").
- That the specimen is transparent rather than opaque, so that light can pass through it.
- That the specimen is thin and flat so that only a single layer of cells is present.
- That some components have been differentially colored (stained) so that they can be clearly distinguished.

Numerical Aperture and Resolution

Numerical Aperture (NA)

The numerical aperture (*NA*), an optical constant, of a lens system is the 'light gathering' capacity of the objective lens of a microscope. The figure of NA is generally engraved on the barrel of an objective. It is an indication of the maximum resolution of which the lens is capable, and it is not a measure of the resolution that the objective will automatically produce.

The ability to reveal details can be expressed in terms of the fraction of the wave front admitted by the objective lens. The lens *AB* can accept a cone of rays from the point P. If 2α is the angle of this cone (Fig. 14.5), the numerical aperture (NA) is defined as:

$$NA = n \sin \alpha$$

Where, *n* is the refractive index of the medium in which the lens is working. Since *n* for air is 1, and since $\sin \alpha$ cannot be greater than one (because, $\sin 90° = 1$); it is clear that no lens working in the air can have a theoretical NA greater than one. The only way of achieving a NA greater than 1, since α cannot rise above 90°, is to increase *n*. This is done by using immersion oil and oil-immersion objective. Immersion oil (with its refractive index 1.56) when substituted for air, *both above and below the slide*, not only increases '*n*' but also permits the lenses involved to operate at greater numerical aperture. *If immersion oil is used only*

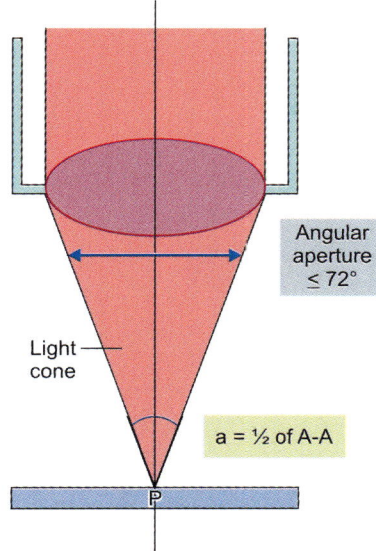

Fig. 14.5: Numerical aperture (NA) of an objective is a measure of its capacity to gather light and resolve fine specimen detail at a fixed object distance.

$$NA = n \sin a$$

Where, n: Refractive of the imaging medium between the front lens of the objective and specimen cover slip, a: One half of AA (angular aperture ≤ 72 degree).

Theoretical maximum NA = 1 (aperture of a lens operating with air as the imaging medium)

between the slide and the objective (a routine but wrong practice followed by those ignorant of correct operation of microscope), the system cannot operate at an NA greater than one.

The numerical aperture can also be expressed by the following equation:

$$NA = R \times \lambda$$

Where, R is the power of resolution and λ is the wavelength of the light which illuminates the optical system. From the above equation, it may be derived that:

$$R = \frac{NA}{\lambda} = \frac{n \sin \alpha}{\lambda}$$

In this expression both n and λ are constant; with usual working the microscope is operated in air (hence $n = 1$) and with white light ($\lambda = 2500 \text{Å}$). **Thus the resolution of any objective in use would be directly propor-**

tional to the sine of half the angle of the entering cone of light.

$$R \propto \sin \alpha$$

The wider the angle of cone of light, the greater is the resolution.

Resolution

The resolution of a microscope can be defined as its capacity to separate clearly two points that are very close together. The objective of any microscope is so designed as to produce a best possible image of a minute object. It must be remembered that the best image is not the largest—it is the clearest. A simple black dot the size of a pinhead is just as understandable as a simple black dot about an inch in diameter. What one wants to know from a microscope to reveal this pattern, if any, is what is termed resolution.

Limit of Resolution

The limit of resolution is measured as the least distance between two points which can be seen as two instead of one. It is the *smallest distance* (d) between two particles when the particles can be discerned as separate entities, and is expressed as:

$$d = 0.612 \, \lambda / NA$$

Where,
- d = limit of resolution
- λ = wavelength of illuminating light source
- NA = numerical aperture of the lens

Resolving Power

The *resolving power* is the ability of a microscope to show two closely lying points as two distinct points. The lesser the distance of limit of resolution more is the resolving power of the microscope. This property of a microscope is one of its most important parameters. It is only necessary to magnify the resolving power to 0.2 mm, the resolving power of the unaided human eye, for all the fine details of an object to be revealed (Fig. 14.6). However, the resolution is dependent primarily on a figure called as the *numerical* length of the light used also affects resolution of the microscope and for an

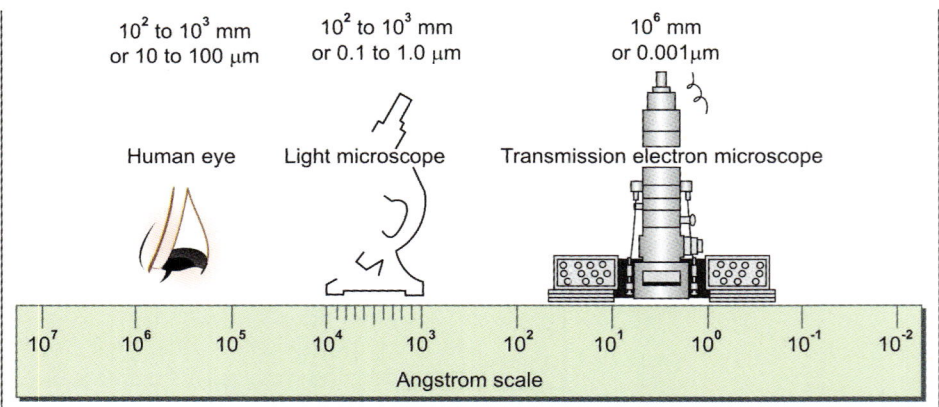

10^2 to 10^3 mm or 10 to 100 μm — Human eye

10^2 to 10^3 mm or 0.1 to 1.0 μm — Light microscope

10^6 mm or 0.001μm — Transmission electron microscope

| 10^7 | 10^6 | 10^5 | 10^4 | 10^3 | 10^2 | 10^1 | 10^0 | 10^{-1} | 10^{-2} |

Angstrom scale

Fig. 14.6: Comparison of limits of resolution of the human eye (A), a light microscope (B), and a transmission electron microscope (C) depicted on an Angstrom scale.

account of relationship between resolution, numerical aperture, and wavelength.

Magnification

The magnification of a microscope is a function of both the objective and the eyepiece (ocular). It is a requirement of an objective entirely secondary to resolution. In addition to the capability of resolution, the objective lens also enlarges and projects image towards eyepiece. Eyepiece (or ocular) further magnifies and projects image towards the detector (retina); it must be remembered that an eyepiece does not resolve the image—only magnifies.

Magnification is the ratio of the size of the image to that of the object. Generally, the magnification (M) of an objective is equal to the tube-length of the microscope divided by the objective's focal length. This relationship is correct only when the given objective is used at the standard tube-length of 160 mm in most of the microscopes.

Magnifying power (M) of a simple microscope is the ratio of the angle subtended by the image at the eye to the angle subtended by the object seen directly, when both lie at the least distance of distinct vision or the near point.

$$M = 1 + D/f$$

Where, D is the least distance of distinct vision and f is the focal length of the lens.

The useful limit of magnification is that which increases the size of the smallest object that can be resolved to the smallest object that can be seen. The total magnification is obtained by multiplying the magnification of objective (*primary magnification*) and that of eyepiece (*secondary magnification*). The values of magnification produced by these lenses are generally engraved on the respective lens barrels.

For large magnifying power the objectives and eyepiece both should be of short focal lengths. On contrary in telescopes, the focal length of objective is large and the focal length of only eyepiece is short. The focal length of any simple lens is the distance from the center of the lens to a point at which parallel rays of light, from a distant source, are focused. The focal length in compound lenses (of most microscope objectives) is the distance between the plane of focus and a point roughly midway between the compound lenses.

The focal length of a lens influences the **angle of view** (Fig. 14.7). The angle of view is wide with the lenses of shorter focal lengths (examples: Microscopes and wide angle lens cameras), and becomes appreciably narrow in cases of long focal length lenses (example, telephoto camera lenses).

A comparison between the resolution and magnification limits is given in Table 14.1.

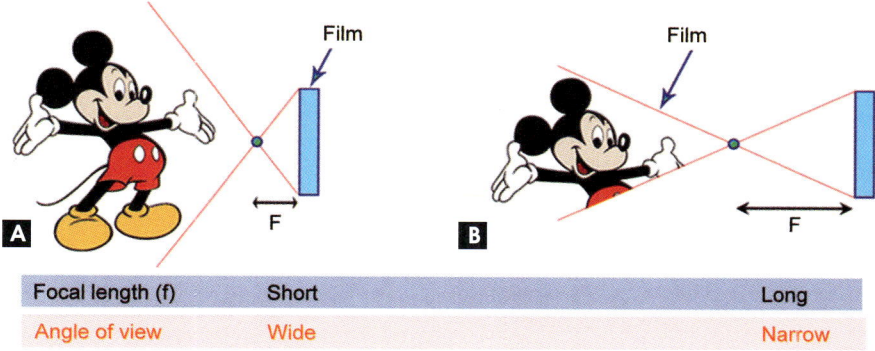

Focal length (f)	Short	Long
Angle of view	Wide	Narrow

Fig. 14.7: Diagrammatic representation to show that the lens focal length is inversely proportional to the angle of view: That means a wider angle of view with lens of shorter focal length as in microscopes and wide angle lenses. As a corollary, the lenses with longer focal lengths have a narrower angle of view as seen in telephoto lenses.

Table 14.1	Comparison of resolution and magnification limits		
Source	**Specimen**	**Resolution**	**Magnification**
Human eye	Human ovum	0.1 mm–100 µm	—
Light microscope	Red blood cell	8.0 µm	Amplitude contrast—white/black
	Bacteria	0.2 µm	Wavelength contrast—color
		Increased by achromatic lenses and oil-immersion lens	**Up to 1000 times**
Electron microscope	Cell organelles	2–5 Ångstrom (Å) $$d = \frac{0.612\,\lambda}{n \sin \alpha} = \frac{0.612\,\lambda}{NA}$$	**Great** (but amplitude contrast only)

Principle of Microscopy

Most tissues are neither small enough nor transparent enough to be examined directly in the microscope. Hence, they are:

- Chemically fixed
- Sectioned into very thin slices called sections
- Subsequently, stained to reveal different components of the cells

There are following *three* things required for viewing cells in a light microscope:

1. Bright light: Must be focused onto the specimen (actually a stained section) by the lenses in the microscope condenser.
2. Specimen preparation: Must be carefully done to allow light to pass through it.
3. Objective and ocular (eyepiece) lenses: Be approximately arranged to focus an image of the specimen in the eye.

Two Main Types of Microscopes

Although there are several kinds of microscopes used for different purposes, in routine use only *two* basic types of microscopes need to be introduced here.

1. Simple/Dissecting Microscope

A **simple microscope** is a microscope that uses only one lens or set of lenses to enlarge an object through angular magnification alone, giving the viewer an erect enlarged virtual image. Simple microscopes are not capable of high magnification. The use of a single convex lens or groups of lenses are still found in simple magnification devices such as the magnifying glass, loupes, and eyepieces for telescopes and microscopes.

As shown in Fig. 14.8, dissecting microscope consists of a biconvex lens which is moved up and down by an adjustment screw to bring the object in sharp focus. The object is placed on the platform and light is focused with the help of a concave mirror fitted below.

In simple microscope, convex lens of short focal length is used to see magnified image of a small object. The object is placed between the optical centre and the focus of a convex lens, its image is virtual, erect and magnified and on the same side as the object. The position of the object is so adjusted that the image is formed at the least distance of distinct vision (D).

2. Compound Microscope

A compound microscope (also referred to as a "light microscope" or "bright field microscope") utilizes two set of convex lenses (Fig. 14.9):

i. Ocular or eyepiece: Set of lens of relatively moderate focal length and large aperture facing the eye (hence the name)

ii. Objective lens set: Of short aperture and short focal length facing the object

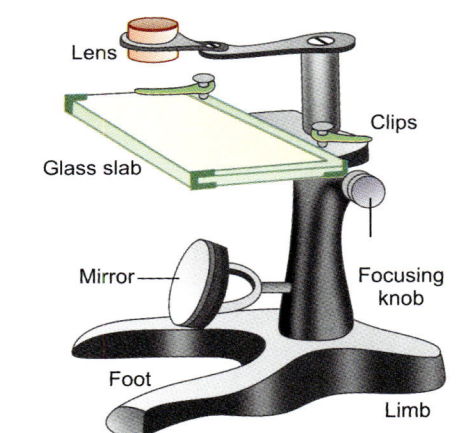

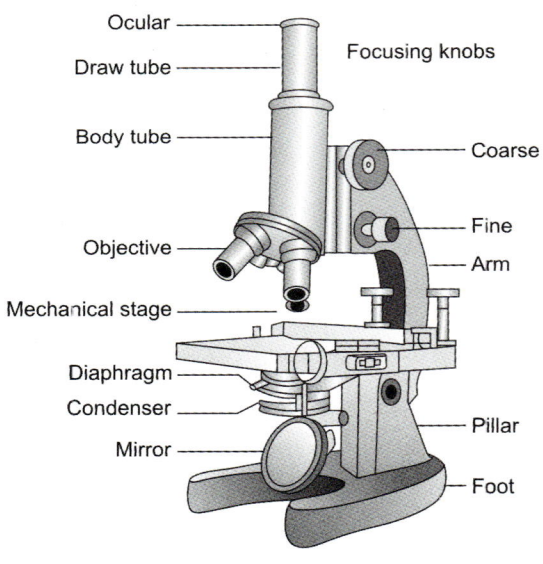

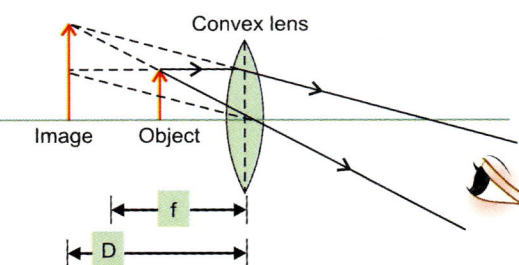

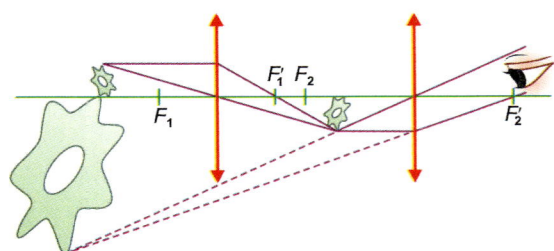

Fig. 14.8: The upper figure shows the parts of a simple microscope. The lower figure shows a convex lens (of 'f' focal length) forming a virtual, erect, and enlarged image of an object at 'D' distance.

Fig. 14.9: A students' model of compound microscope with its parts. On the top is the path of rays forming an image, which is magnified and resolved.

(hence, the set is called objective). In order to achieve higher magnifications we must use the compound microscope, which was originally developed by the Janssen brothers in the Netherlands and Galileo in Italy around the beginning of the 1600s. In its simplest form, the instrument is composed of two convex lenses aligned in series: An object glass (more commonly referred to as an objective) closer to the object or specimen, and an eyepiece (ocular) lens closer to the observer's eye (with means of adjusting the position of the specimen and the microscope lenses). The compound microscope achieves a two-stage magnification where the objective projects a magnified image into the body tube of the microscope and the eyepiece further magnifies the image projected. The total magnification equals the magnification of the objective multiplied by the magnification of the eyepiece.

The earliest compound microscopes were hindered by optical aberrations (both chromatic and spherical). Such defects result from the fact that white light is composed of numerous wavelengths, and when light waves pass through the periphery of a lens, they are not brought into focus with those passing through the center. The images that early microscopes produced were often blurred with colorful halos until lens makers in the mid-1700s discovered that by combining two lenses made of glass with different color dispersions, much of the chromatic aberration could be reduced or eliminated. Modern microscopes are often modular with interchangeable parts for different purposes, and can have several lenses arranged one behind the other, thus allowing magnifications of up to 2000X and higher, and the capability of producing images with remarkable clarity and contrast.

In the microscope beam path (Fig. 14.10A), the object or specimen is recorded by the objective and is first projected at infinity with a parallel bundle of wave-fronts or rays. In effect, the light rays originating from one point of the specimen travel in straight, parallel lines behind the objective. The **tube lens** then functions in a similar way to a camera to focus the parallel ray bundles, producing **second magnifier** a magnified intermediate image located inside the eyepiece at its front focal plane.

The eyepiece, acting as a, translates the dimension of the intermediate image into parallel rays. The resulting viewing angle of the sophisticated compound microscope system is much larger than results from direct observation (Fig. 14.10B), where the object is seen directly from a distance of approximately 25 centimeters. The region where these parallel bundles intersect is termed the **eye point**, and that is where the iris of the eye should be located. The cornea and lens of the eye focus these parallel rays onto the retina. As described above, the total magnification equals the objective magnification multiplied by the

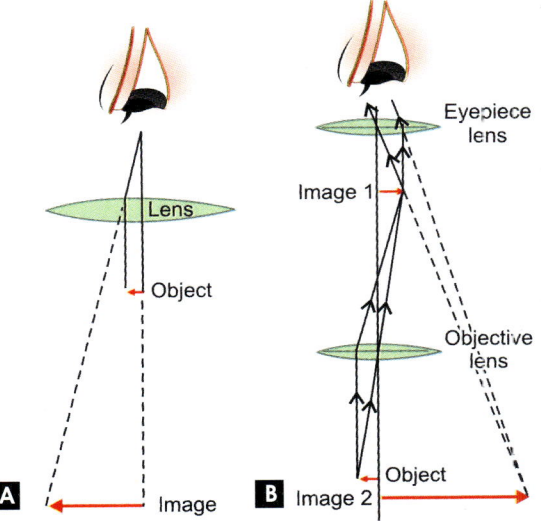

Fig. 14.10: Comparison between light paths in (A) simple microscope, and (B) compound microscope.

eyepiece magnification. In this illustrative example, the overall magnification of the microscope is 100X (10X objective with a 10X eyepiece).

At a selected **numerical aperture** (the sine of the angular aperture of the objective multiplied by the refractive index of the imaging medium) where the microscope presents a magnified image with a magnitude equivalent to the resolution limit of the human eye, further magnification beyond this point does not result in the resolution of even finer specimen detail.

The range of **useful total magnification** for an objective and eyepiece combination is defined by the numerical aperture of the system. There is a minimum magnification necessary for the detail present in an image to be resolved by the eye, and this value is typically set at 500 times the numerical aperture (500 × **NA**). At the other end of the spectrum, the maximum useful chromatic aberration and exchangeable objective lenses to adjust the magnification. A compound microscope also enables more advanced illumination setups, such as phase contrast.

The objective lens of the microscope produces an enlarged primary image of a very tiny object O at I_1. This is called *initial* or *primary magnification* usually engraved on the barrel of the microscope objectives. This magnified image serves as an object for the eyepiece (ocular) which acts as a telescope being at the other end of the body tube. The first lens (field lens) of the eyepiece produces a magnified image of I_1 at I_2 (Fig. 14.11), within the barrel of the eyepiece. The top lens (eye-lens) of the eyepiece produces a small image of I_2 at I_3. This image (I_3) is usually about a millimeter in diameter and is called *Ramsden's disk*. The disk is rather small and it requires the human eye to transform it into illusion of a magnified image of the object (O). This is done by advancing the observer's eye down towards the eyepiece until Ramsden's disk (I_3) is just inside the cornea (between the cornea and lens of observer's eye). The eye-lens then casts an image of

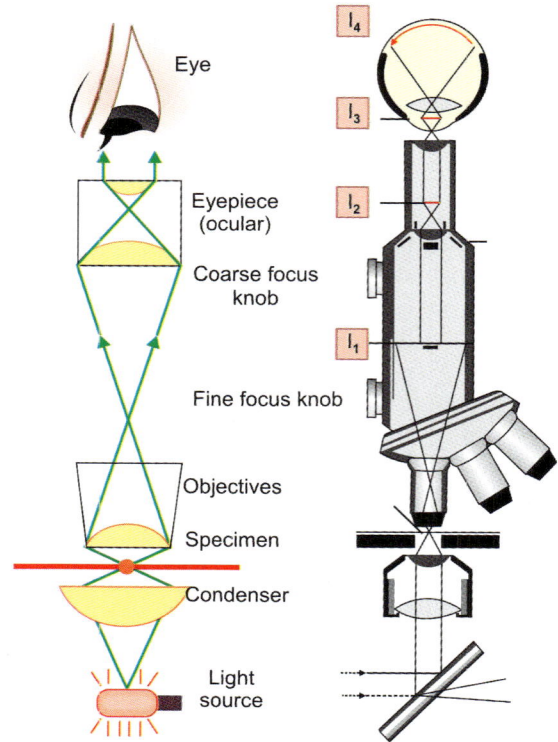

The light path in bright-field light microscope

Fig. 14.11: The light microscopy allows to magnify up to a thousand times, and to resolve details as small as 0.2 μm (a limitation imposed by the wavelike nature of light, not by the quality of lenses). Three things are required for viewing cells in a light microscope: (i) A bright light must be focused onto the specimen by lenses in the condenser. (ii) The specimen must be carefully prepared to allow light to pass through it. (iii) An appropriate set of lenses (objective and eyepiece (or ocular) must be arranged to focus an image of the specimen (usually a section) in the eye.

I_3 on the surface of retina at I_4. It is this fourth image which actually "fills our eye".

The distance between the outer surface of the eye-lens of the ocular and the image (I_3) formed by it is called the *eye-relief* of the ocular (*see* Chapter 16). In most eyepieces it is so stupidly short that wearers of spectacles have to remove them in order to get the Ramsden's disk (I_3) into the cornea and thus fill their eye with image.

SUMMARY

History and Background

- Leeuwenhoek's simple microscopes gave clearer images than his compound microscope.
- The first description of the nucleus was carried out by Leeuwenhoek, in 1700.
- **Robert Brown** (1773–1858), a Scottish botanist, introduced the term nucleus in microscopy.
- **Marie François Bichat** (1771–1802) verified 21 textures compared to pieces of cloth.
- Tissues, composed of cells and extracellular matrix, are of four fundamental types:
  - ◆ Epithelia
  - ◆ Connective tissue
  - ◆ Muscle tissue, and
  - ◆ Nervous tissue

Development of Microscope

- The word *microscope* is derived from the Greek *micros* (small) and *skopeo* (look at).
- In 1645 Rhetia designed the first binocular microscope.
- In 1675, Antonie van Leeuwenhoek developed a remarkably simple microscope by mounting a lens between two flat pieces of metal.
- In twentieth century many improvements in the manufacture and usefulness of various types of microscopes have been made.

Some Terms Related to Microscope

- *Definition:* It is a term to express the capacity of an objective to render outline of the image of an object *clear* (transparent, not clouded, not dim) and *distinct* (easily discernible separate or differing in identity).
- *Numerical Aperture:* The term numerical aperture (NA) of an objective is a measure of its capacity to gather light and resolve fine specimen detail at a fixed object distance.

$$NA = n \sin \alpha$$

Where, n: Refractive of the imaging medium between the front lens of the objective and specimen cover slip, α: One-half of angular aperture <72°.

- *Resolution:* It is a term defined as its capacity to separate clearly two points that are very close together.

The *limit of resolution* is measured as the least distance between two points which can be seen as two instead of one. It is the *smallest distance* (d) between two particles when the particles can be discerned as separate entities, and is expressed as:

$$d = 0.612\ \lambda / NA$$

Where,
- d = limit of resolution
- λ = wavelength of illuminating light source
- NA = numerical aperture of the lens

The *resolving power* is the ability of a microscope to show two closely lying points as two distinct points. The lesser the distance of limit of resolution, more is the resolving power of the microscope.

- **Magnification:** It is a term defined as the ratio of the size of the image to that of the object. It is a function of both the objective and the eyepiece (ocular).

The total magnification is obtained by multiplying the magnification of objective (*primary magnification*) and that of eyepiece (*secondary magnification*).

Principles of Microscopy

There are following *three* things required for viewing cells in a light microscope:

1. *Bright light:* Must be focused onto the specimen (actually a stained section) by the lenses in the microscope condenser.
2. *Specimen preparation:* Must be carefully done to allow light to pass through it.
3. *Objective and ocular (eyepiece) lenses:* Be approximately arranged to focus an image of the specimen in the eye.

Basic Types of Microscopes

Only *two* basic types of microscopes are in routine use (*see* Fig. 14.10):
- *Simple/dissecting microscope*
- *Compound microscope.*

15

Aberrations in Microscope Lenses and their Correction

INTRODUCTION TO ABERRATIONS

The literal meaning of the word *aberration* is 'deviation from the type'. In optics the term is used to imply all such defects in the performance of a lens (or a lens system) on account of which the image of an object, formed either by reflection or by refraction, is imperfectly formed and it tends to become different from the object. However, difference in the size is not considered to be a defect. The lens aberrations may be classified into two groups:

1. *Monochromatic aberrations*:
   - Spherical aberration
   - Coma
   - Astigmatism
   - Curvature of field
2. *Chromatic aberrations*.

MONOCHROMATIC ABERRATIONS

Spherical Aberration

In this defect the image is of curved shape, and appears distorted when cast on flat surface. Because of the varying thickness of lenses at their centres and periphery, the focal lengths at these sites differ. As a result the marginal rays striking a lens surface are bent maximum in comparison to those striking closest to the center. Thus the paraxial rays from an object bend least and are focused at a point different from the one focusing the marginal rays (Fig. 15.1).

The distance between the two points of focus is the measure of *axial* (*central* or *longitudinal*) *spherical aberration*. The axial spherical aberration is of two subtypes: (i) *Positive* for convex or convergent lenses, and (ii) *negative* for concave or divergent lenses. The best possible position of the image is neither at the point where marginal rays converge nor at the point where paraxial rays converge; but somewhere in between these two points where the section XY of the refracted beam by a transverse plane will have minimum area. This is called circle of least confusion. The radius of this circle is the measure of the *lateral spherical aberration*.

Coma

In this monochromatic aberration, there is a comet-like appearance of a pin-point object

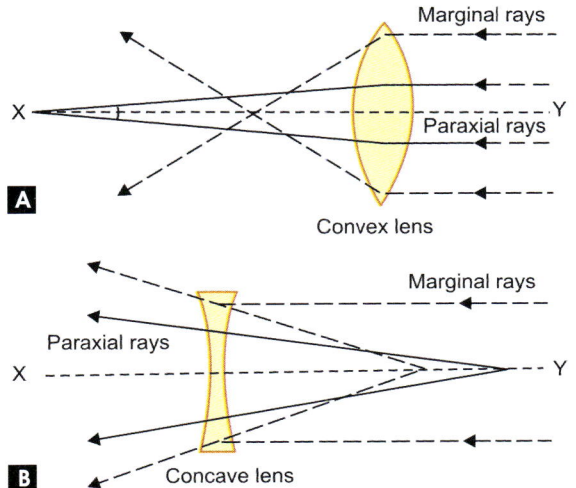

Fig. 15.1: Positive (+) and negative (−) axial spherical in convex (A) and concave (B) lenses. At XY section, the radius of the circle is the measure of lateral spherical aberration.

situated just off the axis of a lens. Overall image of the point object consists of a series of circles (Fig. 15.2) each representing the image from the different zones of the lens. This defect occurs even if a lens has been corrected for central (longitudinal) spherical aberration and produces a sharp image of a particular point object on the axis. Coma is basically of two types: (i) *Positive coma* for concave (divergent) lenses having negative spherical aberration; and (ii) *Negative coma* for convex (convergent) lenses having positive spherical aberration.

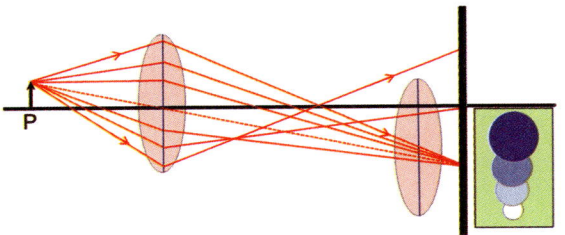

Fig. 15.2: A comet-like appearance of a point object P is to a defect of coma. The comet is the resultant of a series of differently illuminated circles (brightest is smallest and lowest one).

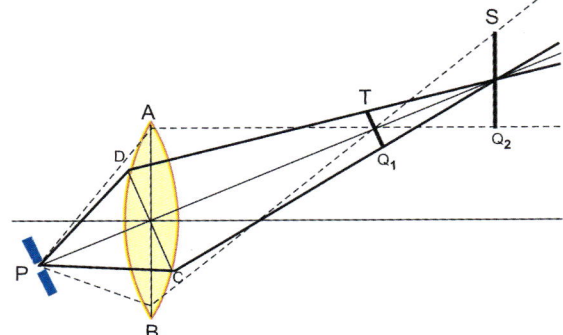

Fig. 15.3: Astigmatic defect of a convex lens showing two focal lines (CD = first focal line; AB = second focal line).

Astigmatism

In astigmatic defect, there is nowhere a sharp image of a pin-point object contained, and the thin pencil of light which diverges from two short lines mutually at right angles is said to be astigmatic. Astigmatism is due to those rays of light which pass obliquely (Fig. 15.3) through a lens.

The light rays passing from a point P in tangential plane give rise to the formation of image at Q_1 on the central ray POQ_1. But the rays from P in sagittal or horizontal plane form a point image at Q_2 farther than Q_1. The convergent beam passing to form the image for Q_2 as it passes Q_1 forms a line-image T parallel to *CD* (called the *first focal line*). The beam making for Q_1 spreads out again and as it crosses Q_2 forms a line-image S which is parallel to *AB* (called the *second focal line*).

Curvature of Field

Due to the curvature of field, the center of field is focused sharply in contrast to the edges which are less sharply focused and are therefore, blurred (Fig. 15.4). It is most obvious when photomicrography is to be performed.

CHROMATIC ABERRATIONS

Chromatic aberrations (*chroma* = color; *aberration* = error) are due to dispersion of light.

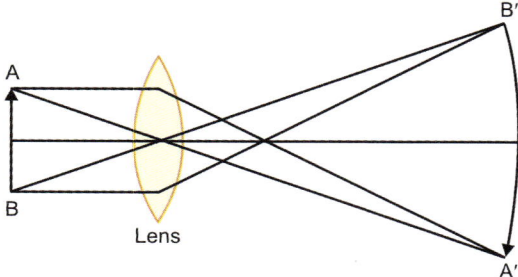

Fig. 15.4: Due to curvature of field defect the image A′ B′ of an object AB is sharply focused at center but less sharply at edges.

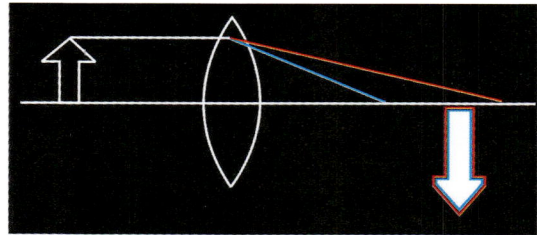

Fig. 15.5: The chromatic aberration defect in a convex lens.

When the white light passes through a lens, the red light is bent least and the violet bent most. due to the phenomenon of dispersion, the red image is larger than the blue image and sticks out around it (Fig. 15.5). The central portion of the image, in which all colors are superimposed, still appears white.

Like the two types of spherical aberrations, the chromatic aberration is also of two varieties. The *axial* or *longitudinal chromatic aberration* is the difference between the axial distance between the red and violet images. The *lateral*

chromatic aberration is due to the difference of image formed by red and violet colors.

CORRECTIONS OF LENS ABERRATIONS

The lens aberrations can be corrected to a great extent (definitely minimised) and the image quality is thus improved.

Spherical Aberration Correction

Spherical aberration can be easily corrected since it is caused by the difference in distance between the center and edge of a lens. The common methods of reducing the spherical aberration are as follows (Fig. 15.6).

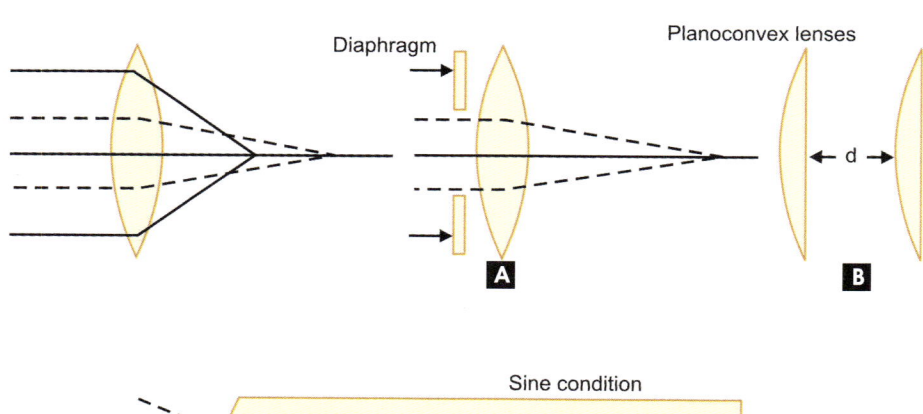

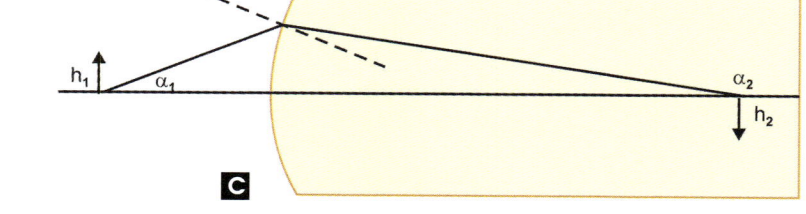

Fig. 15.6: Different methods for correction of spherical aberrations can be done either through a 'stop' A, or by the use of planoconcave lenses (B), or by satisfying the sine condition (C) of lenses.

i. To place a *diaphragm* in front of a lens. This allows only the central portion of the lens being used, and is by far the simplest way of correction.

ii. To use *Coddington lenses*, consisting of a glass sphere round which a deep equatorial groove is ground. The light reaching the eye is limited to the central portion.

iii. To use *planoconvex lenses* separated from each other by a specific distance equal to the difference between the focal lengths of the two lenses. Such is the case of the arrangement of lenses in eyepieces of microscopes.

iv. If a condition satisfying the *sine condition* is fulfilled, the defect of coma is eliminated. The *sine condition* (Fig. 15.6 C), is expressed by the formula:

$$\mu_1 h_1 \sin \alpha_1 = \mu_2 h_2 \sin \alpha_2$$

Where,

μ_1 = Refractive index of the medium in which the object (specimen) lies

μ_2 = Refractive index of the medium in which the image is formed

h_1 = Size of the object

h_2 = Size of the image

α_1 = Angle between the optical axis and the incident ray

α_2 = Angle between the optical axis and the refracted ray

The surface of any lens satisfying the sine condition is called in optics *Aplanatic surface*.

Chromatic Aberration Correction

Chromatic aberration is much more difficult to be corrected than the spherical aberration. The only solution so far discovered makes use of the fact that the relation between the index of refraction and dispersion differs in a different kinds of glasses. A *doublet lens*, consisting of a convex and a concave lens cemented together by means of a transparent medium, is used (Fig. 15.7).

The convex lens, made of *crown glass*, has a high refractive index and low dispersion. It bends the light to a great deal but the colors are separated very little. The concave lens, made up of *flint* glass, has a low refractive index but high dispersion. It bends the light very little but pulls the different colors proportionately close together. As a combined result the chromatic aberration is reduced by combining crown and flint glasses possessing different dispersions for the various colors. Red and violet images are brought approximately to the same focus. With the use of a doublet lens the image will be relatively free of spherical aberration also as the thick edge of the concave lens balances the thick centre of the convex. Unfortunately, this arrangement is a theoretical dream. Glasses of very high refractive index and very low dispersion or *vice versa* do not exist.

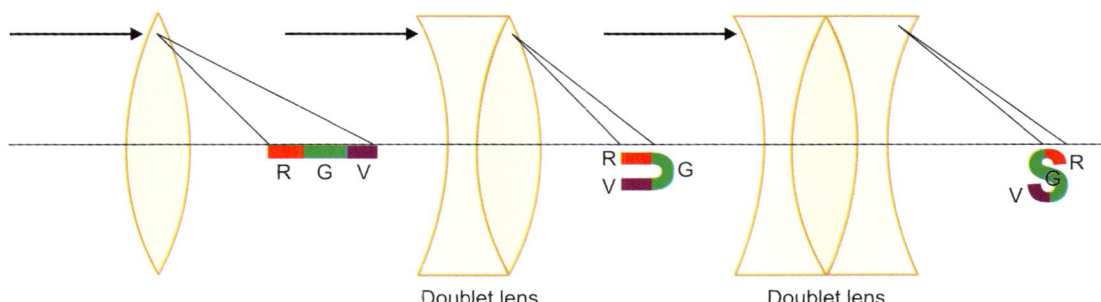

Fig. 15.7: Correction of chromatic aberration achieved by the use of 'doublet' lenses.

Even after perfect correction for both types of lenses aberration is done, the vibratory nature of light itself presents the formation of an absolute point image of a theoretical point source of light. The image possesses a positive circular area often termed *circle of confusion* (Fig. 15.8), or *airy disc* or *antipoint*. The diameter of disc is directly proportional to the wavelength of the light used. *Sharper photographs are produced by using blue light than are obtained by using red*. The lenses with small aperture produce a bigger antipoint in contrast to lenses with large aperture and small antipoint. But, lenses with large aperture have a drawback.

These possess correspondingly less depth of focus, especially with nearby objects. When greater depth of focus is desired, we resort to *stopping down*, i.e. closing the diaphragm until the desired effect is obtained; or in other words, we work with a lower aperture. Any increase in the numerical aperture reduces the depth of focus, although to the benefit of a greater resolution.

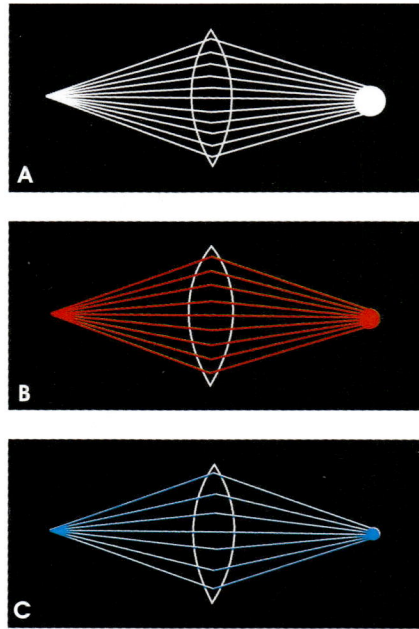

Fig. 15.8: Variable sizes of the circle of confusion (airy-disc or antipoint) with different lights used: Largest (A) with white light, medium-sized (B) with red light, and smallest (C) with blue light.

SUMMARY

Introduction to Aberrations

Aberration is defined as 'deviation from the type'. In optics the term is used to imply all such defects in the performance of a lens (or a lens system) on account of which the image of an object, formed either by reflection or by refraction, is imperfectly formed and it tends to become different from the object.

Types of Aberrations

The lens aberrations may be classified into *two* groups:

1. *Monochromatic aberrations*

   - *Spherical aberration:* The image is curved and appears distorted when cast on flat surface
   - *Coma:* The image has a comet-like appearance of a pin-point object. It is situated just off the axis of a lens. Overall image of the point object consists of a series of circles each representing the image from the different zones of the lens. Coma is basically of *two* types: (i) *Positive coma* for concave (divergent) lenses having negative spherical aberration; and (ii) *Negative coma* for convex (convergent) lenses having positive spherical aberration.
   - *Astigmatism:* The image is nowhere a sharp image of a pin-point object. The thin pencil of light which diverges from two short lines mutually at right angles is said to be astigmatic. Astigmatism is due to those rays of light which pass obliquely through a lens.
   - *Curvature of field:* The image at the center of field is focused sharply in contrast to the edges which are less sharply focused and are therefore, blurred. The defect is more conspicuous when photomicrography is done.

2. *Chromatic aberrations:* Chromatic aberrations are due to dispersion of light. Two varieties of chromatic aberrations may be seen:

   - The *axial* or *longitudinal chromatic aberration* is the difference between the axial distance between the red and violet images. The *lateral chromatic aberration* is due to the difference of image formed by red and violet colors.

Corrections of Lens Aberrations

- *Spherical aberration correction:* May be achieved by selecting any of the following methods (see text for further reading):

  i.   By far the simplest way of correction is to place a *diaphragm* in front of a lens allowing to use only central part of the lens

  ii. Use of *Coddington lenses*

  iii. Use of *planoconvex lenses* separated one another by a distance equal to the difference between the focal lengths of the two lenses.

  iv. The *sine condition* eliminates the coma defect

- *Chromatic aberration correction:* More difficult to correct than the spherical aberration.

  A *doublet lens* consisting of a convex and a concave lenses cemented together by means of a transparent medium is used.

16

Parts of Compound Light Microscope

In most of the histology laboratories, the microscope used by students is referred to as the *light microscope*, the compound microscope, or perhaps more appropriately the *bright-field microscope* (Fig. 16.1). Essentially, the instrument consists of a *light source*, a *condenser* that focuses rays of light on the tissue section, a *stage* on which slide is placed, an *objective lens* and an *ocular lens* or *eyepiece* through which the section may be directly seen.

PARTS OF A LIGHT MICROSCOPE

Parts of light microscopes used in majority of laboratories (Fig. 16.2) are divided into *two* groups: *Mechanical* and *optical*.

Mechanical Parts

The mechanical parts in a light microscope are:
1. The stand with a heavy foot bearing an upright *pillar*, to which is hinged a *limb* which carries the *body-tube* at its upper end, and to which a flat *stage* is attached at the lower end. The body-tube is the outer brass tube of the microscope. This tube is caused to move by the action of focusing knobs. The *coarse* and *fine focusing adjustment knobs* move the body-tube up and down.

2. A body tube is an integral part of the microscope as it holds the eyepiece and connects it to the objective (Fig. 16.3).

   Lower end of the body-tube is threaded to take the objectives, or a *nosepiece*. At its upper end, the body-tube contains a sliding *draw-tube* into which a lens system known as the ocular or eyepiece will slide. The draw-tube should be used to give the correct tube-length for the objective in use, due allowance being made for the thickness of the cover slip. Generally, the tube length is adjusted to 160 mm or 170 mm. The body tube supports the eyepiece and lenses. It also maintains the proper distance between the eyepiece and the objective lenses

   The **mechanical tube length** of an optical microscope is defined as the distance from the nosepiece opening, where the objective is mounted, to the top edge of the observation tubes where the eyepieces (oculars) are inserted. For many years, almost all prominent microscope manufacturers designed their objectives for a **finite tube length**. The designer proceeded under the assumption that the specimen, at focus, was placed at a distance a "little" further than

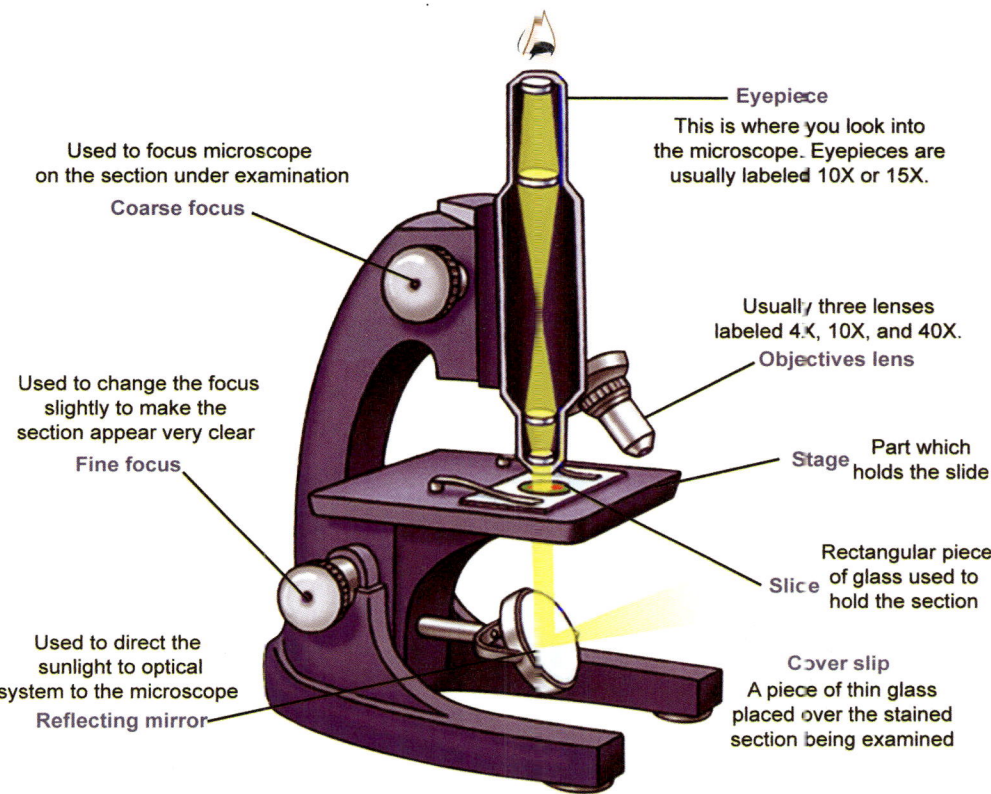

Used to focus microscope on the section under examination
Coarse focus

Used to change the focus slightly to make the section appear very clear
Fine focus

Used to direct the sunlight to optical system to the microscope
Reflecting mirror

Eyepiece
This is where you look into the microscope. Eyepieces are usually labeled 10X or 15X.

Usually three lenses labeled 4X, 10X, and 40X.
Objectives lens

Part which
Stage holds the slide

Rectangular piece
Slice of glass used to hold the section

Cover slip
A piece of thin glass placed over the stained section being examined

Fig. 16.1: A simple model of school microscope showing the basic parts and their role in microscopy.

the front focal plane of the objective. The objective then projects a magnified image of the specimen which converges (is brought into focus) at the level of the eyepiece diaphragm, located ten millimeters below the top edge of the openings of the microscope observation tube, where the eyepieces are inserted.

Tube length has now been standardized to the Royal Microscopical Society (RMS) suggestion of 160 millimeters for finite-corrected transmitted light microscopes. Objectives designed for a 160 millimeter finite tube length microscope bear the inscription "**160**" (mm) on the barrel. The "tube" generally is not a straight line and the light waves are transmitted from the objectives to the eyepieces (oculars) with mirrored beam splitters. This is the case with most modern microscopes, especially those equipped with trinocular heads for photomicrography. The schematic infinity-corrected systems illustrated in Fig. 16.4 indicate how additional optical components can be inserted into the light path. Figure 16.4A is a diagrammatic representation of an infinity-corrected system showing a specimen on a micro slide being illuminated by a substage condenser. The image-forming light rays pass through the objective and form a parallel light beam that is focused by the tube lens into the eyepiece. Accessories can be inserted into the parallel light beam without further optical correction as illustrated in Fig. 16.4B, which shows a Wollaston prism and several polarizers inserted into the pathway. Figure 16.4C illustrates the insertion of a beam splitter

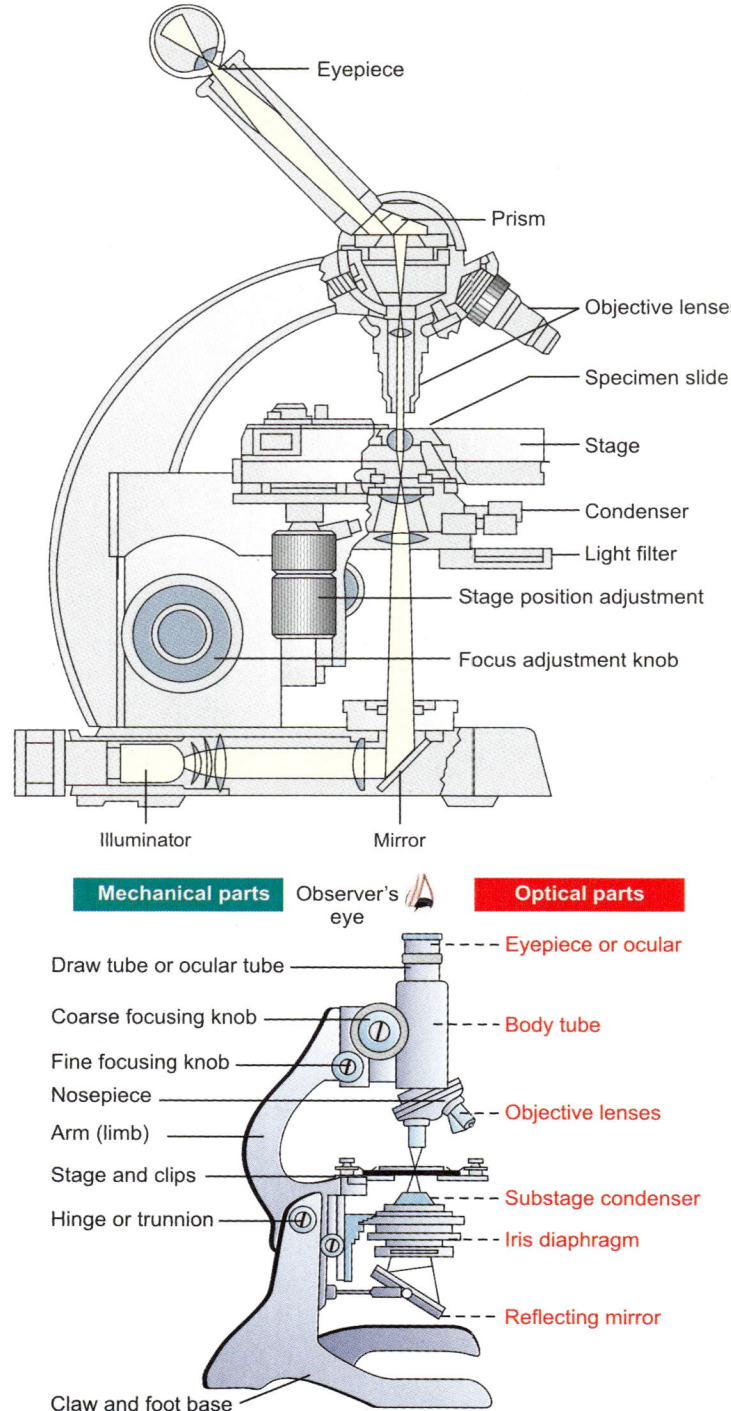

Fig. 16.2: Schematic drawing of a light microscope shows its main components and the pathway of light from the substage lamp to the eye of the observer. On the right side, mechanical and optical parts of a student model light microscope are shown.

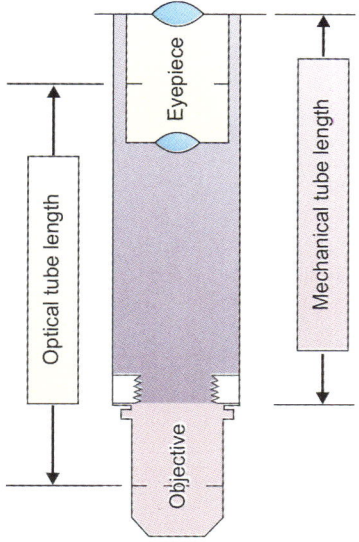

Fig. 16.3: Mechanical and optical tube lengths of a light microscope.

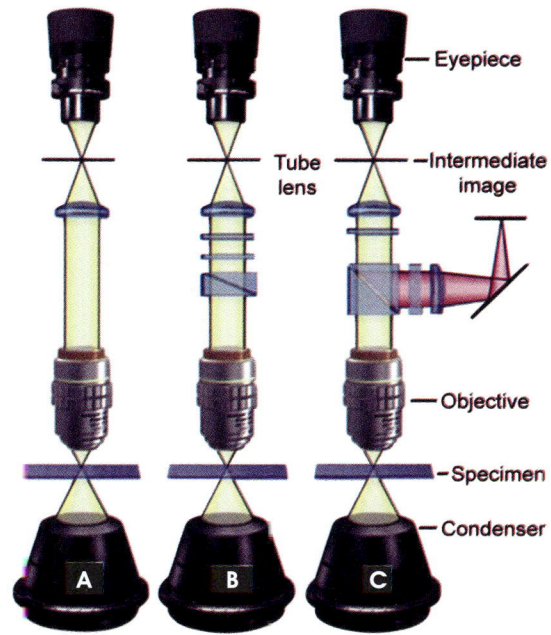

Fig. 16.4: Diagrammatic representation of infinity-corrected systems showing a specimen (section) on a glass slide illuminated by a substage microscope condenser: (A) A parallel light beam focused by the tube lens into the eyepiece, (B) A Wollaston prism and several polarizers inserted into the path of light and (C) With insertion of a beam splitter, which diverts the light to a parallel beam set up (shown in purple).

into the parallel light beam. This beam splitter diverts light to the external accessory positioned on the right of the parallel beam.

3. The microscope stage is a square or circular metal plate with a hole in the center. The stage is provided with 2 clips, which hold the specimen/ slide in place on the stage. A glass slide with section to be examined is placed on the stage over the hole. However, in better models of microscope, the stage is of mechanical type. The mechanical stage is a modified stage with a *rack and pinion* device by means of which a slide can be moved in two directions at right angle (Fig. 16.5). It is also useful because, by means of Vernier readings, it is possible to replace a slide in any given position.

4. The nosepiece with 3 or 4 holes for attachment of objective lenses which can be rapidly interchanged. It is 12 mm to 18 mm thick. The nosepiece holds the objective lenses and can be turned to increase the magnification (Fig. 16.6).

5. The substage with rack and pinion moves the condenser.

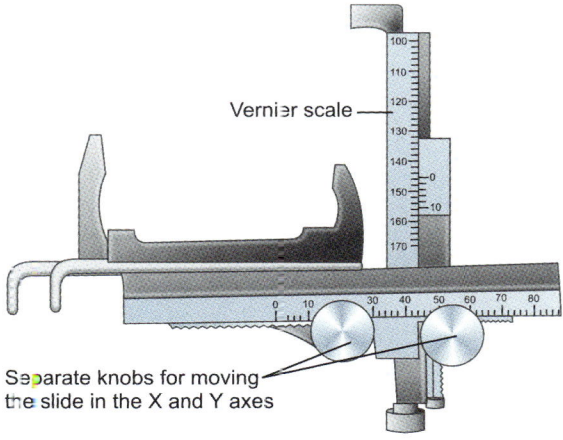

Fig. 16.5: Mechanical stage of a light microscope is movable in two axes at right angles to each other.

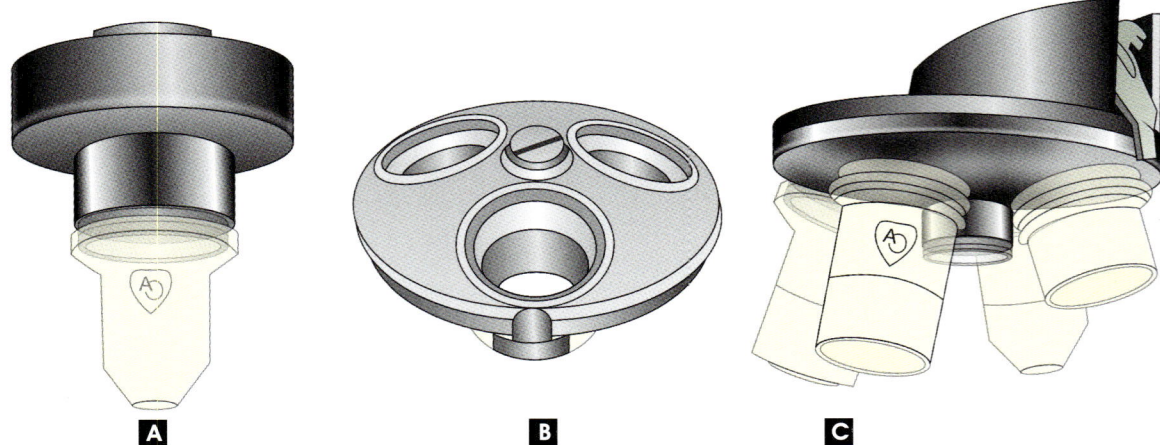

Fig. 16.6: Different types of microscope nosepieces with: (A) Single, (B) Three, and (C) Multiple adapters.

6. The field (iris) diaphragm on the base of the microscope, which determines the diameter size of the field of illumination. It controls the amount of light on the specimen/slide.

7. The substage diaphragm (also called aperture diaphragm) is placed underneath the condenser for regulation of the size or aperture of cone of light passing through the section and entering the objective. Too large an aperture leads to light glare, and too small an aperture leads to unpleasant diffraction effects and loss of microscope's definition.

Optical Parts

The optical parts in a light microscope are:

a. Microscope Lamp

The character of the image obtained with a microscope is as dependent upon proper illumination (Fig. 16.7) as it is upon a good objective, because the formation of the image of an object is dependent upon the light which is transmitted through the material and picked up by the lens system and the observer's eye. The theoretically ideal illumination source is a point of light, but this is difficult to achieve or to utilize. The best possible microscope lamp is not always practical. However, it is impor-

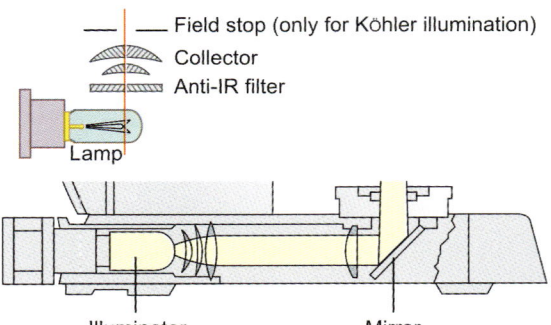

Fig. 16.7: Portion of the microscope in Fig. 16.2 has been cut and enlarged to show illuminator (an in-built part) as a source of light reaching the mirror from which it gets directed towards the microscope condenser placed below the stage.

tant to remember that the proper adjustment of a crudest light system improves the image tremendously. On the other hand, the incorrect use of the most expensive lamp results in a very poor microscope image. When a simple lamp is used without a ground glass it should be provided with a frosted, blue daylight bulb.

b. Mirror

The mirror is provided in those light microscopes which do not possess an integral light system. It reflects light to the lens of the microscope, and is meant to direct the light from its source upwards toward the condenser. In ordinary (student model) microscopes it has a concave surface which converts the beam of light falling on it into a cone of light. In good quality research microscopes the mirror possesses two surfaces—a concave and a plane (Fig. 16.8).

The concave surface of the mirror is used when working is done with low-power objective only; that is when substage condenser is out of the system. When an object is visualized under higher magnification it necessitates the use of substage condenser. Under this condition, only the plane surface of the mirror must be used to redirect the light from its source toward the condenser without altering its form. It must be remembered that the concave surface of the mirror should never be used

when substage condenser is in the optical path. Nevertheless, the concave surface of the mirror may be regarded as a 'crude' condenser.

c. Substage condenser

One of the key components of the optical system of many compound microscopes is a condenser. A condenser is a lens that is substage in position and gathers light from microscope light source. It serves to concentrate light from the illumination source into a cone of light that illuminates the specimen with parallel beams of uniform intensity from all directions over the entire view-field. The light cone is in turn focused through the object and magnified by the objective lens.

The substage condenser is a very short-focus lens combination ($f = 11$ mm) consisting of

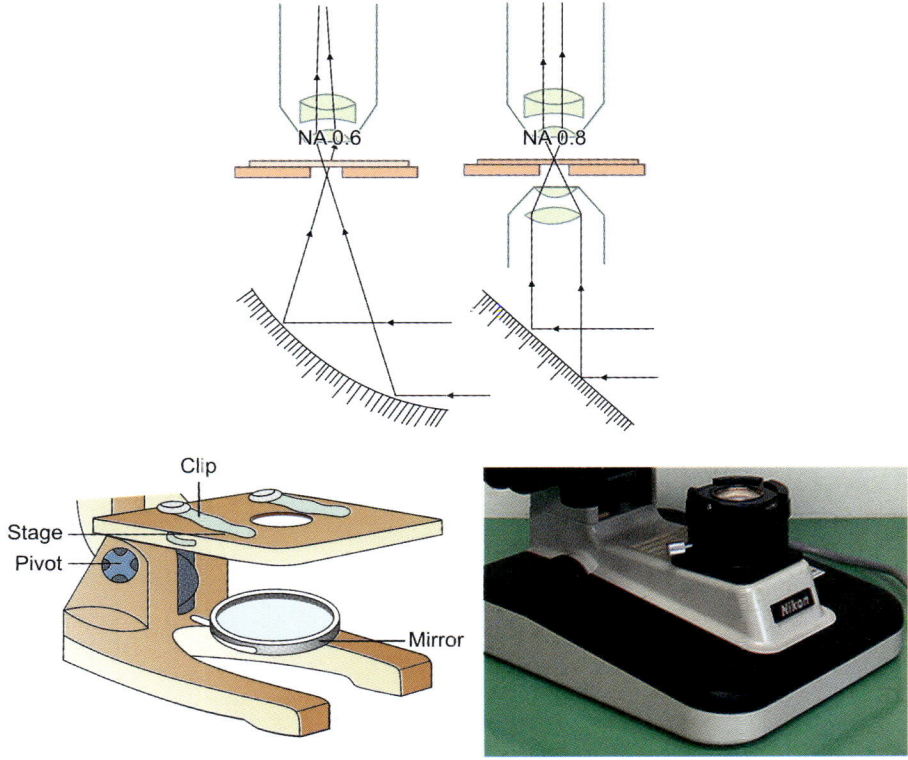

Fig. 16.8: In good quality microscope mirror has both: (A) Concave and (B) Plain surface for reflecting the light to the condenser. The concave mirror casts a narrow cone of light (NA 0.6) in comparison with the plain mirror (NA 0.8). The plain surface of the mirror should be used only if microscopes are provided with condenser.

two simple lenses—an upper hemispherical lens and a lower larger biconvex lens (Fig. 16.9) which are mounted between the microscope mirror and the microscope stage. It is provided not in all, but in only some types of microscopes.

The substage condenser bends the column of light coming from the plane surface of mirror, inwards and forms a uniform cone of light a little above its upper lens. It forms a real image of the light source in the plane of the specimen. It also governs to a large extent the resolution and visibility of semitransparent structures. Thus it can be said that the substage condenser controls the quality of the image which will be formed by the objective.

If somehow (by those ignorant of the proper and correct use of the microscope—*see* Chapter 17) the concave surface of the mirror is used along with the substage condenser, the light rays bisect within the upper lens of the condenser; and it becomes **impossible** to focus them on the specimen. It is due to this drawback that as a general rule whenever substage condenser is in use, only plane surface of the mirror must be used.

For moving the substage condenser up and down, there is rack and pinion arrangement in microscopes. This up and down movement serves as a critical light focusing system particularly when photomicrography (Part 3 of this book) is planned.

Engravings on Microscope Condenser Housing

The majority of condensers have the following *three* engravings (Fig. 16.10):
1. Its type (Abbe, achromatic, aplanatic, etc.)
2. Numerical aperture (NA)
3. A graded scale that indicates the approximate adjustment (size) of the aperture diaphragm

Types of Microscope Condensers

The substage condensers used in microscopes are of *four* types (Fig. 16.11):
1. Abbe condensers
2. Aplanatic condensers

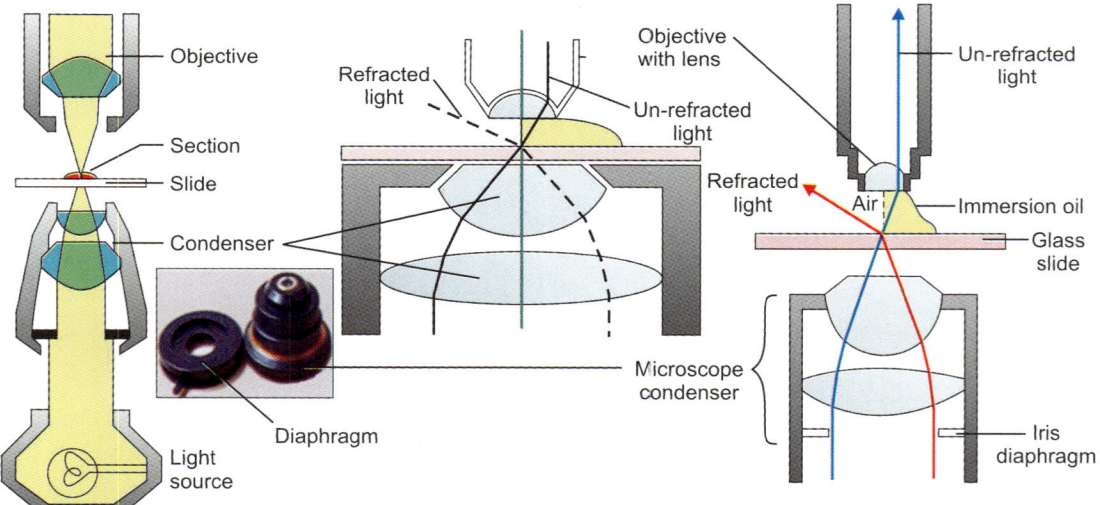

Fig. 16.9: Substage condenser is integral as well as very important part of a good quality light microscope. (A) The optical path shows that cone of light from illuminating source passes through condenser lenses to enter the objective lens; (B) and (C) Show that the light bends (diffracted) in air so much that it misses the small high magnification (40X); but the use of immersion oil over the glass slide (with stained film or smear) keeps light from bending so that the undiffracted light enters through the lens of oil immersion objective (100X) and hence provides better resolution.

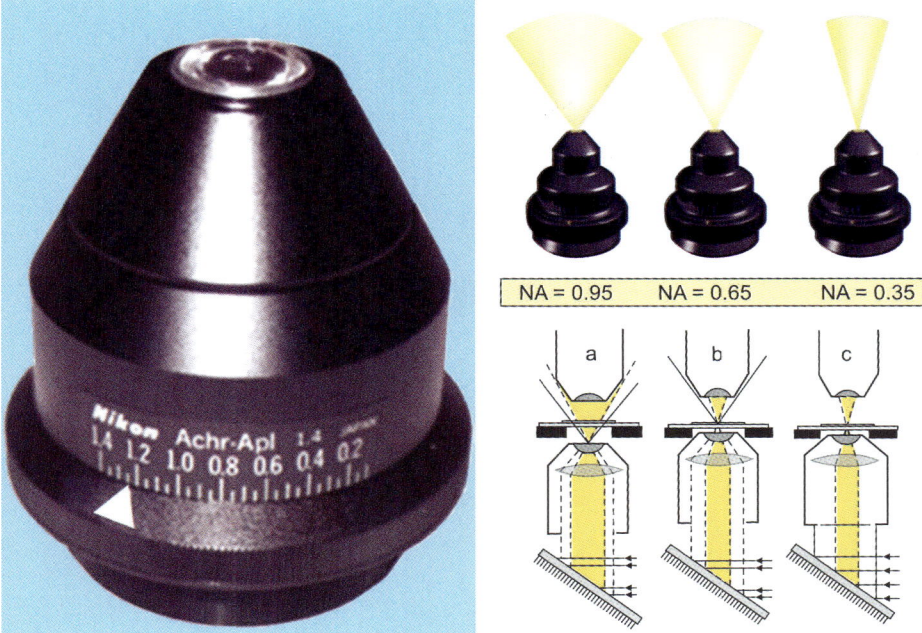

Fig. 16.10: On the left is shown a microscope condenser with engravings on its housing. Drawings on the right are diagrammatic representations of substage condenser's positions and the illuminating cone size versus numerical aperture (NA). The maximum light reaches to the objective lens with the condenser positioned highest (c).

3. Achromatic condensers
4. Aplanatic achromatic (compound) condensers

The condensers are corrected both for spherical and chromatic aberrations (Table 16.1).

Criteria for Selection of Good Microscope Condenser

The condenser in any microscope is a very important component. Two main things should be looked in the condenser before finally purchasing a microscope:

1. Numerical aperture

A critical factor in choosing substage condensers is the numerical aperture performance that will be necessary to provide an illumination cone adequate for the objectives. The condenser numerical aperture should be equal to or slightly less than that of the highest objective numerical aperture. Therefore, if the highest magnification objective is an oil-immersion objective with a numerical aperture of 1.40, then the substage condenser should also have an equivalent numerical aperture to maintain the highest system resolution. In this case, immersion oil would have to be applied between the condenser top lens and the underside of the microscope slide (Fig. 16.12) to achieve the intended numerical aperture (1.40) and resolution. Failure to use oil will restrict the highest numerical aperture of the system to 1.0, the highest obtainable with air as the imaging medium.

2. Thickness of glass slides

Another important consideration is the thickness of the microscope slide, which is as crucial to the condenser as cover slip thickness is to the objective. Most commercial producers offer slides that range in thickness between

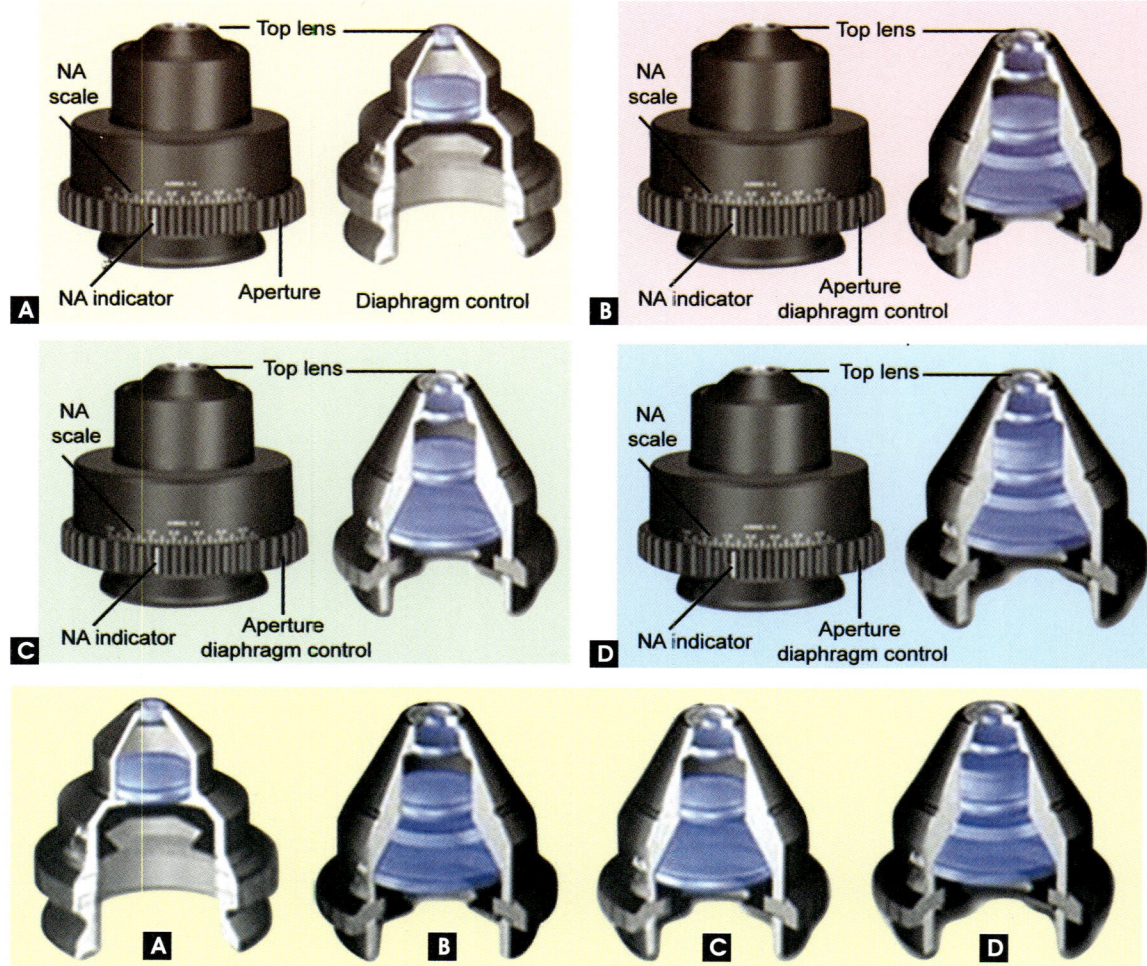

Fig. 16.11: Optical corrections in substage condensers of different kinds: (A) Abbe type, (B) Aplanatic type, (C) Achromatic type, and (D) Aplanatic-achromatic compound type. Upper drawings show together with a schematic sectional view for looking the external appearances of these types in the same sequence together with each type's internal lens arrangements shown in blue.

0.95 and 1.20 mm with the most common being very close to 1.0 mm. A microscope slide of thickness 1.20 mm is too thick to be used with most high numerical aperture condensers that tend to have a very short working distance. While this does not greatly matter for routine specimen observation, the results can be devastating with precision photomicrography. It is recommend that microscope slides be chosen that have a thickness of 1.0 ± 0.05 mm,

and that they be thoroughly cleaned prior to use.

Objectives of Microscope

The objectives are powerful magnifying lenses attached to the revolving nosepiece of a microscope (Fig. 16.13). They are so designed as to produce the best possible image of an object. A set of objectives, commonly three being provided in a good quality microscope,

Table 16.1	Different types of microscope condensers with their features and use			
	Abbe	**Aplanatic**	**Achromatic**	**Aplanatic achromatic**
Diagram				
Features and aberration corrections	• Simplest and **least corrected** • **Least expensive** • Capable of passing bright light, but • **Not** corrected for either chromatic or spherical optical aberrations • Image is **not sharp** as it is surrounded by blue and red colors at the edges	• **Well corrected** for **spherical aberration** (green wavelengths) • **Not corrected** for chromatic aberration	• Usually contain three to four lens elements • Corrected in two wavelengths (red and blue) for chromatic aberration	• **Highest level correction** for both spherical and chromatic aberration • **Condenser of choice** for use in critical color photomicrography with white light
Lens elements	2 in simplest form	5 lens elements	4 lens elements	8 internal lens elements (2 doublets and 4 single lenses)
NA	**1.25**; up to 1.4 in high-end models (with 3 or 4 internal lens elements)	Typical aplanatic condenser with a numerical aperture of 1.40	0.95 the highest attainable without requiring immersion oil	1.38
Suited for	Mainly for routine observation with objectives of modest numerical aperture and magnification wide cone of illumination is able to work with long working distance objectives	Capable of producing excellent black and white micrographs when used with green light generated by either a laser source or by use of an interference filter with tungsten-halogen illumination	Useful for both routine and critical laboratory analysis with "dry" objectives also for black and white or color photomicrography	

Note: Abbreviations in diagrams are: 1=condenser top lens, 2=internal lens element, 3=mounting flange, and *=aperture diaphragm

are at least **parfocal** so that when the nosepiece is rotated to change the objective only a small adjustment of focus is necessary. Ideally as the magnification increases, the focusing movement required should always be upwards away from the specimen to avoid damage to the lower lens of the objective in use and the cover slip. Infinity-corrected objectives come in a wide range of magnifications, from 1.5X to 200X, and in various qualities of chromatic and spherical correction—from simple achromats to planachromats and precision

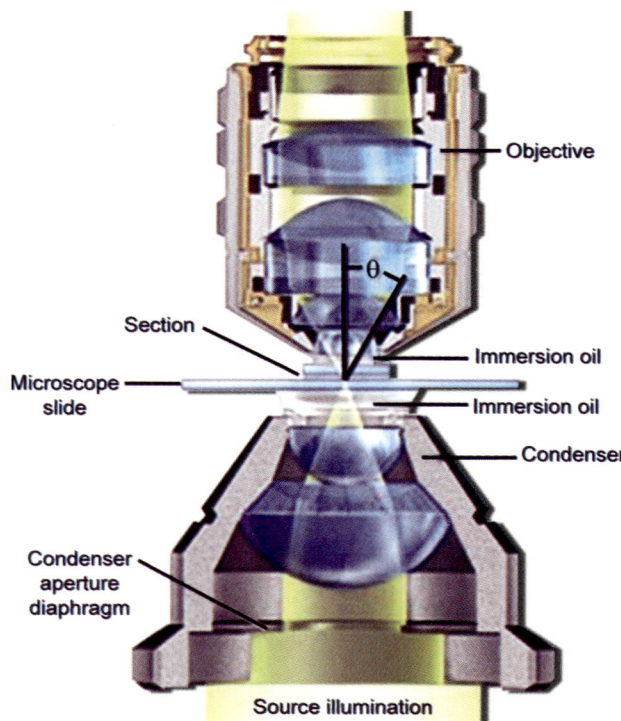

Fig. 16.12: Abbe condenser (most commonly used condenser in standard microscopes) and its optical pathway. Abbe condensers have two lenses (*Refer to* Table 16.1)

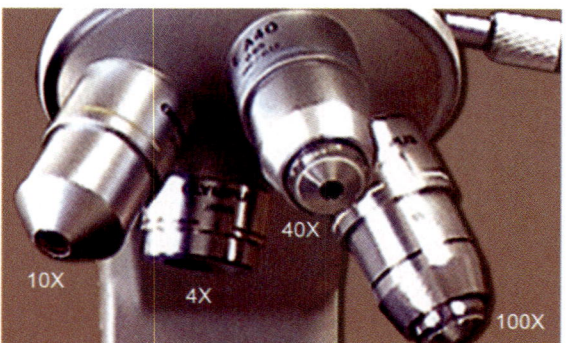

Fig. 16.13: The **objective lenses** are the most important components of a light microscope. There are four objectives in a good quality make, and are responsible for image formation, magnification, the quality of images and the resolution of the microscope. The typical numerical aperture (NA) are: 0.1 for 4X; 0.25 for 10X; 0.65 for 40X and 1.25 for 100X lenses.

plan-apochromats. Most, but not all, are designed to be used dry that is with air in the space between the objective and the specimen. The bright-field series have the customary microscope thread for screwing into the nosepiece. The objectives which are used for bright-field/dark-field observation usually have wider diameter threads and require a nosepiece with wider openings for attaching.

The compound microscope achieves a two-stage magnification (Fig. 16.14). The objective projects a magnified image into the body tube of the microscope and the eyepiece further magnifies the image projected by the objective. The amount of total magnification depends on the focal length of the eyepiece, which is placed near the focal point of the objective to magnify this image.

Sometimes, a very low power objective (3X or 4X) is also provided in a microscope. This is meant just for scanning of sections.

b. High-power Objectives

A high-power objective (40X) generally consists of a number of lenses mounted together. These are made up of different kinds of flint and crown glasses (sometimes of Jena glass). The lowest lens in the objective is made up of crown glass and it consists of a glass hemisphere, the plane surface of which faces towards the object. Of remaining lenses, those which are concave are made up of flint glasses, while the convex lenses are made up of different kinds of crown glasses (Fig. 16.16).

c. Oil-immersion Objectives

The oil-immersion objectives (90X or 100X) are like the high-power objectives. They are used only when very high magnification work requiring finest detail resolution is needed. A special difficulty is encountered with the use of an oil-immersion objective, which will be described when setting of microscope is dealt. The low-power, high-power, and very-low-power objectives are sometimes designated as of *dry* type objectives, because a film of air is present between the top of cover slip and the bottom lens of the objective. In that case an oil-immersion objective is regarded as of *wet* type because of thin layer of immersion oil replaces the air. The best quality oil immersion objectives have a NA of 1.4, but routinely, those with NA 1.0 or even less are used for most of the histological work.

Objectives are generally available with somewhat confusing array of characteristics. *Three* kinds of objectives are in ascending scale of performance and cost (Fig. 16.16).

i. Achromatic objectives

These objectives manufactured by Amici (1827) and Lister (1827), are the cheapest ones and adequate for general laboratory use. The achromatic objectives signify objectives 'without color'. They are corrected for one

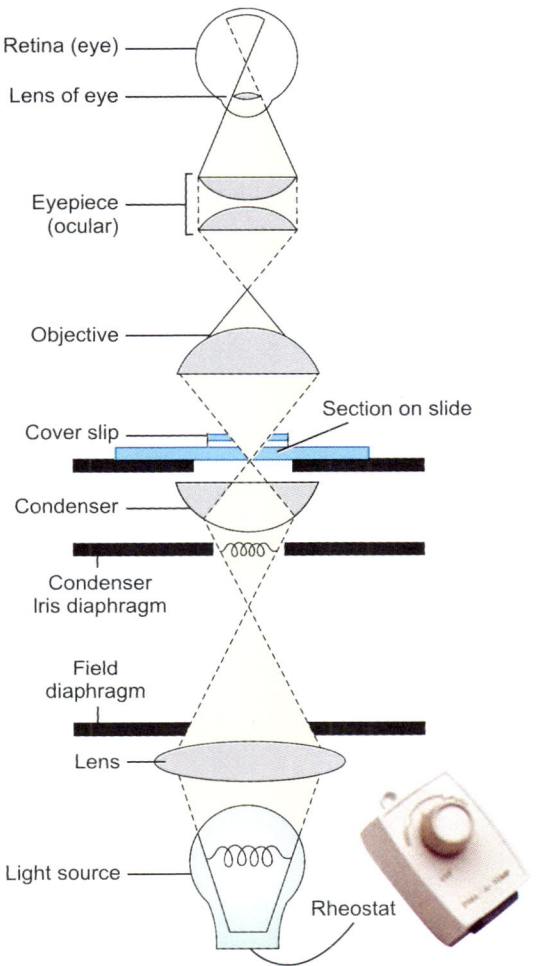

Fig. 16.14: Two-step magnification in a microscope.

Types of Objectives

Commonly three types of objectives are attached in the revolving nosepiece of a microscope (Fig. 16.15).

a. Low-power Objectives

A low-power objective (10X) consists of a lower plano-convex lens of crown glass, and a divergent (concave) lens of flint glass. It is the plane surface of plano-convex lens which faces towards the specimen. Whereas the plano-convex lens diminishes the spherical aberration, the concave lens (flint-glass) corrects the chromatic aberration.

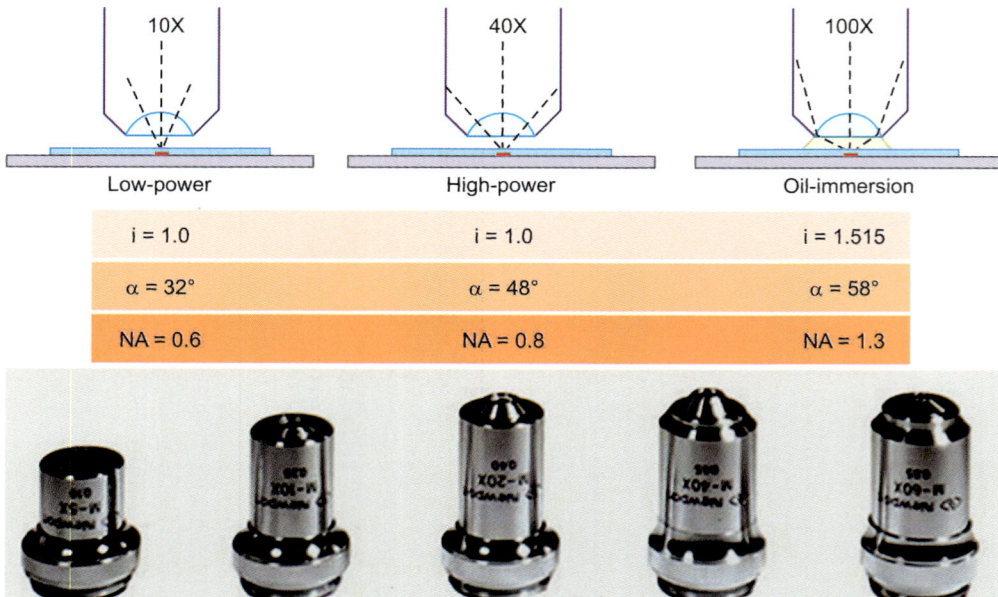

Fig. 16.15: The low and high power objectives (both high dry lenses) work in air which has a refractive index of 1.0.; and numerical apertures 0.6 and 0.8 respectively. Notice that in the extreme right diagram, Immersion oil objective has a considerably higher numerical aperture, sometimes even up to 1.56 because of the oil's refractive index 1.515 (higher than 1). Using an immersion oil bends more light into the lens capturing more orders of diffraction from the object. (Keep in mind that finer details or more closely spaced objects will give much higher angles of diffraction than will larger objects with less fine details).

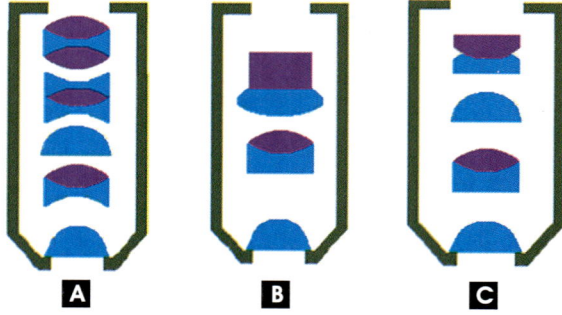

Fig. 16.16: Number of lenses (of crown-glass in blue and of flint-glass in purple shade) mounted together in: (A) High-power, (B) Achromatic, and (C) Apochromatic objectives.

color (called the preferred color) of the spectrum. The color corrected lenses of achromatic objectives are designed to bring the images of all colors to the same focus. In practice, however, only the yellow and green-blue images are coincident. Because these objectives are constructed entirely of glass they exhibit a residual chromatic fault known as the *secondary spectrum* (and this cannot be excluded). The correction for spherical aberration is usually done in central part (yellow-green region) of the spectrum. The faults on either side of this region are, therefore, diminished. But these faults become more apparent as the color of light used differs from that of the preferred color (for which correction has been done).

Achromatic objectives are also corrected for chromatic aberration by superimposing two spectral color regions so that they also come to a focus at the same point.

ii. Semi-apochromatic (fluorite) objectives

These objectives are constructed partly of fluorite or some other special type of glass which reduces the secondary spectrum. They are intermediate between the achromatic and apochromatic types; being superior to achromatic ones but not quite equal to apochromatic objectives. Fluorite objectives require compensating eyepieces for best performance.

iii. *Apochromatic objectives*

Abbe (1886) constructed apochromatic objectives using newly discovered Jena glasses. Apochromatic lenses are relatively costlier than achromatic objectives, have a still more perfect high correction in the center of field, and are best for resolving fine details. Spherical and chromatic aberrations have been corrected as far as possible. Sometimes the apochromatic objectives are called *with separated colors* objectives. The high color correction is possible through the use of a mineral fluorite, which possesses characteristics not obtainable in any known glass. The spherical aberration is corrected for two regions in the spectrum instead of one only, as in achromatic objectives, and three colors (red, blue, and green) are superimposed in correcting chromatic aberration, thus making the lenses of the objective almost ideal. Like fluorite or semiapochromatic objectives (Fig. 16.17), apochromatic

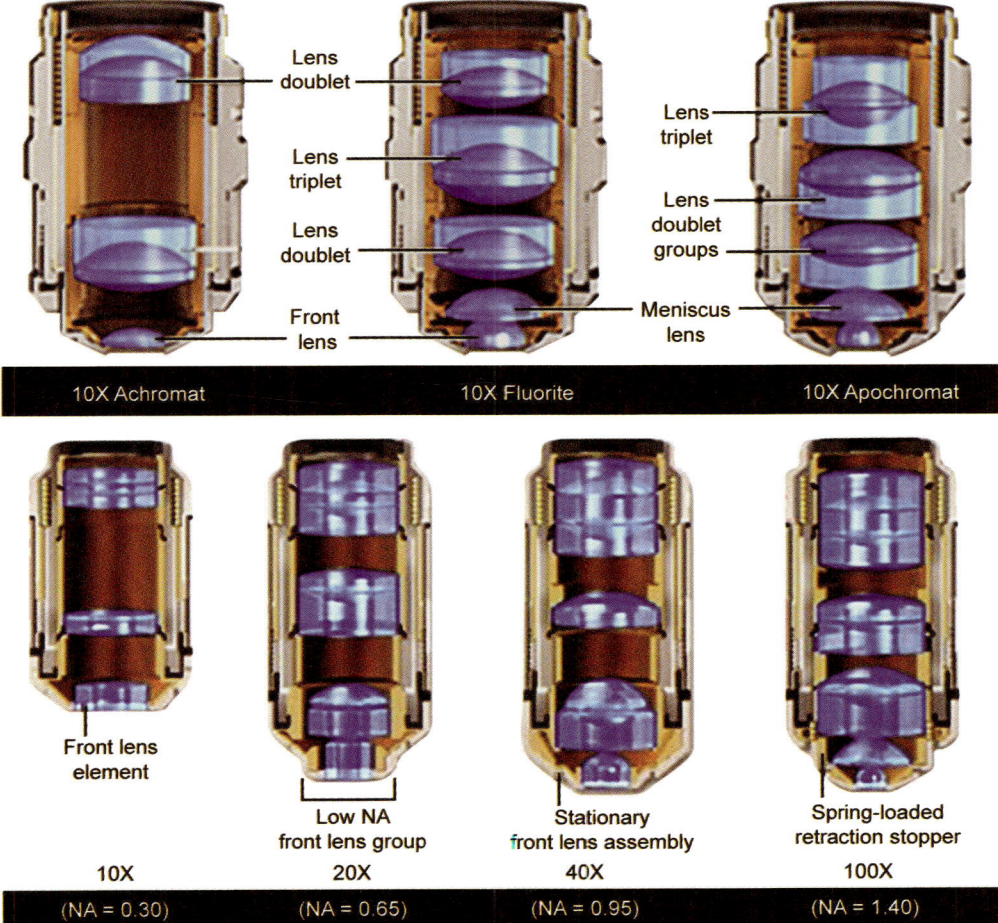

Fig. 16.17: Upper drawings represent lens-system designs in common objective optical correction factors. Lower drawings are shown apcchromat objectives of different numerical apertures (NA) with front lens group, front lens assembly, and spring-loaded retraction stopper

ones also require the use of compensating eyepieces because a part of correction of aberrations is accomplished in the eyepiece. It may be taken that all colored images lie in the same plane, hence apochromatic objectives are of great use in accurate photomicrography. Flat-field apochromatic (plan-apochromatic) objectives are preferable for photomicrography and for scanning wide fields.

Interpretation of Different Engravings Over Barrels of Objectives

Often microscope manufacturers mark certain parameters engraved on the outer side of objective barrels. These engravings differ in different makes. Knowledge of the meaning of each engraving requires a little understanding for the proper use of the microscope (Fig. 16.18).

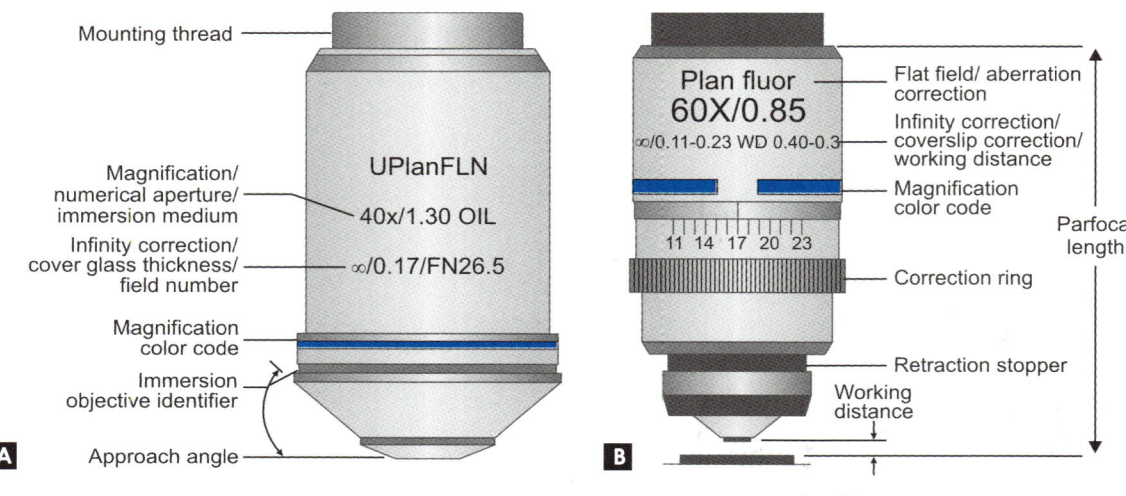

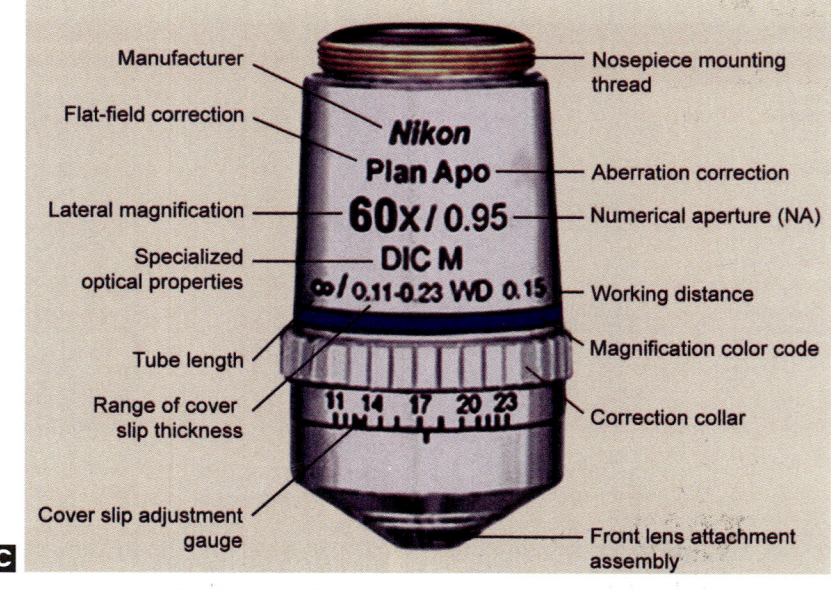

Fig. 16.18: The engravings on the barrels of microscope objectives shown here serve only as examples. The format of the engraved specifications will vary between objectives and manufacturers. (A) Oil-immersion objective; (B) 'Dry' objective; and (C) Engravings on the barrel of plan apochromat objective.

Two methods of designating the primary magnification of objectives are still in use.

i. The *older method* is to engrave the equivalent focus (EF) on the barrel of the objective. The equivalence is to the magnification of a simple lens at a distance of ten inches. A simple lens of one inch focal length will cast an image magnified ten times at a distance of ten inches. Thus if ½" is engraved on the barrel, it would mean that the lens will give a 20X image at the same distance. Similarly, engraving of 2/3 inch will mean a lens with 15X image. However, if EF is engraved in millimeters (which is a stupidity that should never have originated), it is first necessary to convert millimeters to inches. For example, if 2 mm is engraved, it will mean 1/12" and it would mean a magnification of 120X.

ii. The *second (modern) method*, obviously intelligent, and followed by high grade manufacturers, is to engrave the actual figure of magnification on the objective barrel. Some objectives are engraved also with values of the tube length (in millimeters) and thickness (in millimeters) of the cover slip for which a particular objective lens is corrected (Fig. 16.10). However, the cover slip thickness is important only when dry field examination (with low or high-power objectives) is being carried. When oil immersion is in use there is no importance in the optical system of the cover slip thickness.

The approximate values of focal lengths, in millimeters and inches, with their numerical aperture (NA) marked over the barrel and magnifications for achromatic objectives in common use are given in Table 16.2.

Numerical Aperture

The numerical aperture (NA), an optical constant, of a lens system is the 'light gathering' capacity of the objective lens of a microscope. The figure of NA is generally

Table 16.2	Comparison of magnificatio and viewfield diameter	
Magnification	**Viewfield diameter (mm)**	
½X	42.4	
1X	21.2	
2X	10.6	
4X	5.3	
10X	2.12	
20X	1.06	
40X	0.53	
50X	0.42	
60X	0.35	
100X	0.21	
150X	0.14	
250X	0.085	

engraved on the barrel of an objective. It is an indication of the maximum resolution of which the lens is capable, and it is not a measure of the resolution that the objective will automatically produce.

The ability to reveal details can be expressed in terms of the fraction of the wavefront admitted by the objective lens. The lens *AB* can accept a cone of rays from the point *P*. If 2θ is the angle of this cone (Fig. 16.19), the numerical aperture (*NA*) is defined as:

$$NA = n \sin \theta$$

Where *n* is the refractive index of the medium in which the lens is working. Since *n* for air is 1, and since sin θ cannot be greater than one (because, sin 90° = 1); it is clear that no lens working in the air can have a theoretical NA greater than one.

The only way of achieving a NA greater than 1, since θ cannot rise above 90°, is to increase *n*. This is done by using immersion oil and oil-immersion objective. Immersion oil (with its refractive index 1.56) when substituted for air, *both above and below the slide*, not only increases '*n*' but also permits the lenses involved to operate at greater numerical aperture. *If immersion oil is used only between the slide and the objective* (a routine but wrong

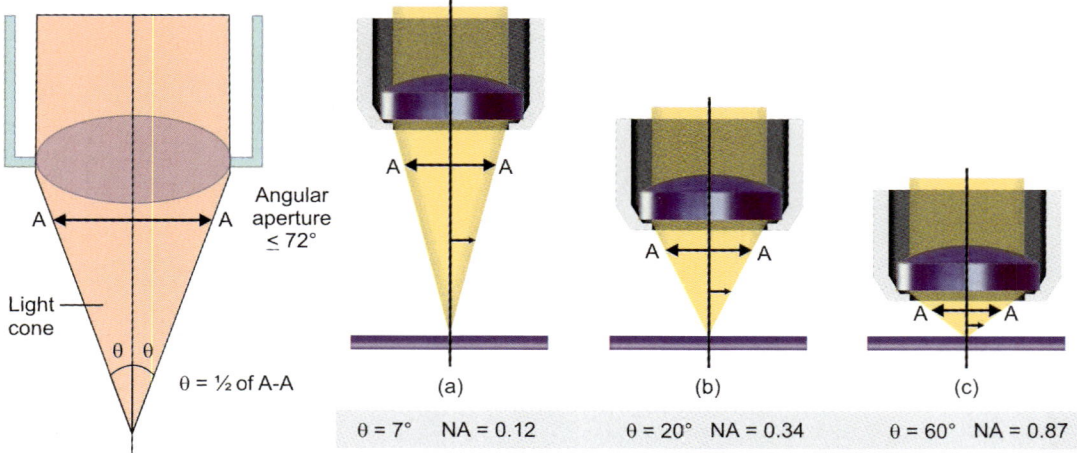

Fig. 16.19: Numerical aperture (NA) of an objective is a measure of its capacity to gather light and resolve fine specimen detail at a fixed object distance. NA = n (sin θ), n: refractive of the imaging medium between the front lens of the objective and specimen cover slip, θ one half of AA (Angular aperture ≤72 degrees. Theoretical maximum NA = 1 (aperture of a lens operating with air as the imaging medium).

practice followed by those ignorant of correct operation of microscope), the system cannot operate at an NA greater than one.

The numerical aperture can also be expressed by the following equation:

$$NA = R \times \lambda$$

Where R is the power of resolution and λ is the wavelength of the light which illuminates the optical system. From the above equation, it may be derived that:

$$R = \frac{NA}{\lambda} = \frac{n \sin \theta}{\lambda}$$

In this expression both n and λ are constant; with usual working the microscope is operated in air (hence $n=1$) and with white light ($\lambda = 2500$Å). Thus the resolution of any objective in use would be directly proportional to the sine of half the angle of the entering cone of light.

$$R \propto \sin \theta$$

The wider the angle of cone of light, the greater is the resolution.

Cover Slip Correction Collars

High magnification objectives designed to be used with air as the immersion medium between the front lens and the cover glass are prone to aberration artifacts due to variations in cover glass thickness (Fig. 16.20) and dispersion. The internal lens elements in a high numerical aperture dry objective may be adjusted to correct for these fluctuations.

Most objectives are designed to be used with a cover glass that has a standard thickness of 0.17 millimeters and a refractive index of 1.515, which is satisfactory when the objective numerical aperture is 0.4 or less. However, when using high numerical aperture dry objectives (numerical aperture of 0.8 or greater), cover glass thickness variations of only a few micrometers result in dramatic image degradation due to aberration, which grow worse with increasing cover glass thickness. To compensate for this error, the more highly corrected objectives are equipped with a correction collar to allow adjustment of the central lens group position to coincide with fluctuations in cover glass thickness. This type of correction is also used on objectives

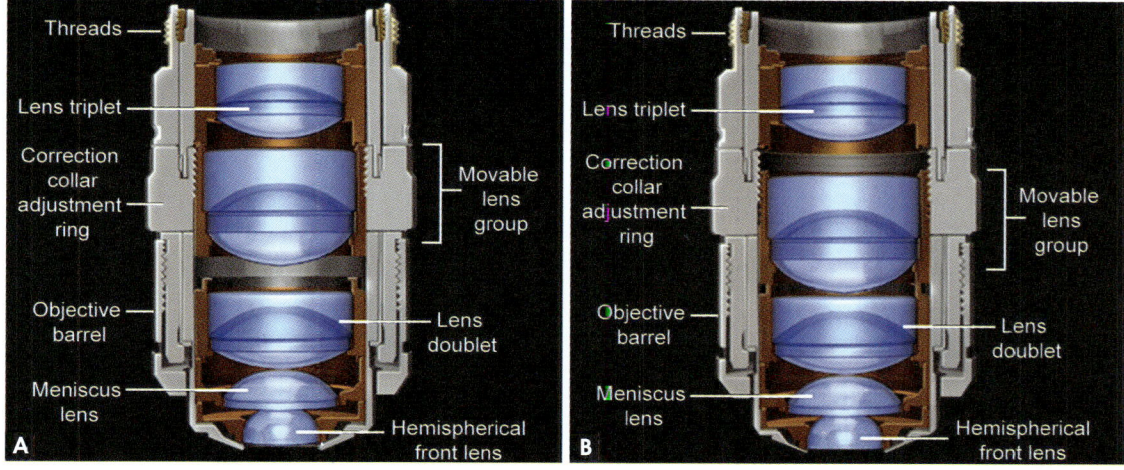

Fig. 16.20: Cover slip correction collars for variable thickness: (A) 0.11 mm thickness, (B) 0.23 mm thickness.

intended to be utilized with glycerin and water as the imaging medium.

Using an objective having a correction collar requires a considerable amount of practice and careful attention. When the collar is adjusted, focus tends to shift and the image often wanders, which can lead to focus errors and an increase in aberration artefacts.

Eyepieces (Or Oculars) Of Microscope

The eyepieces are the optical components that are fitted on the top of body tube of the microscope merely by a sliding fit. They can be removed at will and one of different magnifying power substituted. Practically all reputable manufacturers of microscopes make their series of eyepieces to the same standard diameter (23 mm). An eyepiece simply enlarges the image of an object produced by a particular objective. It neither improves the quality of the image nor adds anything to the resolution of the objective.

The basic design of a microscope eyepiece consists of *two* lenses (or lens doublets) separated by an air space the length of half the sum of the focal lengths of the two lenses. The lens closest to the eye is called the *eye lens* and the other lens (closest to the objective) is called the *field lens* (Fig. 16.21).

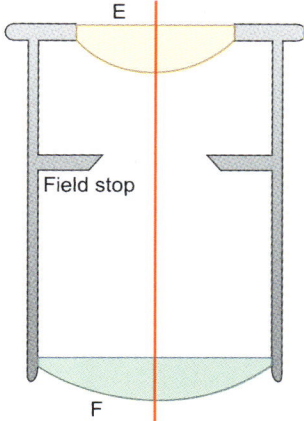

Fig. 16.21: The eyepiece is basically formed by two lenses: An upper eye lens (E) and a lower field lens (F) separated by an air space the length of half the sum of the focal lengths of the two lenses.

The result of having two lenses within the eyepiece is to reduce the size of the intermediate image and thus allow the observation of a larger field of view. The eyepiece focuses the real image (from the objective) outside of the eye lens at a position called the *eyepoint* or *exit pupil*, which corresponds to the position of the focal point of the eye (Figs 16.22 and 16.23).

The *eye relief* of an optical instrument (such as a telescope, or a microscope) is the distance

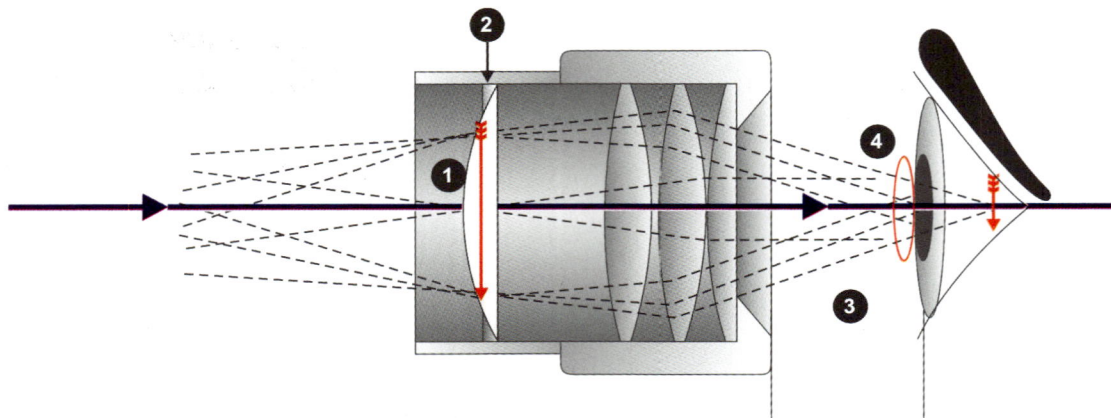

Fig. 16.22: Optics of a high eye point eyepiece showing eye relief and exit pupil: 1=Real image, 2= Field diaphragm, 3=Eye relief, and 4=Exit pupil/ Ramsden disc.

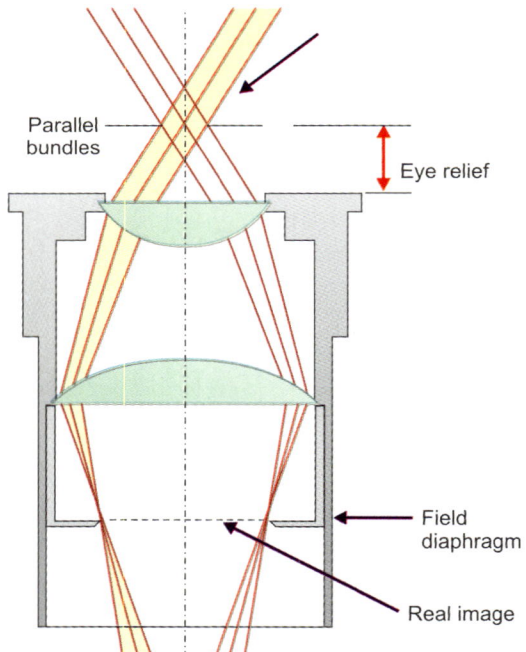

Fig. 16.23: Formation of the exit pupil in an eyepiece.

greater than 5–7 mm to avoid observer difficulties and is usually 7–13 mm. In most of the microscopes the eye relief is so stupidly short that wearers of spectacles have to remove them in order to get Ramsden's disk inside their cornea and thus 'fill' their eye with the image.

Exit Pupil of Eyepiece

One of the first things one needs to know when taking picture with the microscope is position of the "exit pupil" of the eyepiece being used (Fig. 16.23). The exit pupil is a small circular area, located at the vertex of a cone of light that comes out from the eyepiece, through which all the light rays that form the image pass. If the diaphragm of the camera is not placed in the plane of the exit pupil, but instead, intercepts the cone and the corners of the image will be dark.

Often, the camera is prevented from attaining the right position by the front surface of the camera lens, which touches the ocular before reaching the proper position. If the exit pupil is too close or too far away from the eyepiece only the corners of the picture will be black; if things work less well your image will be reduced to a little disk. The ideal condition is when the circular field of the image is just outside the corners of the field of the camera. To obtain this condition, you can

between the outer surface of the eye lens of the eyepiece and Ramsden's disk within which the user's eye can obtain the full viewing angle. If a viewer's eye is outside this distance, a reduced field of view will be obtained. Generally, **the higher the magnification and the larger the intended field-of-view, the shorter is the eye relief. Eye relief** must be

take advantage of the zoom lens provided on almost all digital cameras.

Figure 16.23 illustrates the path of the marginal rays in an eyepiece. The objective of a microscope forms the real (or primary) image on the plane of focus of the eyepiece, which usually is located at its field diaphragm. It can be seen in this figure the light coming from a point of this image passes through the lenses of the ocular and emerges from it as a narrow bundle of parallel light rays. This diagram shows the paths of the rays coming from two marginal points of the image, but the same is true for all other points. So, the light bundles that come from all image-points pass through the same circular area, which is also called the *Ramsden circle*. This is the exit pupil of the eyepiece, or the eyepoint. Therefore, the whole of these parallel bundles forms a cone of light that passes through the pupil. As we have seen, the vertex of this cone is not a point, but a small disk called the air-disc (Fig. 16.24).

Types of Eyepieces (oculars)

There are a handful of eyepiece types which have evolved over time, each trying to address the various problems of amplification of the prime image from the objective. Standard types have been assigned the names of the men who first designed these amongst them are names of scientists/astronomers... Huygens, Kellner, Ramsden, etc. There are different types of eyepieces available (Fig. 16.25 and Table 16.3). A brief account of some of the types is as follows:

i. *Regular* or *Huygenian eyepieces* are the most common type of eyepieces used in routine microscopes. These eyepieces without or with pointer (for indicating the given area or cell in the microscope field) are available in market. Huygens (1864) introduced compound eyepieces consisting of two lenses. The lower lens represents the objective of an optical instrument. It forms a real image of the image formed by the objective; and since it also enlarges the field of view—it is called *field lens* (Fig. 16.25 A). The second lens lies on the top of the ocular and is termed *eye lens*. Both the field and eye lenses are generally plano-convex with their **convex surfaces facing down; flat sides face the observer**. Approximately middle of these lenses, there is a fixed circular opening which, by its size, defines the circular field of view that is

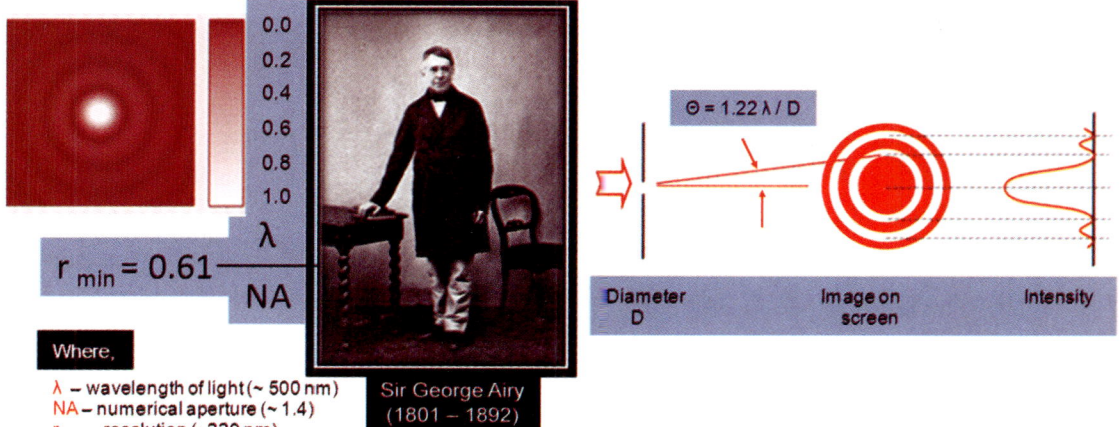

$$r_{min} = 0.61 \frac{\lambda}{NA}$$

Where,

λ – wavelength of light (~ 500 nm)
NA – numerical aperture (~ 1.4)
r_{min} – resolution (~220 nm)

$\Theta = 1.22 \lambda / D$

Diameter D   Image on screen   Intensity

Sir George Airy (1801 – 1892)

Fig. 16.24: Airy disc and resolving power: British Astronomer George Airy discovered and described the diffraction pattern of a circular aperture. The lower picture depicts diffraction through a circular aperture of diameter D. Notice that because of diffraction, images formed by 'perfect" optics are fuzzy.

observed when looking into the micro-scope.

ii. *Ramsden eyepieces* have the field diaphragm below the field lens (Fig. 16.25B). These types of eyepieces are mostly used for measurement purposes in conjunction with a scale placed on the diaphragm. In Ramsden type of eyepieces, the two lenses are of equal focal lengths, separated by a distance equal to two-thirds of the numerical value of the focal length of either. Like Huygen's type the eye lens in Ramsden type has its plane surface facing towards top; but the field lens has its plane surface facing down. Thus the convex surfaces of both eye and field lenses face each other. These eyepieces are termed *positive eyepieces.*

iii. *Kellner eyepieces* are basically like Ramsden type of eyepieces, except that their eye and field lenses are separated by a distance numerically equal to the focal length of either (Fig. 16.25D). These eyepieces have a very wide field of view; but due to a disproportionate magnification of the peripheral part of the field,

there appears a noticeable spherical aberration.

iv. Devised by Abbe (1886) *compensating eyepieces* are specially designed to be over corrected for a final compensation in the degree of magnification of the red and blue images of apochromatic objectives, which are not of equal size otherwise. These eyepieces may, however, be used with advantage for high-power achromatic objectives, as well. The barrel of compensating eyepieces is commonly marked to distinguish them from ordinary eyepieces. If not so, a practical way to distinguish the two types is by means of holding an eyepiece to the eye and observing an extended light source. Whereas an ordinary type of eyepiece will show a blue fringe to the edge of the diaphragm, a compensating eyepiece will show a red fringe.

Apochromatic and fluorite objectives need compensating eyepieces, whose properties are essential to the making of the final image, such that these objectives cannot produce a fully corrected image

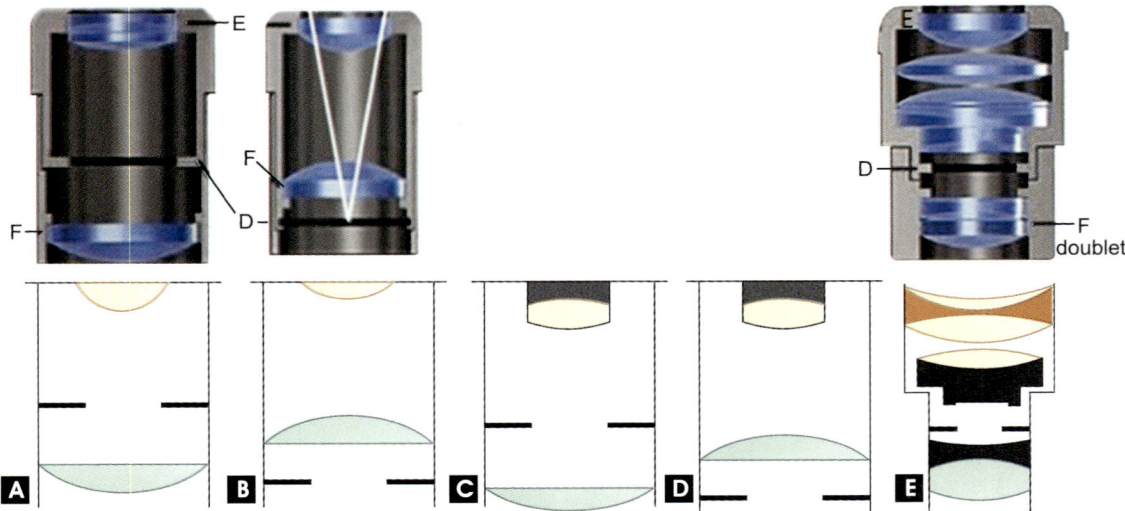

Fig. 16.25: Types of eyepieces: (A) Huygenian; (B) Ramsden; (C) Achromatised Huygenian; (D) Achromatised Ramsden (Kellner); (E) Modern 'Periplan' eyepiece designed to reduce the curvature of field often associated with simple Kellner eyepiece. D=Diaphragm, E=Eye lens, and F=Field lens.

by themselves. In the achromatic system it is generally understood that the objective is designed to produce a corrected image, which is then magnified by the (non- compensating) eyepiece. Usually, but not always, these eyepieces are marked with a 'K' or similar to denote the compensating property.

v. *Comparison eyepieces* are parts of a device which can be fitted over two microscopes side by side. The images from these are brought together into one eye lens in such a manner that one-half of the field is the view through one microscope while the other half of the field is that through the other microscope (*see* Chapter 18).

vi. *Projection eyepieces* are of regular Huygens' type, but the eye lens (which is generally achromatized) is fitted in a separate tube which has a spiral adjustment for focusing the diaphragm on a photographic plate and for giving a better state of correction of the final image. Projection eyepieces are better suited for photomicrography; although most of the ordinary type eyepieces can be used fairly well (Fig. 16.26).

To produce an ideal photomicrograph, a flat field is obtained with apochromatic objective and a special kind of eyepiece called *Homal eyepiece* (Fig. 16.27). These

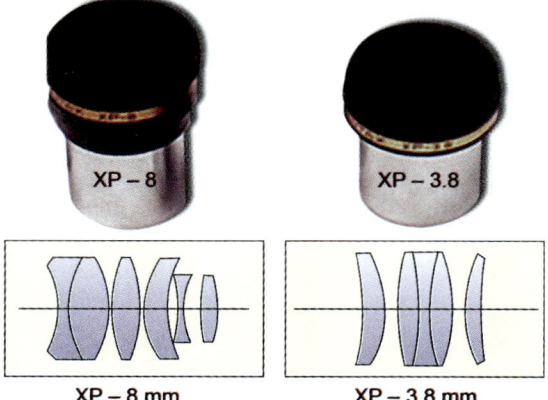

Fig. 16.26: Pentax SMC XP lenses and their lens arrangements.

Fig. 16.27: Homal eyepieces.

are negative eyepieces, with the eye-point lying (Ramsden disk) within the lens; hence they cannot be used for visual purposes. The Homal eyepieces being of larger diameter than standard oculars, adapters are required for fitting them to the microscope. **Homal** is a special projecting system used to take pictures in the visible light. Optical system of homals partially corrects field of view curvature and magnification chromatic aberrations of achromatic and apochromatic objectives. But using Homals eyepieces lose about 1/3 of the field of view.

Homal-IV eyepieces should be used with objectives for 160 mm and 190 mm tube length if their powers is 60X or higher, and with 90X infinity tube objectives (using f=250 mm tube lens).

Focal length: minus 20.28 mm; Linear field of view: 8 mm; Mount diameter: 27 mm; Coated glass.

vii. *Ultraplane eyepieces* are available as a set of three eyepieces; and possess the same diameter as standard oculars. Hence special adapters are not needed. The ultraplane eyepieces are considerably costlier; but they will fully repay all their prices when ideal results are desired in a photomicrograph.

viii. *Demonstration eyepiece,* which can be used by *two observers* at a time (Fig. 16.28), is a

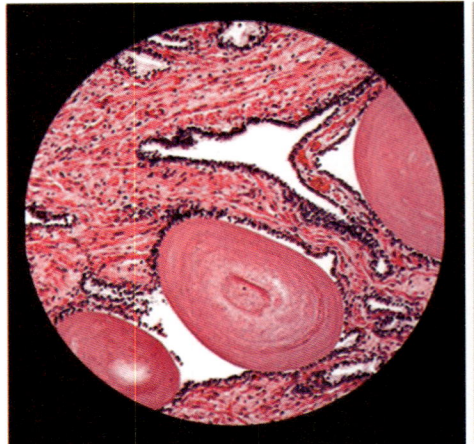
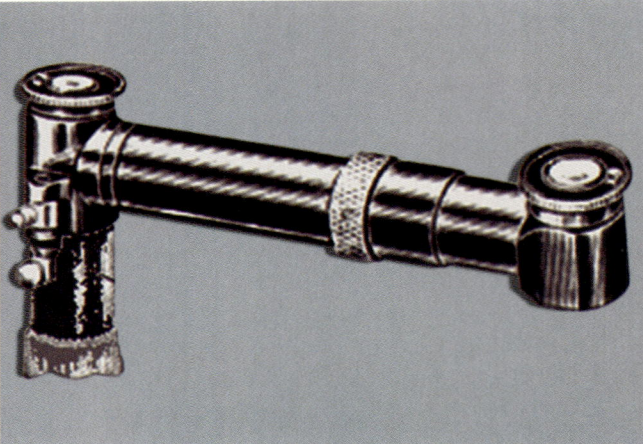

Fig. 16.28: Demonstration eyepieces are used for looking by two observers at a time.

device for attachment to any monocular microscope which takes the standard eyepiece. The side tube extends horizontally, with the auxiliary eyepiece extending at right angles in a removable mount which can be focused independently, so that the second eyepiece can be tilted and adjusted for convenient observation. In demonstration eyepiece a pointer, visible in both eyepieces, is useful for directing attention to any desired area of the slide being viewed.

High eye-point oculars, designed for glasses wearers, have an eye relief of 15–20 mm (Fig. 16.29). The diameter of the eyepiece aperture (in mm) is called the **Field-of-view-number**, or **Field Number** (FN). Remember that the eyepiece aperture is located at the intermediate image plane. The FN is often embossed on the barrel of the eyepiece. Power [P, magnification of an eyepiece is defined as D/focal length; where D = the closest distance of distinct vision, or 250 mm (average in humans)].

The rays of light leaving the eyepiece intersect at the exit pupil, or aperture, of the eyepiece, sometimes called by its

Fig. 16.29: Eyepiece with long eye-relief is convenient for those microscopists who want to see in microscope without taking away their spectacles.

older name of Ramsden disc. This eye-point is the place where the eye is placed in order to see the whole field of view, and is indicated by the letter 'R' in Fig. 16.30. Beginners in microscopy often find difficulty in holding their head sufficiently steady at the correct height at the

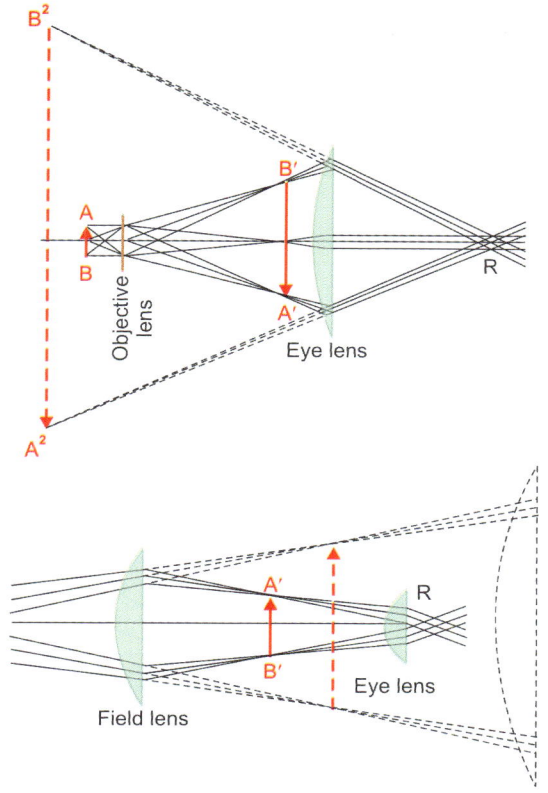

Fig. 16.30: Function of the eye lens in the eyepiece.

plane of the exit pupil for the field(s) of view not to move uncontrollably. With higher magnification eyepieces, the exit pupil is formed closer to the top eye-lens of the eyepiece. This sometimes makes the entire field of view difficult to see, particularly for spectacle wearers, and so manufacturers often make 'high-eyepoint' eyepieces, and include rubber cups around the eyepiece to assist the observer in holding their head at the correct height. These rubber cups can be folded down out of the way if desired. As with objectives, the higher the magnification of the eyepiece, the smaller is the field of view.

ix. *Zoom eyepieces* (Fig. 16.31) are also available in microscopes marketed by reputed manufacturers.

x. *Digital eyepieces* for microscopes with 1/3 inch CMOS image sensor and an impressive 1600×1200 resolution is a great way to experience viewing things in close up on your computer via USB connection (Fig. 16.32) on your computer and gives you a real-time images directly

Fig. 16.31: Zoom eyepieces are special kind of eyepieces.

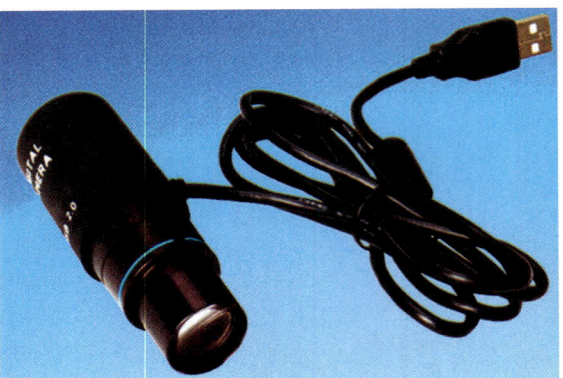

Fig. 16.32: A digital eyepiece is used viewing things in close up on your computer via USB connection.

on to your computer screen. This eyepiece is the best way to view high resolution images of everything in the microscopic world without having to bend over your microscope and squint with one eye?

The image quality of digital eyepiece is purely amazing as well because digital eyepiece uses a 1/3 inch 2 megapixel image sensor for delivering high image quality and remarkable detail like you have never seen before. The digital eyepiece can capture images and recordings in a beautiful 1600×1200

Table 16.3	Salient features of some important kinds of microscope eyepieces (oculars)		
Huygenian or negative eyepiece	**Ramsden or positive eyepiece**	**Achromatized Ramsden (Kellner eyepiece)**	**Periplan eyepieces**
• Called the internal diaphragm or regular eyepiece • Due to formation of the primary image by the objective, on the negative side of the field lens, these eyepieces are called *negative eyepieces*.	• Design created by **Jesse Ramsden** (1782); now called the external diaphragm eyepiece • 2 plano-convex lenses of same glass and similar focal lengths placed less than one eye-lens focal (7/8 or 7/10)	• Has the single eye-lens replaced by doublets • Useful in micrometry	**Advantages** • Better correction for lateral chromatic aberration • Increased flatness of field • Overall better performance when used with higher power objective
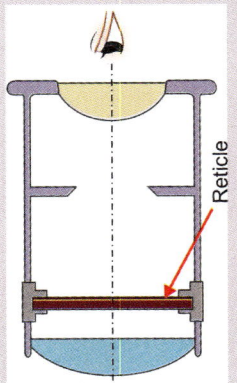	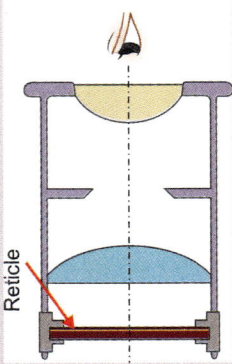	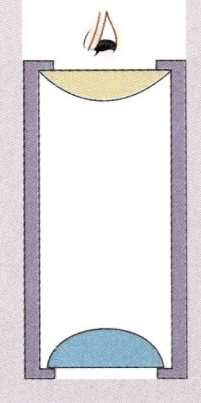	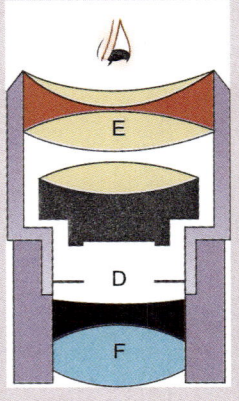
• Distance between the two lenses is numerically equal to twice the focal length of eye lens	• Eliminates distortion from the image of any scale in the eyepiece, but is not fully achromatised		**Disadvantages** • Produces **internal reflection** (minimises by internal silver coating)

Contd...

Table 16.3	Salient features of some important kinds of microscope eyepieces (oculars) (Contd...)		
Huygenian or negative eyepiece	**Ramsden or positive eyepiece**	**Achromatized Ramsden (Kellner eyepiece)**	**Periplan eyepieces**
• Field diaphragm lies at the focus of the eye lens and thus limits the field of view			• Most microscopes now feature such **wider body tubes** that have greatly increased the size of intermediate images • Wide-eye field eye-pieces that increase viewable area of the section by 40%
Simplest type of eye-piece, and works well with achromatic objec-tives, but it will cause distortion of image points off the optical axis	Highly suitable for monochromatic light source		

resolution for both video and photos. With the easy setup and user-friendly software, even children can get in on the action. Simply attach the eyepiece to any microscope then attach the USB cable to your computer to start capturing the wonders.

The eyepieces must be corrected for optical errors in the same way that objectives are corrected. To correct for this distortion, the Ramsden (external diaphragm or positive) eyepiece is also more easily accessed and fitted. In some cases the eyepiece of the microscope is also used to correct the chromatic difference of magnification found in semi-apochromat and apochromat objectives.

Choosing the Correct Eyepiece

The eyepiece is an integral part of any com-pound visual system. Its purpose is to magnify the real image formed by an objective or intermediate relay system. For this reason, the eyepiece design affects the magnification, image brightness, field flatness, and aberra-tions of these systems. As in all optical systems, the designer trades one advantage for another,

in an effort to optimize the eyepiece for specific applications. In choosing an appropriate eyepiece, one needs to consider the particular system being assembled. Since each eyepiece has individual advantages, a specific system may favor one eyepiece over another.

Anyone looking to buy a microscope knows, or quickly learns, that the total magnification of a microscope is arrived at through the simple expedient of multiplying the power of the objective lens by that of the eyepiece. So,

• 10X objective plus a 10X eyepiece = 100X magnification
• 100X objective lens with 20X eyepiece = 2,000X magnification

False Magnification

False magnification is when the power of the eyepieces employed pushes the **maximum useful magnification** above 1,000 times the numerical aperture (NA). For example, you can achieve 1,000X magnification by using a 40X/0.65 NA with 25X eyepieces. However, the total magnification of 1,000X exceeds the

Table 16.4 A comparison between Huyghens and Ramsden type eyepieces		
Parameters	Huyghens eyepieces	Ramsden eyepieces
Diagram		
	Often used in economical systems	Lateral color is not fully corrected because the focal plane is external
Focal lengths of Field lens Eye lens	3f f	f f
Distances between the field and eye lenses	2f	$\frac{2}{3}f$
Focal length of eyepiece	$\frac{3}{2}f\ (-)$	$\frac{3}{4}f\ (+)$
First principal focus	Between the field and eye lenses	In front of the field lens
Principal planes	Crossed	Crossed
First principal point	Distance behind the eye lens = f	Distance behind the eye lens = f/2
Second principal point	Distance in front of the eye lens = f	Distance in front of the eye lens = f/2
Angular extent of field	Sharp up to 25° Fair up to 40°	Sharp up to 25° Fair up to 40°
Nature of image	Convex towards eye	Flat
Lateral chromatic aberration	Small	Small
Distortion	Small; Field-stop and final image not equally distorted	Small; Field-stop and final image equally distorted

value of 0.65 NA multiplied by 1,000 (1000 × 0.65 = 650).

As a result, while your image will be magnified 1,000 times, it will yield no further useful information or finer resolution of detail. Quite the contrary, you will likely experience significant to severe degradation in resolution. The image becomes blurry in much the same way as when you try to zoom in on a webpage. It gets bigger, but there is no improvement in the resolution; no improvement in the amount of detail you can see. In fact, excessive magnification introduces artifacts, diffraction boundaries, and halos into the image that obscure specimen features and complicate the interpretation of visual observations.

With a 100X objective lens with an NA of 1.25 and 10X eyepiece, one achieves the same level of 1,000X magnification. The difference is that not only do you achieve higher magnification, but you also benefit from improved resolution. In other words, you can see materially better details in the image. Why? Because you have not exceeded the maximum useful magnification of 1000 × NA, which in this example is 1000 times 1.25 NA (1.25 × 1,000 = 1,250).

If you were to use a microscope camera with the two different objective lens/eyepiece combinations, you would see an even greater difference in clarity and detail between the two solutions.

Having read this far, it should come as no surprise to know that every compound microscope is designed and sold with 10X eyepieces as the standard benchmark. There is minimal benefit in using higher power eyepieces and considerable disadvantages. As if this were not enough, there are very few and rarefied applications in light microscopy that actually warrant higher magnification than 1,000X. By the same token, over 90% of stereo or low power applications employ magnifications of less than 45X.

So why do you need higher power eyepieces. Typically, you do not and you should not be bamboozled into paying extra for microscopes with additional eyepieces unless you have a specific requirement to isolate something in the field of view. For example, you may wish to measure an specific element in a smaller field of view using a reticule. In other words, there are some applications that may warrant higher power eyepieces. For general purpose use, however, they are at best not required and at worst, a waste of money. With that in mind, we sell just one microscope, the OM88, with additional 16X eyepieces. This

Table 16.5	Summary of the common objective/eyepiece combinations that lie in the range of useful magnification			
Objective (NA)	10X	15X	20X	25X
4X (0.12)	✓*	✓	✓	✓
10X (0.25)	✓*	✓	✓	✓
40X (0.65)	✓	✓	✗	✗
100X (1.25)	✓	✗	✗	✗

*These two objective/eyepiece combinations fall below the minimum useful magnification range. This is usually set at 500 times NA. However, it is highly arbitrary and all our microscopes operate effectively with 10X eyepieces and both 4X and 10X objective lenses.

is a popular microscope with doctors and clinics and the 16X eyepieces operate at the maximum useful magnification for the 40X objective lens.

The important thing to note about Table 16.5 is that the higher power eyepieces (15X, 20X and 25X) do not operate effectively in combination with the higher power objective lenses. In other words, it is meaningless to advertise a high power, compound microscope as "2,000X Magnification" since it does not work. In reality, standard, light microscopes are designed for a maximum of 1,000X magnification... so, please, do not waste your money on claims of higher magnification via higher power eyepieces.

SUMMARY

Parts of a Compound Light Microscope

Mechanical

1. Base or metal stand
2. Pillars
3. Inclination joint (hinge)
4. Arm (curved)
5. Body tube
6. Draw tube
7. Coarse focusing knob
8. Fine focusing knob
9. Stage (simple or mechanical)
10. Nosepiece (revolving)

Optical

1. Light source
2. Diaphragm
3. Condenser
4. Substage condenser
5. Objective lenses
6. Eyepiece or ocular

Mechanical Parts

1. Base—supports the microscope and/or electronics
2. Arm (limb)—supports the lenses, mirrors, and body tube
3. Body tube—passes light from the head to the eyepiece (or ocular)
4. Rotating head—contains mirrors and allows the body tube to rotate 360°
5. Focusing knobs—these are two in number:
   - Coarse focus knob moves the stage up and down quickly

- Fine focus knob makes small focus adjustments, particularly when using high power objective
6. Stage—holds the slide with the clips. It contains a circular aperture that allows light to pass through the section placed on the slide on its way to the objective lens being used.

Optical Parts

1. Light source—it illuminates the section placed on the glass side brought in a position that section lies against the hole in the stage. A shining bright light falls on the section from below. A power switch turns the light "on" and "off". The diameter of light put into the optical system of the microscope is controlled by an iris (or field) diaphragm placed near the base.
2. Diaphragm rotating dial—it controls the passage of light through the stage. Numbers on the dial indicate the relative amount of light passing, with, in majority of microscopes, "5" being the most and "1" being the least. This diaphragm is placed underneath the condenser of the microscope, hence called the *substage diaphragm* or *aperture diaphragm.*
3. Objective lens system—it produces most of the magnification (primary) as well as resolution. The high-power objective (blue stripe) magnifies 40X; the low-power (yellow stripe) magnifies 10X.
4. Eyepiece (or ocular)—it is placed where one views the enlarged image of a given section. It contains a lens called ocular that further magnifies the section's image by 10X.

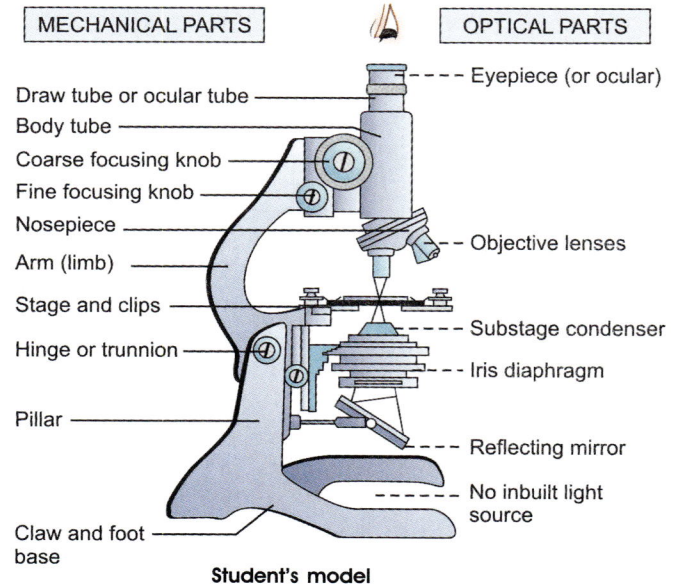

Student's model

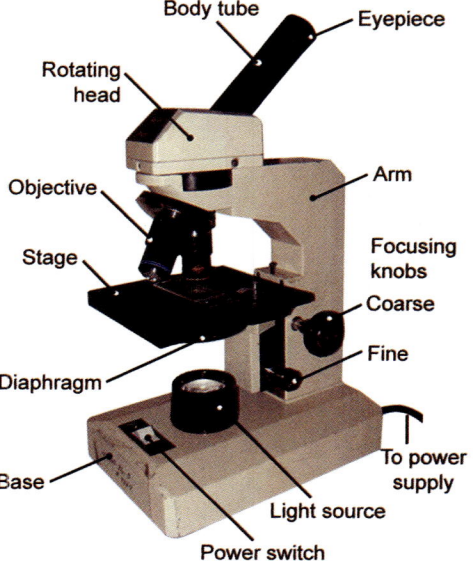

Research model

17

Best Performance of Light Microscope

SUITABLE MATERIAL TO BE PHOTOGRAPHED

The optical light microscope is one of the most extensively used instruments in medical and biological laboratories engaged both for the research and teaching purposes. Of all the instruments it is probably the least understood and the most indifferently used. The reason seems to be that no matter how poorly a microscope is adjusted, some kind of an image is obtained, whereas with practically every other scientific instrument a precise and correct setting up and handling are essential if the apparatus is to work at all. The problem is made worse by the fact that many users of the microscope are themselves unaware that a poor quality image is obtained and that they are only using their microscope to a small fraction of its potential capabilities. The object of this book is to help a student of histology get the best possible results in terms of image quality and in terms of the efficient visualisation and interpretation of the structures which are revealed.

The following points are essential for obtaining best results in microscopy whether stained sections are to either be visualized for interpretation of the histological details in the sample visualized or photomicrographed subsequently:

i. Suitably prepared section (including its thickness and staining)—thicker sections (8 to 10 µm) are to be obtained for low to medium power work; while thinner sections (4 to 5 µm) are recommended if higher magnification examination is anticipated

ii. Optimum thickness of the glass slide and cover slip—slides.

iii. Mountant of proper refractive index—the refractive index should be almost nearest possible to that of the tissue under view—or else Becke line (*see* Chapter 12) might interfere the field of view.

iv. 'Good' quality light microscope and its proper use (including photomicrographic camera if required)

It is, therefore, advisable to familiarise with the position and mode of operation of various controls in a microscope. It is a good idea to spend a few minutes with the manufacturer's instruction book supplied at the time of purchase of the microscope. This is very important as microscopes available in most of the laboratories have a separate light source, and allow, because of their very simplicity, the student to be introduced to the principles and practice of good microscopy. The very latest models of microscope are very highly priced but are much convenient to use and provide a superb image of very high quality across a large field of view. In newer microscopes the stage moves up and down for focusing, whereas in older models the body tube moves and carries the objective and the eyepiece.

SELECTION CRITERIA FOR A GOOD MICROSCOPE

The quality of a microscope depends mainly upon its objective lens, the system of lenses in the objective is so designed as to produce the 'best possible' image of an object. It is noteworthy that the best image is not the largest, it is the clearest. However, the points to be present in any good microscope are summarised below:

1. The *stand* should have the maximum of steadiness with the minimum of weight, and it should remain firm even when a microscope is placed in a horizontal position.

2. The *tube* of the body of a microscope should be made up of large diameter and should also be painted black in order to prevent internal reflection.

3. The *draw-tube* should not work too stiffly.

4. The *nose-piece* should be revolving for the objectives of different resolution may be swung into position.

5. The *coarse adjustment* should not work too easily.

6. The *fine adjustment* should be a very slow movement.

7. The *stage* should be mechanical type (preferably detachable) with at least a movement of 4 to 5 cm in the horizontal and of about 2 cm in the vertical direction. There should be divisions reading by Verniers so as to note the exact position of an object viewed. It should also have a central aperture of not less than 4 to 5 cm to prevent the immersion oil getting to the surface of the stage when an oil immersion condenser is used. A rotating stage is almost useless except that it is a little easier to clean.

8. The *substage* should be provided with a rack-and-pinion focusing system, and should possess screws for the centering of the condenser.

9. The *condenser* should be a good quality Abbe condenser. A poor condenser ruins the performance of the remainder of the optical system of the microscope.

10. The *mirror* should have flat and concave surfaces both, and must be attached to the tail-piece in such a manner that its centre lies in the optic axis of the microscope (Fig. 17.1). If one looks down through the instrument, after removal of all the lenses from the body and the substage, the central line of vision should strike the center of the flat surface of the mirror, and not the space on the brass mount intermediate between the curved and the flat surfaces of the mirror.

11. The *eye-relief* should not be too small so that it necessitates to practically touch one's eye to the eye-lens of the ocular in order to fill up the eye with the image of an object.

12. The *objective lenses* should be apochromatic. The set of these should be *parfocal* so that when the nosepiece is rotated, to bring a particular objective into the system, only a little adjustment in focusing is required to focus the section.

13. The *eyepieces* and objectives should be preferably of the same make so that their optical corrections match as designed.

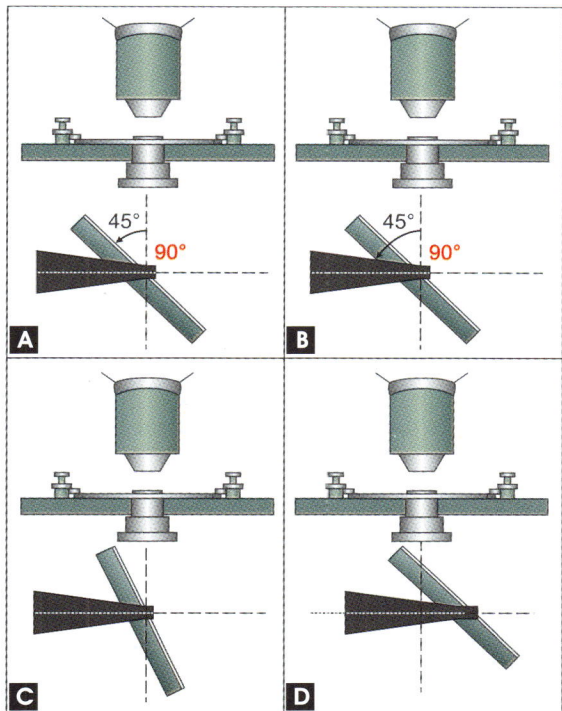

Fig. 17.1: Alignment of two optic axes with common point at the center of the microscope mirror. The vertical interrupted lines represent the optical axis of the microscope. The horizontal dotted lines represent light-beam at 90° to the microscope axis. The ideal condition (A) is achieved in best models of research microscopes where the turnnion pivots are in the plane of mirror surface. In standard equipments, best condition (B) can be achieved with fibres in the axis of the microscope but below the axis of light-beam, to locate silvered surface of the mirror at the intersection of the two axes. The standard microscope with poor conditions (C) generally has pivots at the intersection of the two axes—the reflected beam cannot lie in the optic axis. In still poorly manufactured microscope design (D) the pivots do not lie in the optic axis.

FURTHER CONSIDERATIONS

Microscope types vary so much in complexity and in the arrangement of their component parts that a user of the instrument has to select it as per his requirements. The most important criterion in the selection of an appropriate type of microscope is the uses to which the instrument will be put and the cost. One should consider whether or not the microscope be required for other than routine bright-field microscopy. Will an oil-immersion objective be required? If the answer is No, expense is saved and the microscope condenser need not have a numerical aperture above 0.9. If scanning of wider areas is expected, one should obtain eyepieces with a wide field. There are currently several reputable manufacturers of microscopes, and an instrument from any of the standard firm would be probably suitable for general use. The microscope head is built to adapt different kinds of body tubes (Fig. 17.2). However, if much photomicrography is intended, then investment in plan-achromats with their flatter field and trinocular heads would certainly be advantageous if funds are not prohibitive.

The modern binocular microscopes have great advantage, over the routinely used monocular, of *less eye fatigue* since both the eyes are simultaneously used in the former. Binoculars are provided with four prisms in the optical system that enable light, emerging from the objective in use, divide equally between two eyepieces. In addition, eyepieces may be set closer or apart, thus *adjusting the interocular distance* to suit the need of the microscope user. Furthermore, binocular microscopes have increased *tube length* from 160 mm (in monocular microscope) to 240 mm. This necessitates use of *compensating eyepieces* for refocussing the image for the new tube length. The larger tube length also has an effect on *magnification increased 1.5 times* than with an ordinary tube length.

PROPER SETTING OF MICROSCOPES FOR THE BEST PERFORMANCE

For obtaining the best possible results and highest quality image from a given optical microscope (Fig. 17.3), it is essential to ensure that the optical parts (condenser, objective, and eyepiece) of the microscope are clean

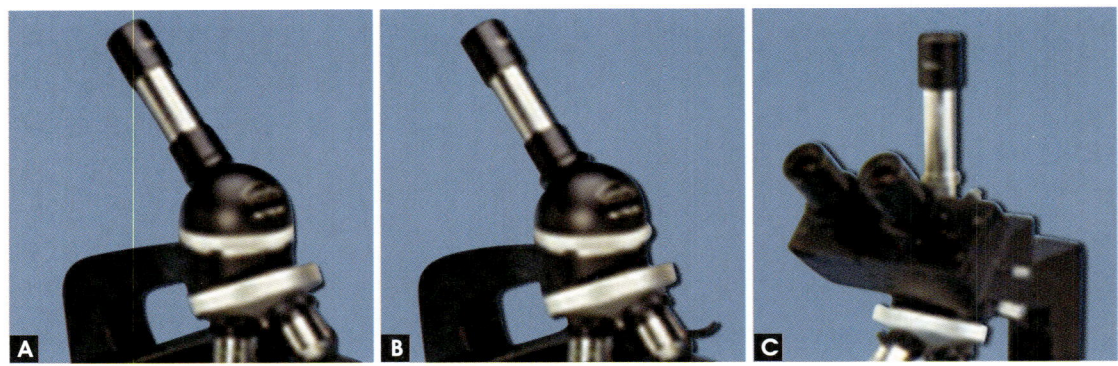

Fig. 17.2: Different types of microscope heads: (A) Monocular, (B) Binocular, and (C) Trinocular.

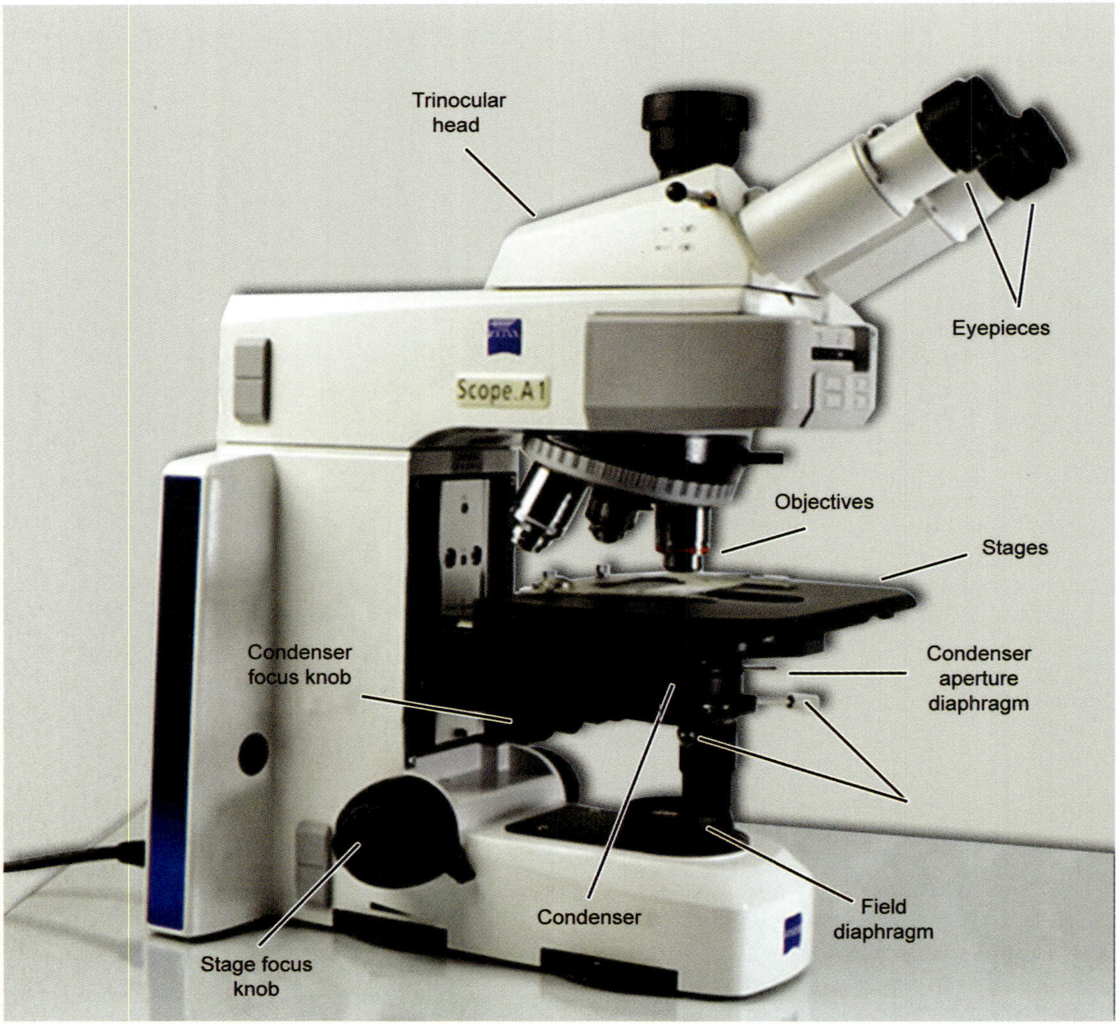

Trinocular head

Eyepieces

Scope.A1

Objectives

Stages

Condenser focus knob

Condenser aperture diaphragm

Condenser

Field diaphragm

Stage focus knob

Fig. 17.3: A trinocular light microscope used for research and photomicrography.

absolutely. Poor quality images (fuzziness and/or loss of contrast) usually result due to dirt or greases on one or more of these optical components. If the image is very severely degraded, it is likely that the front lens of the objective is contaminated. It may be cleaned gently with the lens-tissue paper or a clean soft cotton cloth. Sometimes the cleaning tissue may require moistening with xylene in order to remove the contamination. *It must be well remembered that alcohol of any strength should never be used for cleaning lenses, as it damages the cementing material by which lenses are placed in position.* Fuzzy spots in the field of view may be located by rotating or moving various optical components, in turn, whilst viewing the image. If the dirty spots move in the field of view as the eyepiece is rotated, the eyepiece needs cleaning.

The following steps may be taken for a proper setting-up of the microscope for visual use:

1. Plug in and switch on the power supply attached to the microscope. If the light is not on—there may be a separate switch on the microscope.

   If the microscope has no built in light; position the lamp nearly 6" opposite the microscope so that the light strikes the centre of the mirror and is directed upwards into the condenser. If there is no condenser provided, use the concave surface of the mirror to direct the light into the condenser. But if condenser is provided, only the plane surface of the mirror must be used.

2. Rack the condenser up fully.

3. Open the aperture diaphragm to its full extent.

4. Check whether a frosted filter is present in the filter-holder; if so remove it temporarily or replace it with a clear one.

5. Swing the lowest objective (10X) into place confirming that the nosepiece clicks into position.

6. Lower the stage of microscope; place a slide, with a stained section of tissue, flat on the stage. Ensure that the cover slip is uppermost.

7. Bring the section into the center over the top lens of the condenser.

8. Move the stage up with coarse focus knobs. Be careful that the specimen does not strike the front lens of the objective. Continue until the slide comes closer to the objective than the distance one expects it to focus at (nearly 10 mm for a 4X objective). Focus the section sharp.

   One cardinal rule for the microscopist is always to rack the objective *down* near the object before looking through the eyepiece and then to focus on the object by racking the objective up and away from the object. *This will avoid damaging the object, or the front lens of the objective,* and is particularly important when using the oil-immersion lenses which have as short working distances as 0.3 mm.

9. Close the field diaphragm until its image is smaller than the field of view. Focus the image into the same plane as the specimen.

10. Centre the diaphragm image by adjustment of centering screws in the condenser mount.

11. Open the field diaphragm until its image is just clear of the field of view.

12. Put back the frosted filter previously removed if it was not replaced by a clear blue one.

13. Remove the eyepiece and look down the body tube of the microscope. The rear lens of the objective should appear well illuminated.

14. Adjust the aperture of the substage diaphragm until about two-thirds of the objective aperture is illuminated.

15. Replace the eyepiece and view the slide.

The microscope is now properly set up for getting the best out of the instrument as far as resolution and image quality is concerned. However, the following points may be noted:

i. If a microscope has a binocular head one or both eyepieces should be adjusted for focus. First set with one eye the steps described above and focus the section sharp. With the other eye focus the section by adjusting the second eyepiece. The section should then be in focus for both eyes. When adjustments with eyepieces have been made, the inter-ocular distance may be adjusted until clear binocular vision is obtained.

ii. Ideally the adjustments described for the set up of any given microscope must be repeated each time a different objective is swung into position. But in practice the aperture of the substage diaphragm can be adjusted to suit the highest power dry objective (40) and may be utilized in the same position for all the dry objectives. It needs to be readjusted when an oil-immersion objective is put into the system. Similarly, the field diaphragm set up for the 4 objective can be used for all other objectives but the oil-immersion. Therefore as routine practice for working with all dry objectives, one should set up the substage diaphragm for 40X and the field diaphragm for a 4X objective. Only for oil-immersion both the diaphragms of the microscope need to be reset.

iii. Sometimes if the section has insufficient contrast, it is, of course, often possible to improve visibility by reducing the working aperture of the system by closing the condenser diaphragm (or by lowering the condenser). This trick, although useful at times, is not a solution to be recommended for general use because it is at the expense of loss in resolution.

iv. If any slide needs oil immersion examination, the use of only low viscosity oil is recommended. The most practical oil for routine use is a mixture of 4 volumes of light mineral oil with one volume of α-bromonaphthelene. With use of such oil there is great advantage when rapid motion of the slide is required during examination, or when fresh wet-preparation (like blood films and smears) are being viewed under a cover slip.

Difficulty in Setting up the Microscope

Sometimes viewing a given slide with a stained tissue section(s) with a light microscope gives rise to eyestrain and headache for the user even after short period examinations. The following errors are the most common causes of difficulty in general microscopy even with microscopes from reputable manufacturers.

a. The microscopic glass slides and cover slips are too thick.

b. The cover slip covering the section is not uppermost over the central hole in the microscope stage.

c. The mountant used has too many air-bubbles present.

d. The substage condenser is not properly focused or centered.

e. The source of light (in instruments not provided with integrated illumination device) is not sufficiently intense.

METHODS OF ILLUMINATION

Köhler Illumination for Transmitted Light and Incident Light

The Köhler illumination devised about 120 years ago, based on sound optical principles, is the most common form of illumination used by the majority of microscopists. It is, however, by no means the only method available to the microscopist. In this article I shall first cover source-focused and Köhler illumination, before proceeding to the alternatives.

Early microscopists used natural light, and subsequently oil or gas lamps which gave a structure less source of illumination. This method is thus called 'source-focused' (also 'critical' or Nelsonian) illumination.

This kind of illumination (Fig. 17.4) merely requires the *lamp source* to be focused by the

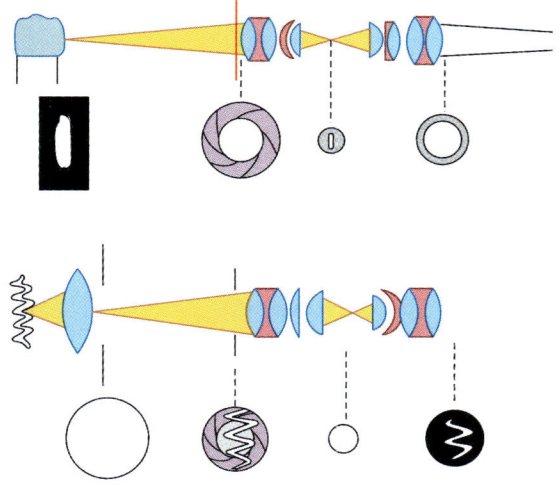

Fig. 17.4: Light distribution in object and aperture planes for source-focused (top) and Köhler illumination (bottom).

condenser into the plane of the specimen. It is essential to restrict the field of view that is lit by closing down the **illuminating field diaphragm** (IFD) in order to reduce glare arising from stray light. The IFD is found either on the front of the lamp housing, or is built into the front focal plane of the condenser. The image of lamp source sometimes has to be magnified to fill the field of view. It is possible to remove the image of the filament from the field of view by using an opal light bulb, or a fine-grain diffuse glass screen in front of a clear bulb. With the passing of oil-lamps and the advent of the much more reliable and less messy electric lighting, a considerable drawback arose because the structure of the small source lamp filament was conjugate with the specimen plane, and thus was visible (Fig. 17.5).

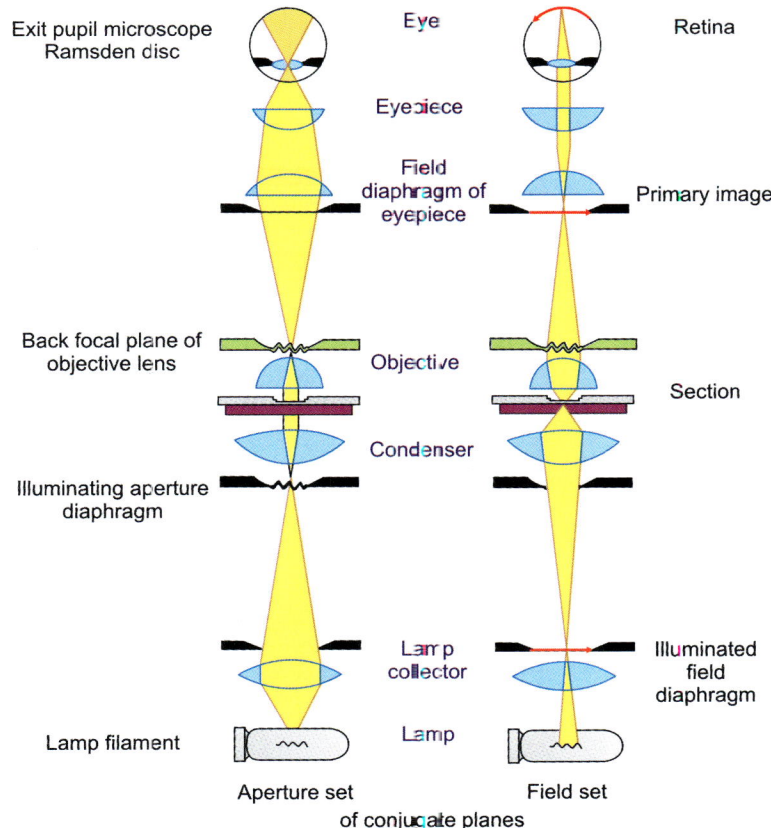

Fig. 17.5: Illustration shows ray paths for each set of conjugate planes for Köhler illumination for a transmitted-light microscope. In practice, these ray paths are superimposed upon one another.

Modern microscope stands are built to contain the lamp built into the stand. This obviates the need to align an independent free-standing lamp. If such a lamp is used, it is helpful to manufacture (or acquire) a wooden board on which to mount the lamp and stand in fixed positions relative to one another, so that the amount of adjustment and alignment required each time is kept to a minimum.

To align the microscope for Köhler illumination on a stand with built-in illumination, proceed as follows:

- Place a well-stained stained specimen on the stage and focus a medium power (10X magnification) objective onto it.
- Close down the illuminating field diaphragm (IFD; usually in the base of the microscope) down to a pinhole.
- Rack the condenser up and down until the image of the iris leaves are in focus at the specimen plane. Open up the IFD to just fill the field of view.
- Remove an eyepiece and observe the back focal plane of the objective.
- Adjust the condenser iris (which can now be seen), so that it is open about 80% of the diameter of the circle seen in the back focal plane of the objective.
- Replace the eyepiece and proceed to use the microscope.

Köhler Illumination on a Monocular Microscope
(With an external light source have an illuminated field diaphragm and condensing lens)

a. Place the lamp horizontally, light it and direct it towards a wall. Focus the filament on the wall, and open the field diaphragm fully. Ensure the color fringes and flare surrounding the filament images are symmetrical, otherwise center the filament to the field condenser lens in the lamp.

b. Ensure centration is maintained as the illuminated field diaphragm is closed down. Place the lamp about 8 inches from the mirror.

c. Adjust the microscope mirror, plane side towards the lamp, so that the lamp beam falls across its center, and focus the filament image sharply.

d. Use a piece of white paper as a screen to make the image visible.

e. Remove the objective, eyepiece and condenser, and white paper screen.

f. Put a piece of tracing paper over the top of the tube, and ensure it is strongly and evenly illuminated.

g. Replace the optical components, and rack up the condenser.

h. Place a *temporary* tracing paper diffuser in front of the lamp, focus on a specimen and centre the condenser. Remove the diffuser and close the field diaphragm; manipulate the mirror until the image of the field diaphragm is central.

i. Focus the condenser until the image of the field diaphragm is sharp.

j. Open up the field diaphragm to fill the field of view.

k. Look under the substage to check that the filament image falls on the condenser iris. Adjust the condenser iris (as above), so that it is open about 80% of the diameter of the circle seen in the back focal plane of the objective.

l. Replace the eyepiece and proceed to use the microscope.

To Align the Microscope for Source-focused Illumination
(Using a mirror stand and independent lamp)

1. Set up the lamp about 8 inches (20 cm) from the mirror on the microscope stand.

2. Remove all lenses (condenser, objective and eyepiece) and also remove any diffusing screen.

3. Adjust the mirror with the plane side of the mirror uppermost to direct light up the microscope tube.

4. Replace the glassware, place a well-stained specimen on the stage and focus a medium power (10X magnification) objective onto the specimen.
5. Rack the condenser up and down until the lamp is focused onto the specimen.
6. Centre the lamp to the optical axis either by using the condenser centration controls, or by closing down the illuminated field diaphragm in front of the lamp and tilting the mirror as appropriate.
7. Open the diaphragm until it just fills the field of view. If needs be, the lamp does not need to have an iris diaphragm in front of it. Provided the lamp is fitted with some sort of filter holder, it is possible to construct a series of aperture holes cut out of stiff card. If the diaphragm is non-centrable and/or the opal bulb has no surface markings, inscribe a cross very lightly on the center of the bulb to aid alignment.

These days the most usual form of electric lamp used is the tungsten lamp ranging from 15 watts as integral illuminators built into the microscope stand, through 30, 50 and 60 Watts to 100 or up to 250 Watts. The voltages of these lamps are either 6 volts or 12 volts direct current stepped down from the 230 volts alternating current supply. An improvement, which reduced blackening of the glass envelope with ageing, came with the introduction of the quartz-iodide (QI) or tungsten-halogen lamp. Iodine vapour, contained within the glass envelope, combines with tungsten evaporated from the electrode onto the glass to form tungsten iodide which, in turn, dissociates at high temperature on contact with the filament to renew the tungsten in the filament and iodine in the lamp.

KÖHLER ILLUMINATION

The Köhler illumination is one key to ideal microscopy. It is incorporated in the design of practically all modern laboratory and research microscopes.

How to Set Köhler Illumination

Köhler illumination is a microscope technique that provides superior control over the light rays during brightfield microscopy by aligning and focusing the microscope, ensuring the best resolution and contrast, as well as a bright, evenly illuminated background for your images. Köhler illumination is critical for advanced techniques and must be aligned.

Required Microscope Parts

In order to set up a microscope for Köhler illumination, the microscope has to have a vertically adjustable, centerable condenser and field diaphragm as illustrated in Fig. 17.6.

Köhler Illumination Procedure

Follow these six simple steps to set up your microscope for Köhler illumination (Fig. 17.6).

The last step is to adjust the condenser aperture iris to match the NA (numerical aperture) indicated on the objective. The aperture iris can also be stopped down to 60% of the NA setting (trading image resolution for contrast.) An alternative way to set the condenser aperture is to remove the eyepiece and look down the eye tube at the back of the objective and adjust the aperture diaphragm until approximately 90% of the area is illuminated (a setting that visually looks very near the edge of the objective but is not quite fully to the edge).

It is to be thoroughly realized that the fullest advantage of a light microscope, for routine examination of tissues, for optimal performance (despite instrument's expensive and highly corrected optics) can be achieved by understanding a few and simple steps for Köhler illumination, introduced in 1893 by August Köhler of the Carl Zeiss corporation.

Necessary Instructions to Provide Appropriate Illumination

1. Focus the section
2. Close the field diaphragm

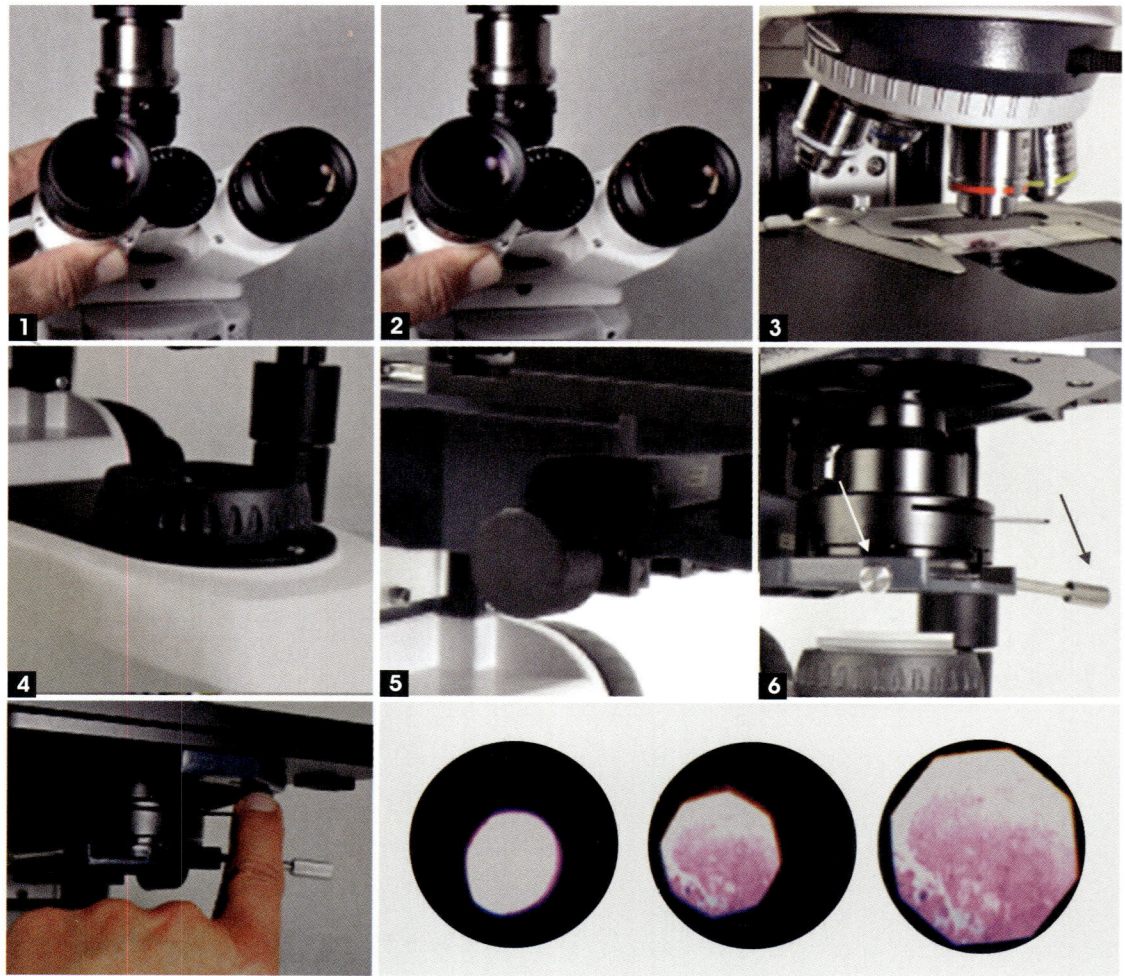

Fig. 17.6: Steps (1 to 6) for obtaining Köhler illumination detailed in the text.

3. Move the microscope condenser up or down until the outline of its field diaphragm appears in sharp focus

4. Center the field diaphragm with the centering controls on the condenser substage

5. Open the field diaphragm until the light beam covers the full field observed

6. Remove the eyepiece and observe the *exit pupil of the objective.* An illuminated circular field, whose radius is directly proportional to the numeric aperture of the objective [as the condenser diaphragm is closed, its outline will appear in this circular field].

7. For most stained sections, set the condenser diaphragm to cover approximately 2/3rd of the objective aperture. [This setting results in the best compromise between resolution and contrast (contrast simply being the intensity difference between dark and light areas in the section)].

CLEANING AND MAINTENANCE OF MICROSCOPE

A regular and proper cleaning of the microscope is essential for its perfect working since this is an exceedingly complicated and delicate

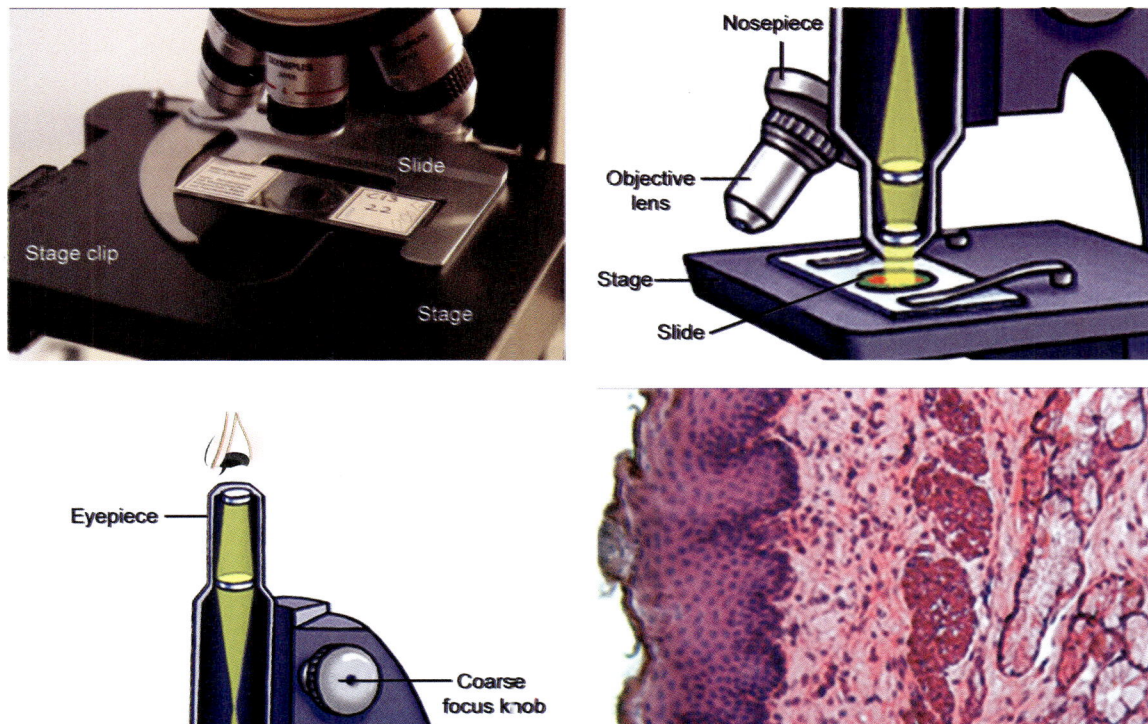

Fig. 17.7: Steps in set-up a microscope to examine a stained section: 1. Place the prepared slide on the stage secured with the stage clip; 2. Click nosepiece to the lowest power (shortest) objective lens; 3. Look into eyepiece. Use coarse focus to bring close in focus; 4. Then use fine focus knob; 5. Adjust iris diaphragm; 6. Click the nosepiece to a longer objective lens. Do NOT use the coarse focus; and 7. Use the fine focus knob to bring the slide into focus.

instrument. A little attention to the cleaning of the microscope daily will, by the removal of the chemically-active and sharp pieces of grit and foreign matter, prolong the life of the instrument and make cleaning less time-taking and easier.

Daily Cleaning Schedule

The microscope should be dusted daily to keep it as clean as possible. The instrument should be covered with a cover provided for the purpose. The stage should always be wiped off spills immediately with an old linen handkerchief. The linen used for cleaning the stage and table should, as a rule, be never used to clean the lenses of the microscope— otherwise the scratch lenses will result.

It should be remembered that although it is impossible to see with dirty lenses, at no cost any one should temper with the microscope.

For the proper cleaning of lens system of the microscope, it should be correctly known which lens is dirty with the dust or finger marks. First look into the microscope—if dust is obvious on the top lens of the eyepiece, it will move with the rotation of the eyepiece. If the dust particles do not move with the rotation of the eyepiece, the condenser, mirror, and/or objective lens may be dirty. Never use an ordinary rough cloth for cleaning these lenses. The lenses should be cleaned by breathing on the glass and wiping them with either best quality chamois leather or lens paper (soft type) or if these things are not available with a soft well-washed silk cloth. Dust on the condenser will be apparent when this is racked up and down, since it will come in and out of focus. If the dust is obstinate (intractable), the cloth or lens paper for cleaning the microscope lenses should be moistened with absolute alcohol. If some Canada balsam has found its way on to the objective, wipe it off gently with a cloth moistened with xylene, and finish off with a dry silk cloth. If the lenses have acquired a greasy deposit, remove it with a piece of clean paraffin oil applied with a piece of soft paper or cloth. A little cleaning of the microscope daily will make the weekly cleaning task a short and simple one. Removal of immersion oil from the microscope system is an absolute necessity before replacing the instrument back in its case.

Weekly Cleaning Schedule

After the daily cleaning has been performed look down into the microscope. If some dust still appears, rotate the eyepiece and if dust particles also move with the rotation of eyepiece, carefully dismantle the eyepiece and clean its both lenses in the usual way. Check the lens systems. The slides of the coarse and fine adjustment, the mechanical stage (if present) and the substage condenser should be wiped with a cloth damped with xylene to remove dust which would otherwise damage the slides. A little light-oil (as supplied for lubricating microscopes) should be applied to slides and joint-points.

If the immersion objective had been less carefully cleaned, after earlier use with Canada balsam or cedar wood oil it will be seen that balsam or oil has dried and then it should be best removed by carefully chipping off the outer portion with a knife, avoiding contact with either metal or glass, and then, removing the remainder with lens paper or a soft cloth moistened with benzene or xylene. Use of alcohol, for cleaning the microscope lenses, specially the objectives, should be avoided since it softens the cement in which the front lenses or the older immersion objectives were mounted. The lenses should not be fingered, but should be held by the lens-casing. They should be kept clean with thin pieces of cloth. When the lenses become greasy, a soft lint-free cloth damp with a little xylene should be used to wipe them clean. Spare lenses should be kept in their cases until needed for use.

SUMMARY

Suitable Material to be Photographed

- Suitably prepared section (including its thickness and staining)—thicker sections (8 to 10 μm) are to be obtained for low to medium power work
- Optimum thickness of the glass slide and cover slip—slides
- Mountant of proper refractive index—the refractive index should be almost nearest possible to that of the tissue under view.

Selection Criteria for a Good Microscope

- The *stand* should have maximum steadiness with minimum of weight
- The *tube* should have large diameter and painted black to prevent internal reflection
- The *draw tube* should not work too stiff
- The *nosepiece* should be of revolving type—properly clicking at position of each objective
- The objective *lenses* should be apochromatic
- The *eyepieces* and objectives must be of the same make to match their optical corrections
- The *coarse focusing knob* should not work too easily
- The *fine focusing knob* be provided with a very slow movement
- The *stage* should be of mechanical type
- The *substage* must be provided with rack-and-pinion focusing system
- The *condenser* should be of a good quality Abbe condenser
- The *condenser numerical aperture* should be equal to or slightly less than that of the highest objective numerical aperture.
- The *mirror* must consist of both plain and concave surfaces
- The eye-relief should be very small.

Proper Set-up for Best Performance of Microscope

It is essential to ensure that the optical parts (condenser, objective, and eyepiece) of the microscope are clean absolutely. Poor quality images (fuzziness and/or loss of contrast) usually result due to dirt or greases on one or more of these optical components.

1. Plug in and switch on the power supply attached to the microscope.
2. Rack the condenser up fully.
3. Open the aperture diaphragm to its full extent.
4. If the filter-holder contains a frosted filter, remove it temporarily by replacing it with a clear one.
5. Swing the lowest objective (10X) into place confirming that nosepiece clicks into position.
6. Lower the stage of microscope confirming the cover slip is towards the objective lens.

7. Bring the section into the center over the top lens of the condenser.
8. Move slowly and carefully the stage up with coarse focus knob until the slide comes closer to the objective than the distance one expects it to focus at (nearly 10 mm for a 4X objective).
9. Focus the section sharp
10. Close the field diaphragm until its image is smaller than the field of view. Focus the image into the same plane as the specimen.
11. Center the diaphragm image by adjustment of centering screws in the condenser mount.
12. Open the field diaphragm until its image is just clear of the field of view.
13. Put back the frosted filter previously removed if it was not replaced by a clear blue one.
14. Remove the eyepiece and look down the body tube of the microscope. The rear lens of the objective should appear well illuminated.
15. Adjust the aperture of the substage diaphragm until about two-thirds of the objective aperture is illuminated.
16. Replace the eyepiece and view the slide.
17. **The above steps produce** Köhler illumination which provides the best possible image of the given section

Cleaning and Maintenance of Microscope

- The microscope should always be carried away with one hand under the base and the other holding the arm.
- The dusting should be done daily.
- The microscope should be always kept covered when not in use with the cover supplied
- The stage should be wiped off any spills immediately.
- Only the coarse focus knob should be used (and **never** the fine focus knob) to find a section when using the low-power objective.
- The microscope lenses should normally be cleaned by breathing on the glass and wiping them either with best quality chamois leather or with soft lens paper.
- If some Canada balsam or DPX has found its way on to the objective lens, it should be wiped off gently with a fine cloth moistened with xylene. Finish with a dry silk cloth piece.
- If lenses have acquired greasy deposit(s) remove grease with a piece of clean paraffin oil applied with a piece of soft paper.
- Always keep the spare microscope lenses in their cases until needed for use.

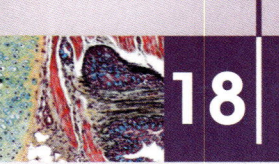

18

Special Types of Microscopes and the Principles for their Use

Besides the bright-field light microscope which has been described in detail (since it is the one most commonly dealt instrument in most of the histology laboratories), there are some other varieties of microscopes. Basically these may be classified by the type of light source used. Whereas an optical microscope utilises visible light; there are certain modifications of this, namely the dark-field, phase contrast, interference, and polarization microscopes. A brief description about each of them is given.

DARK-GROUND (DARK-FIELD) MICROSCOPE

Dark-ground microscope (Fig. 18.1) is a simple modification of the light microscope useful for seeing extremely minute particles of matter (usually a colloid suspension) and for studying large transparent object (e.g. living protozoa, crystals, and spicules, etc.) which are ordinarily invisible with a light microscope using the transmitted light. The dark-ground microscope utilizes a strong, oblique light that does not enter the objective lens. Hence it requires only a central patch stop (annulus) which fits into the substage condenser blocking out the central rays of a light that would enter the objective while allowing the peripheral rays only to illuminate the object.

Principle of Dark-ground Microscopy

Dark-field microscopy is one of the simplest and cheapest contrast enhancing techniques well suited for uses involving live and unstained biological samples, such as a smear from a tissue culture or individual, water-borne, single-celled organisms. It works well for specimens that have a refractive index which is different from its surrounding medium, but which are difficult to see because they lack color. A bright-field microscope can be adapted as a dark-field microscope by adding a special disc called a **stop to the condenser**. The stop blocks all light from entering the objective lens except peripheral light that is reflected off the sides of the specimen itself. The resulting image is a

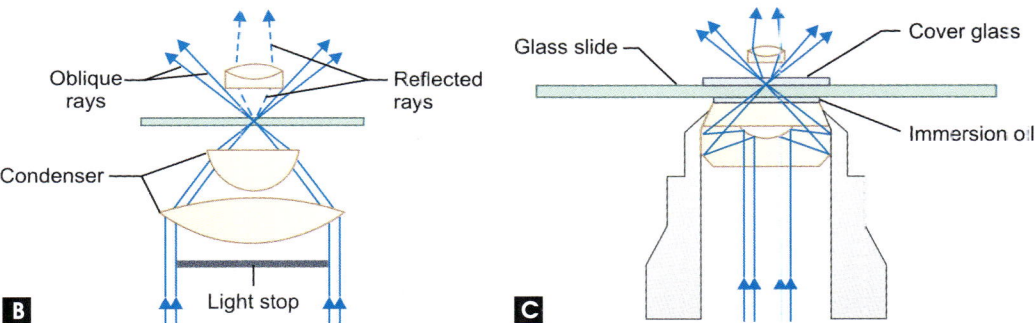

Fig. 18.1: (A) Dark-field microscope with a dark-ground Illuminator, (B) the star diaphragm allows only peripheral light rays to pass up through the condenser (requires maximum illumination), (C) a cardioid condenser provides greater light concentration for oblique illumination than the star diaphragm.

brightly illuminated image surrounded by a dark or black background (Fig. 18.2). Considering the simplicity of the setup, the quality of images obtained from this technique is impressive.

Bright-field images, but the light that reflects off the sides of the feature will be visible in the dark-field images (Fig. 18.2).

To achieve a dark-field image, it is necessary to place a dark-field filter (a "patch stop") into the filter holder of the condenser. This filter prevents light to directly enter the objective (therefore the background appears dark). The

specimen will be illuminated from the side and will scatter some of the light to enter the objective. The specimen will appear bright on dark background.

It can be compared to dust floating in the air with sun shining in from the side through a window. The dust is illuminated by the sun and appears bright on dark background.

There are *two* possibilities to achieve a dark-field image:
• By using specialized dark-field condensers: This is the best but also the most expensive solution.

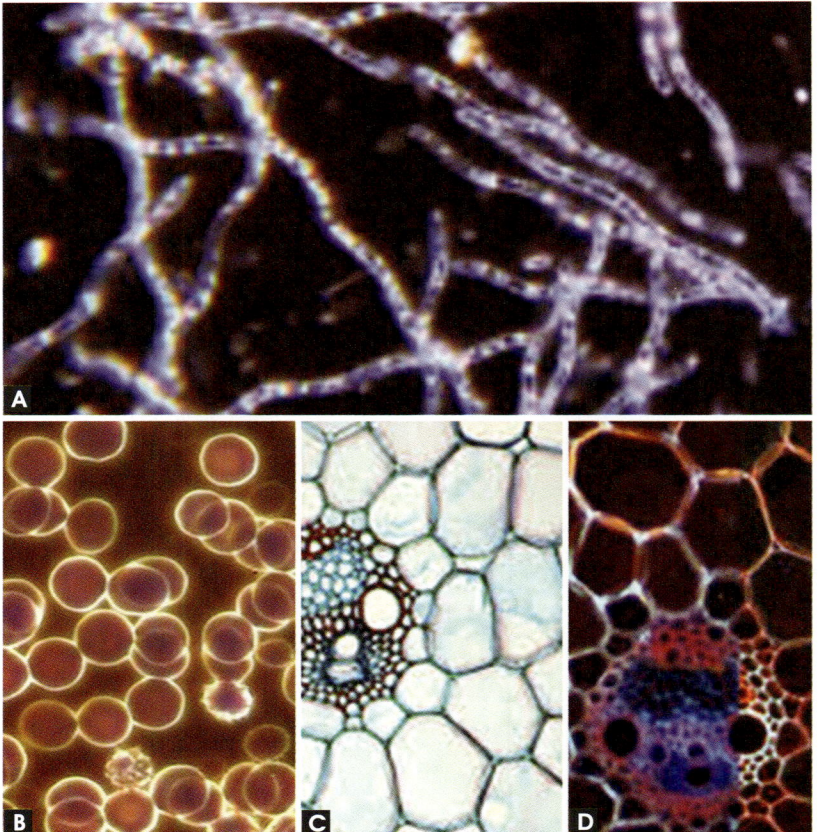

Fig. 18.2: (A) Bacteria Spirillum—dark-field microscopy, (B) A thin blood smear, (C) Section of maize seen with bright-field microscopy, and (D) A dark-field image of the maize showing a less contrast due to the opened aperture diaphragm and a different color representation.

- By using a dark-field filter (a "patch stop") which is placed into the filter holder of the condenser. It is possible to make the patch stop out of cardboard or a tin can using a cutting knife and scissors.

  Dark-field microscopy (dark-ground microscopy) describes microscopy methods, in both light and electron microscopy, which exclude the unscattered beam from the image. As a result, the field around the specimen (i.e. where there is no specimen to scatter the beam) is generally dark.

  A given colloidal suspension with numerous minute particles is taken in a container with an optically worked glass window at one side and the top (Fig. 18.3). A fine beam of intense light from an arc is allowed to enter the suspension horizontally through the side glass window of the container. It is focused with the help of convex lens which is placed between the light source and the side window. The particles are lighted up only by oblique rays which do not enter the objective lens of the microscope. Thus, the particles are seen as bright against a black background. The brightly illuminated appearance of the particles is due to the difference in their refractive indices with that of the medium in which they are suspended.

  The required stop disc can be a built-in component supplied by manufacturer as an accessory dark-field condenser, or it may be

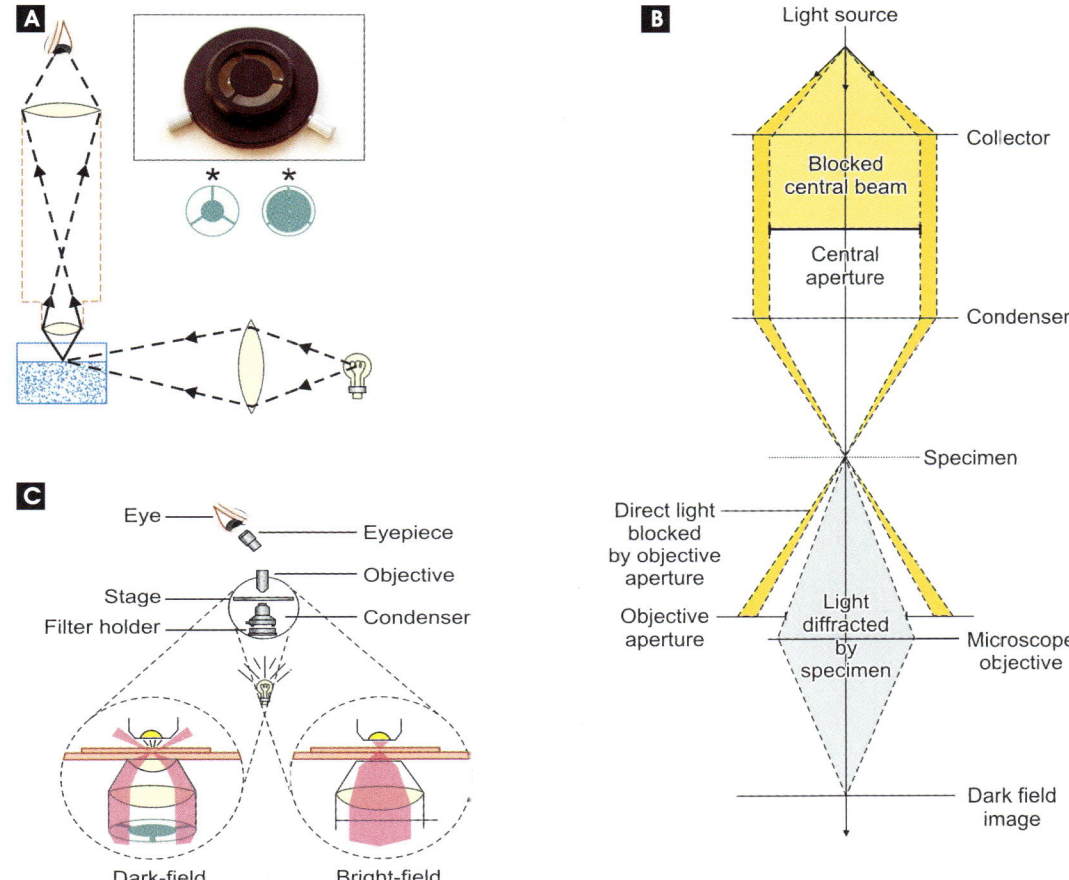

Fig. 18.3: Principle of dark-ground microscopy (A) Two *stops* (*) each supported with three arms are shown; Inset at right upper corner is a dark-field filter (patch stop) placed into the filter holder of the condenser. To the left and the right are the centering screws; (B) Diagram illustrating the light path through a dark-field microscope; (C) Comparison of light passing in a dark-field and a bright field set up of microscopy.

constructed from black cardboard or thin metal with three support arms (Fig. 18.3) following the steps given below:

1. Focus any stained slide with the objective lens which you intend to use for dark-ground. Usually, a low-power objective should be used.
2. Adjust the microscope for source-focused illumination.
3. Remove the eyepiece and look down at the back lens of the objective.
4. Close the iris diaphragm of the condenser until its image just enters into view in the back lens.

5. Remove the condenser and measure the diameter of the opening in the iris diaphragm. Add 15 per cent to this figure. This gives the diameter of the central opaque disc which would be required to get a dark-ground effect. The disc may also be constructed by trial and error method.

The microscope is now set for source-focused direct illumination. When dark-ground is required swing in the stop and open the iris diaphragm fully. Only those rays refracted or diffracted by the object enter the objective and illuminate the object as bright against a black background. The tiny particles in the suspension appear as glistening spots similar to the phenomenon of dust particles "seen" in a beam of sunlight entering a darkened room.

Advantages of Dark-field Microscopy

- It is a simple procedure which can be used on live transparent specimens, specimens which normally need to be stained (and therefore killed).
- The images appear spectacular and are visually impressive.
- Dark-field microscopy even allows for the visualization of objects that are *below* the resolution of the microscope. These objects will appear as bright spots on a dark background. It is, however, not possible to see the shape of these objects.

Disadvantages of Dark-field Microscopy

The main limitation of dark-field microscopy is the low light levels seen in the final image. This means that the sample must be very strongly illuminated, which can cause damage to the sample. Dark-field microscopy techniques are almost entirely free of artifacts, due to the nature of the process. However, the interpretation of dark-field images must be done with great care, as common dark features of bright-field microscopy images may be invisible, and vice versa.

- Dark-field microscopy is very sensitive to dirt and dust located in the light path.
- It is not suitable for all specimens. If the refractive index of a transparent specimen is similar to the surrounding medium, then the specimen light will pass right through the specimen and it will not be scattered into the objective.
- The intensity of the illumination system must be high so see the specimen properly.
- It is necessary to open the condenser aperture diaphragm, and this limits the effective use of the diaphragm.
- One patch stop is generally sufficient for low magnification work, but at a higher magnification the quality of the image drops. It may be necessary to experiment with different patch stop sizes for the different objectives.

PHASE-CONTRAST MICROSCOPE

A phase contrast microscope (Fig. 18.4). Phase contrast microscopy is based on the fact that

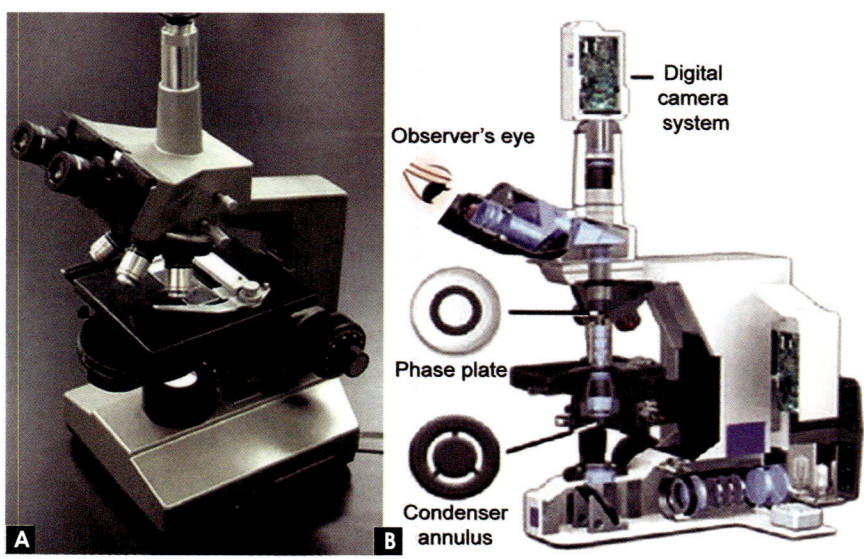

Fig. 18.4: Phase contrast microscope.

light passing through media with different refractive indices slows down and changes direction. Light travels fastest in air (refractive index nearly 1.0), more slowly in water (refractive index about 1.3), and slower still in glass (refractive index about 1.5). This microscope also contains special condensers that throw light *out-of-phase* and cause it to pass through the object at different speeds (Fig. 18.5). Live, unstained organisms are seen clearly with this microscope, and internal cell parts such as mitochondria, lysosomes, and the

Golgi body can be seen with this instrument. Thus light waves traversing equal distances through media with different indices; will not emerge at the same time. In other words, they will emerge *out-of-phase* with each other. The formation of image is a result of *interference* between the direct light (not passed through the object) and the diffracted light (passed through the object). The different components in unstained tissue alter the phase of light waves that pass through them to different extent. These phase differences are also, as such, not discernible to the eye. In phase contrast microscopy these phase differences are converted into differences of amplitude. Because the amplitude is responsible for the contrast, the differences of amplitude make different tissue components to be seen with different degrees of brightness (*greater amplitude, greater brightness; less amplitude, less brightness*). Thus the different constituents are distinguishable from one another. In Fig. 18.5A, the two waves are perfectly *in-phase*. If they are made to combine, there results a production of waves of greater amplitude (hence greater brightness). This is called *constructive interference*. In Fig. 18.5B these waves are out-of-phase with one another, the combination results in what is called *destructive interference*, since the waves obliterate each other It is evident that by variously combining the light waves which are out-of-phase with one another it is possible to produce combined waves of greater or lesser amplitude and hence of greater or lesser brightness respectively. Maximum destructive interference will result if, by optical means, the phase difference between the two sets of waves is increased from $\frac{1}{4}\lambda$ to $\frac{1}{2}\lambda$. The phase change is brought by interposing a phase plate annulus (a film of glass of sufficient thickness to alter the phase of green light by $\frac{1}{4}$ wavelengths) at the back focal plane of the objective lens of the phase contrast microscope (Fig. 18.6) and other optical plate is placed within the condenser lens system.

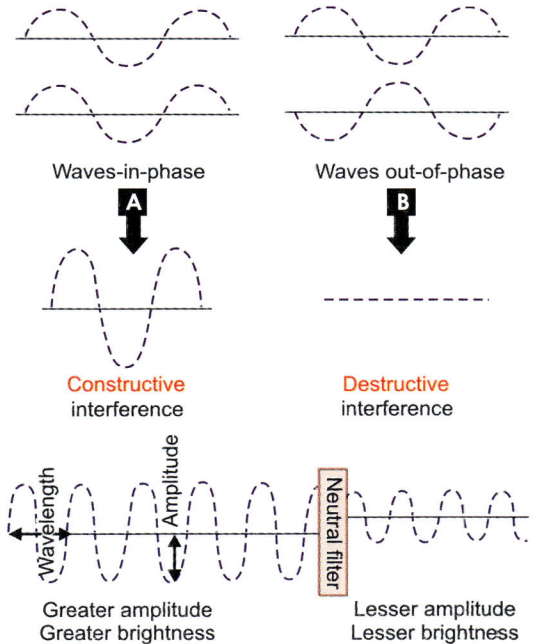

Fig. 18.5: (A) Waves-in-phase and (B) Waves out-of-phase. The different constituents are distinguishable from one another. In (A) the two waves are perfectly *in-phase*. If they are made to combine, there results a production of waves of greater amplitude (hence greater brightness). This is called *constructive interference*. In (B) these waves are out-of-phase with one another, the combination results in what is called *destructive interference*, since the waves obliterate each other. It is evident that by variously combining the light waves which are out-of-phase with one another it is possible to produce combined waves of greater or lesser amplitude and hence of greater or lesser brightness respectively.

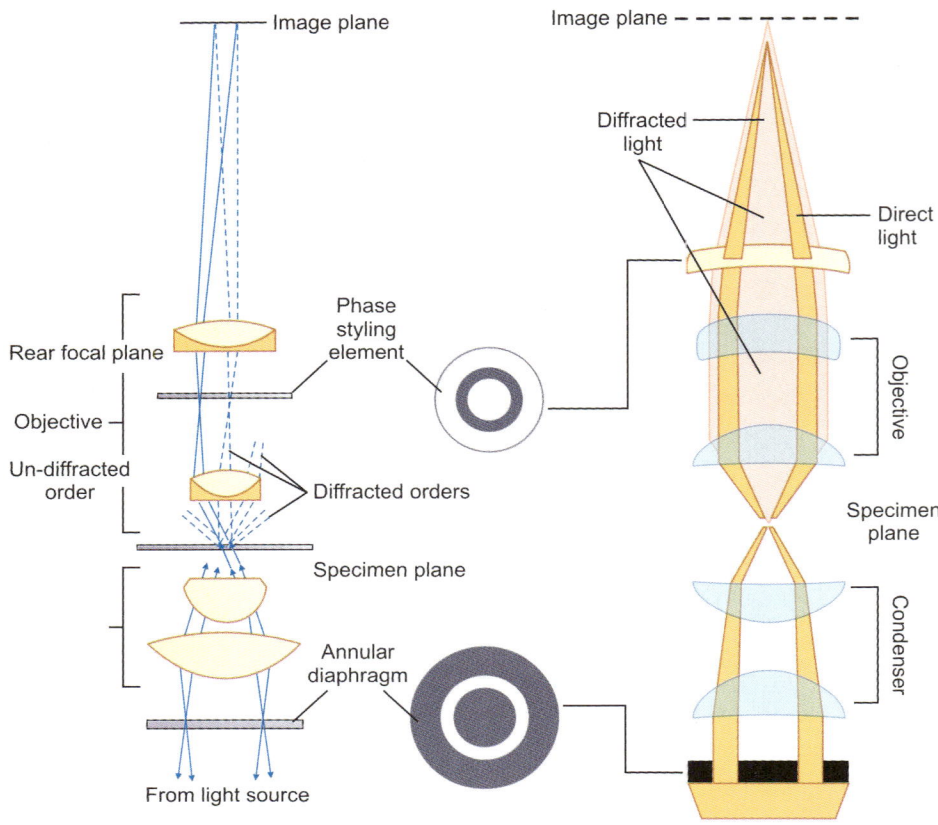

Fig. 18.6: Principle of phase contrast microscope.

Phase Contrast Microscopy

It is a contrast-enhancing optical technique that can be utilized to produce high-contrast images of transparent specimens (Fig. 18.7), such as living cells (usually in culture), microorganisms, thin tissue slices, fibers, glass

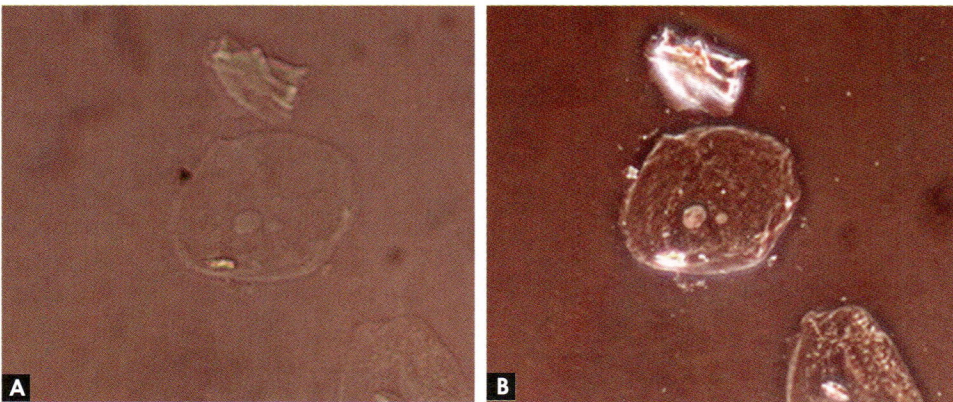

Fig. 18.7: The same cells imaged (A) With traditional bright-field microscopy, and (B) With phase contrast microscopy.

fragments, and sub-cellular particles (including nuclei and other organelles). Used for viewing unstained cells and tissues. Light changes speed when passing through tissue components with different refractive indices. These components then appear lighter or darker relative to each other. This is used to study the behavior of living cells, observe the nuclear and cytoplasmic changes taking place during mitosis and the effect of different chemicals inside the living cells. By using the phase-contrast microscope, an image of strong contrast of the object is obtained. In effect, the phase contrast technique employs an optical mechanism to translate minute variations in phase into corresponding changes in amplitude, which can be visualized as differences in image contrast.

One of the major advantages of phase contrast microscopy is that living cells can be examined in their natural state without previously being killed, fixed, and stained. As a result, the dynamics of ongoing biological processes can be observed and recorded in high contrast with sharp clarity of minute specimen detail.

INTERFERENCE MICROSCOPE

Interference microscopy is a *variation of phase-contrast microscopy* that uses a prism to split a light beam in *two*. The interference microscope sends through the specimen two separate beams of light which then are combined in the image plane. After recombination, difference in retardation of light causes *interference* utilized to measure the refractive index or thickness of the sample under examination. The phase differences cause *interference* used qualitatively to provide contrast in transparent object and quantitatively to obtain information about the concentration of substances in cells and in addition to measure dry cell mass. It may be used also for quantitative studies of macromolecules of the cell components, for example it is used for determination of lipid, nucleic acids and protein contents of the cell.

In fact, the interference microscope is a bright field light microscope with the addition of the following elements:
1. A **polarizer** between the light source and the condenser.
2. DIC **beam-splitting** and **beam-combining prisms**. A beam of light from a single source (*coherent beam*) is split in such a way that one part of the beam passes through the specimen and the other passes to the side of it. The latter is called the *reference beam*. Before entering the objective lens the two beams are recombined by a special system called a *beam recombiner*.

   Manipulating the prism changes the beam separation, which alters the contrast of the image. When the two beams pass through the same material across the specimen they produce no interference. When the two beams pass through different material across the specimen such as on the edges, they produce alteration when combined.
3. **An analyzer**. The incoming light is split inside an **interferometer**, one beam going to an internal reference surface and the other to the sample. After reflection, the beams recombine inside the interferometer, undergoing constructive and destructive interference and producing the light and dark fringe pattern. **Differential interference contrast (DIC) microscopy** enhances contrast by creating artificial shadows (Fig. 18.8), as if the object is illuminated from the side. But DIC microscopy and phase contrast microscopy both use polarized light, which is unsuitable when the object or its container alter polarization. Traditional phase-contrast methods enhance contrast optically, blending brightness and phase information in single image (Fig. 18.9).

POLARIZATION MICROSCOPE

Polarization is a phenomenon in which a single ray of light, after passing through certain

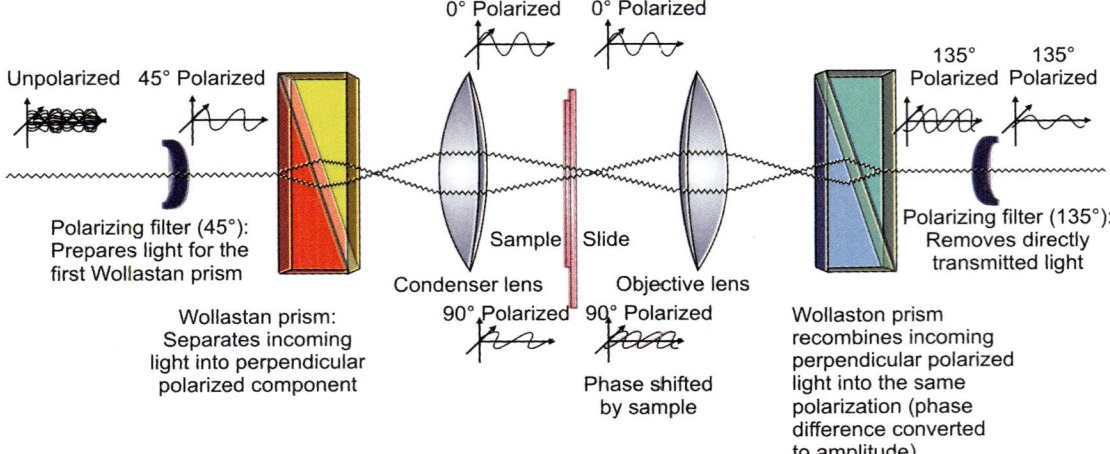

Fig. 18.8: Showing path of light in differential interface contrast microscopy (DIC). Two parallel light beams pass through the specimen and combine to produce an image.

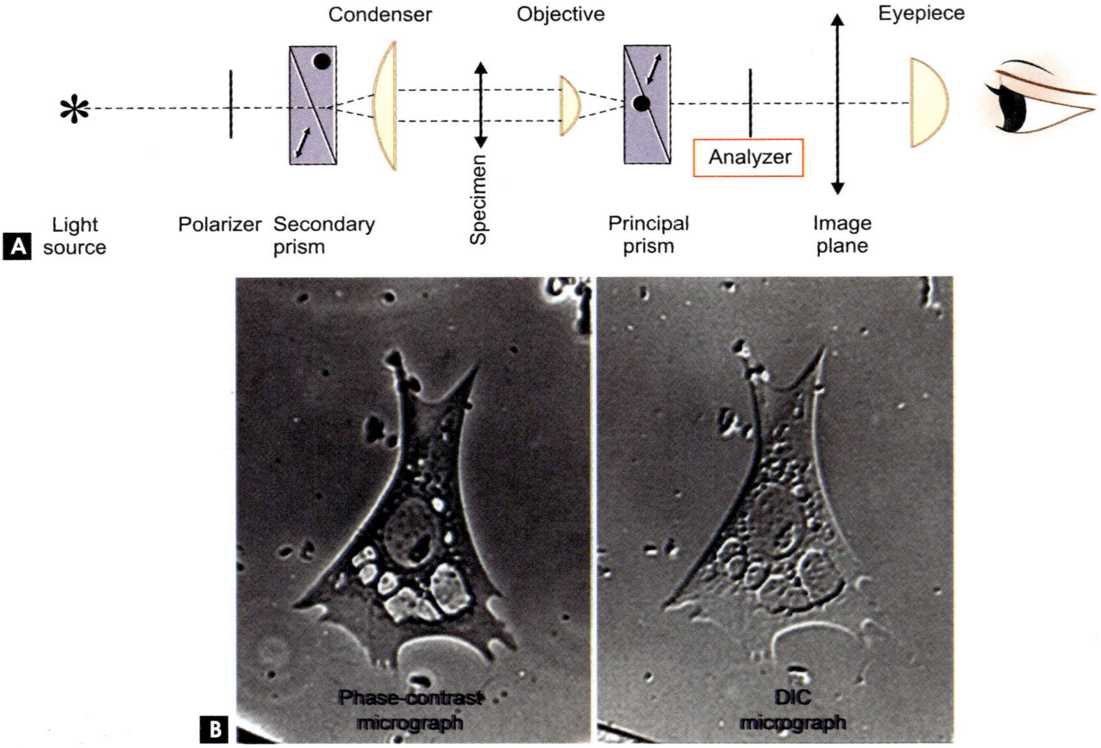

Fig. 18.9: (A) Path of light in differential interface contrast microscopy. (B) Two micrographs are shown. Objects with high refractive index appear high and those with low refractive index appear depressed.

substances or body tissues, is divided into 2 refracted light rays. Such substances are *crystalline* or *birefringent* (doubly refractile).

Tissues not falling into the crystalline group are *amorphous* or *monorefringent*. Birefringence is caused by the orientation of particles of biological materials too small to be resolved even by the best microscope lenses.

In 1669 Erasmus Bartholinus discovered that a ray of light, when incident on a crystal of calcite, is not refracted according to the ordinary law, but forms *two* refracted rays, one of which does not necessarily lie in the same plane as the incident ray and the normal to the refractive surface. Calcite is a transparent crystalline form of calcium carbonate. It was at one time found in great quantities in *Iceland*, and hence known as *Iceland spar*. Calcite crystallizes in many forms, each of which may be reduced, by cleavage, to rhombohedra, bounded by six similar parallelograms with equal to 101° 55' and 78° 5' (Fig. 18.10). Isometric projection of a long crystal of calcite forms a Nicol's prism.

Crystalline or semi-crystalline body tissues like bones and structures with linear symmetry (muscle fibers, nerve fibers, cilia, flagella, collagen fibers, and rods and cones of the retina), and structures with radial symmetry (lipid droplets within the adrenal cortex, and starch granules) can be studied by polarizing microscopy.

A polarizing microscope (Fig. 18.10) is a conventional light microscope, containing a rotating stage with 2 polarizing elements such as a Nicol prism made from calcite (crystalline calcium carbonate) and balsam. In place of Nicol prisms, sheets of Polaroid films are most often used at present. These contain special organic compounds which totally absorb vibrations of light. In a polarizing microscope, the first polarizing element—the *polarizer* is interposed in the light path below condenser under the stage in the filter tray, and the second element-the *analyzer* is located above it being placed within the barrel of the microscope above the objective lens. The analyzer can be constructed by putting a small Polaroid disc in the centre of a small brass or cardboard box with a central hole formed by a sharp cork borer. The analyzer is then fitted over the eyepiece of the microscope and turned until the main axes of polarizer and analyzer become perpendicular to each other. This

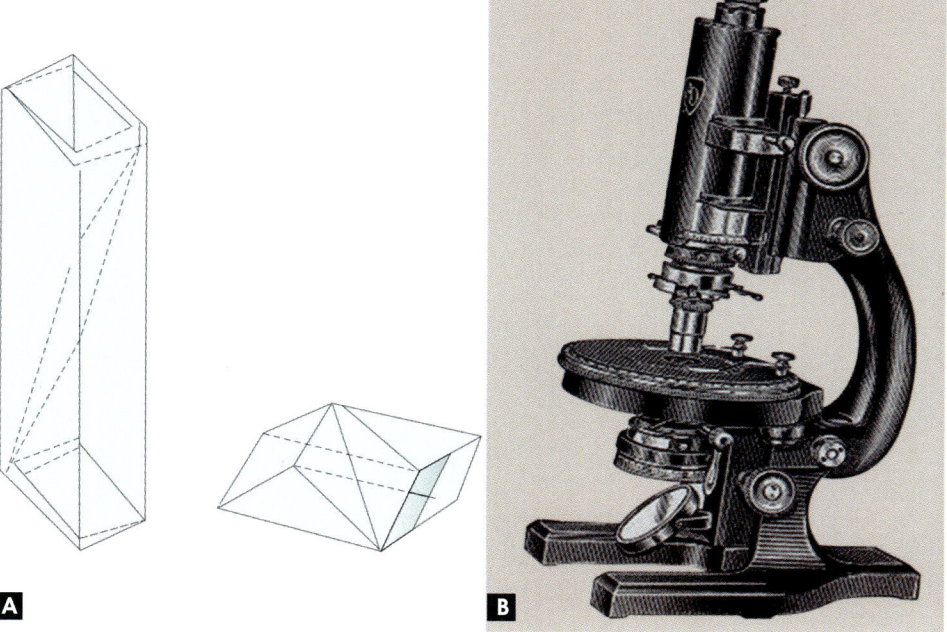

Fig. 18.10: (A) Nicol's prism used in (B) A polarizing microscope.

prevents any light seen in the eyepiece, and the field appears black. The field remains black even if some amorphous (or *isotropic*) object is placed on the stage, microscope field shows light with greater or lesser intensity, due to the passage of one of the 2 refracted light rays polarized rectilinearly. Thus the polarizing microscope can distinguish between amorphous and crystalline biological tissues. Also, their internal arrangement can be visualized at the submicroscopic level.

Polarizing microscopy: Light passes through a polarizing filter and exits vibrating in only one direction. A second filter is placed perpendicular to the first filter, i.e. no light will pass through. However, if a sample containing oriented molecules like collagen or microtubules their ordered structure rotate the axis of the light and appear as bright structures on a dark background.

METALLURGICAL AND INVERTED MICROSCOPES

A metallurgical microscope is a device in which an opaque object is examined through a reflecting device. The instrument has no mirror and no substage condenser. Its stand is modified and the stage is without any apertures. The metallurgical microscope is used in engineering for examination of the structure of metals, and the alteration of that structure with service, and for examination of defects arising in manufacture.

Recent designs in metallurgical microscope tend towards the so-called 'inverted microscopes' (Fig. 18.11). One of the chief points in the design of this instrument is the arrangement of the stage in an uppermost and easily accessible position so that the metal specimen can be conveniently brought into its correct position for examination (perpendicular to the long axis of the microscope) with minimum amount of trouble.

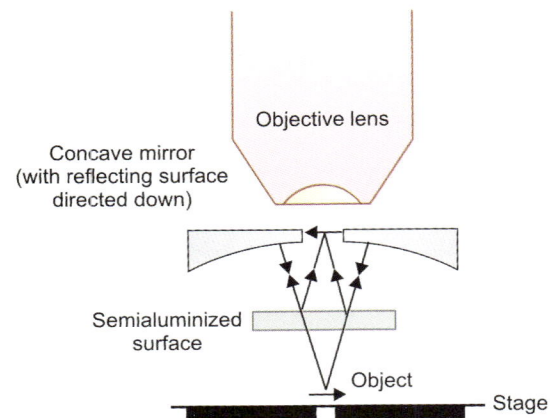

Fig. 18.11: Optical path in an inverted microscope.

McARTHUR MICROSCOPE

This microscope measures only $4 \times 2.5 \times 2$ inches yet it is capable of magnification from 30X to 1500X and covers the same area of the specimen and with the same detail as a conventional microscope. The radical difference is the light path. Light enters the instrument from the top (Fig. 18.12) passes through a conventional condenser to the specimen and then through the objective lens beneath. It is then reflected through 90° along a light tube at the base of the instrument before being reflected again through 90° to pass upwards through the eyepiece to the eye.

Advantages of McArthur Microscope

a. The specimen is always in focus whatever the thickness of the slide.

b. The image in the eyepiece is erect rather than inverted.

c. A satisfactory examination can be made at the highest magnification even under conditions of extreme vibration.

d. The objective lenses themselves are mounted three at a time on a slide which is simply pushed in and out to change the lens, with alternative sets of lenses mounted on different slides which can be changed in a few seconds.

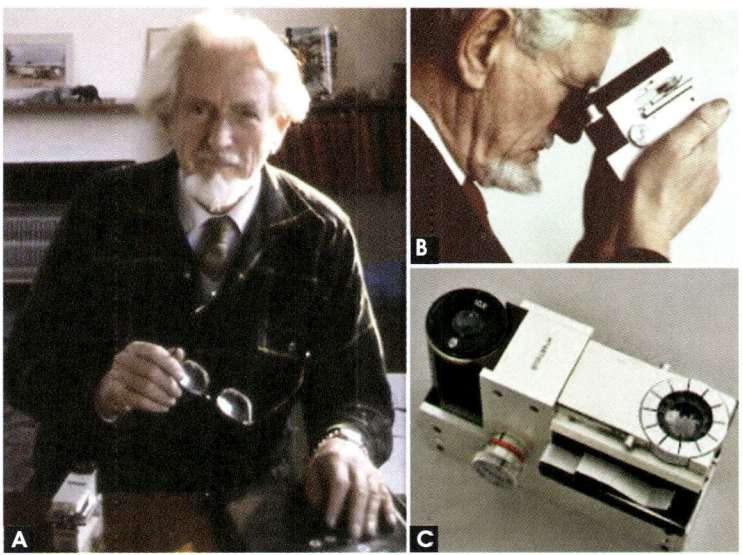

Fig. 18.12: (A) John McArthur at his work room, (October 1977), (B) Using the microscope, (C) McArthur microscope.

e. Each microscope can easily be adapted for any of a multitude of specialist uses.

FLUORESCENCE MICROSCOPE

A fluorescent microscope (Fig. 18.13) requires a strong light source (high mercury arc) which is rich in ultraviolet rays. Fluorescence is a phenomenon in which when ultraviolet light (invisible due to its very short wavelength up to 136 Å) is incident on certain substances, they emit visible light. This is due to the fact that a substance absorbs a part of the incident light of shorter wavelength and then immediately re-emits an appreciable part of it with light of longer wavelengths (visible) than those absorbed. This is called *Stoke's law*.

There are a large number of substances; like uranium glass, kerosene oil, various lubricating oils, certain dyes called *fluorochromes*, paraffin, Canada balsam, and cedar-wood oil; which exhibit fluorescence. In fluorescence microscopy, tissues treated with a fluorescent dye are examined with ultraviolet light. They become luminous and are seen as bright objects

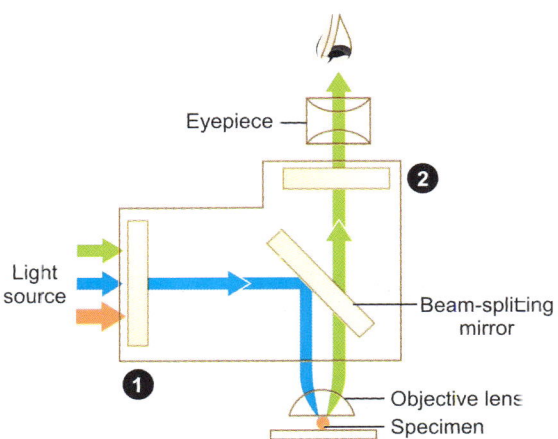

Fig. 18.13: The fluorescence microscope is similar to an ordinary light microscope except that the illuminating light is passed through two sets of filters. The first filter v' filters the light before it reaches the specimen, passing only those wavelengths that excite the fluorescent dye (that has been used for staining cells, detected with the aid of a fluorescence microscope). The second filter w' blocks out this light and passes only those wavelengths emitted when the dye fluoresces. Dyed specimens show up on a dark background.

against a dark background. The technique also enables to demonstrate some constituents of cells (e.g. vitamin A and B_2) which have a characteristic *primary fluorescence* in ultraviolet light.

As the fluorescence emitted by the specimen consists of wavelengths in the visible range of spectrum, it may be examined and photographed. The shorter wavelength of ultraviolet light gives better definition and greater resolution. However, a glass ultraviolet absorbing barrier filter (yellow filter) should be fitted in the optical system of the microscope in order to screen out extra ultraviolet light and prevent ocular damage. Of much greater value is the practice of demonstrating *secondary fluorescence* by treatment of tissues with certain fluorochromes. Since the secondary fluorescence is much more intense than the primary form, by a careful choice of fluorochromes valuable diagnostic information may be obtained.

Fluorescence Microscopy

Some substances which when illuminated at a particular wavelength emit light at a longer wavelength, i.e. fluoresce. Cells and tissues are usually illuminated with UV light with emission monitored in the visible spectrum. Some fluorescent molecules have an affinity for certain macromolecules, e.g. acridine orange binds DNA and RNA.

Fluorescent compounds can be coupled to marker molecules that specifically bind to different tissue or cell components. Fluorescent dyes absorb light at one wavelength and emit it at another, longer wavelength. Some such dyes bind specifically to particular molecules in cells and can reveal their location when examined with a fluorescent microscope. An example is a stain for DNA (green in Fig. 18.14). Other dyes can be coupled to antibody molecules, which then serve as highly specific and versatile staining reagents that bind selectively to particular macromolecules, allowing seeing their distribution in the cell. Figure 18.10 depicts a microtubule protein in the mitotic spindle stained red with a fluorescent antibody.

CONFOCAL MICROSCOPE

A *confocal microscope* is a fluorescence microscope with a laser as its source of illumination. This is focused onto a single point at a specific depth in a specimen, and a pinhole aperture in the detector allows only fluorescence emitted from the exact point of focus to be

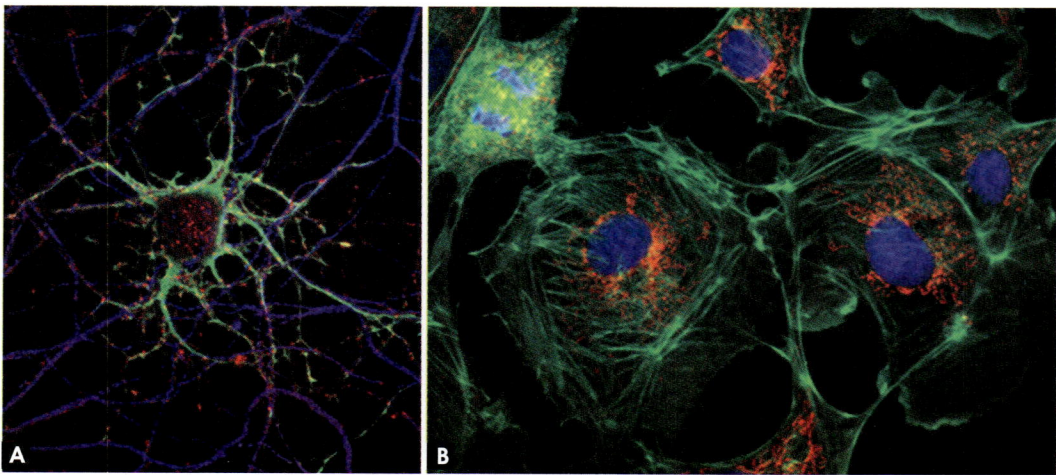

Fig. 18.14: (A) Neuron as seen with confocal fluorescence microscopy; (B) Cells showing Mitochondria (red), actin filaments (green) and nuclei (blue).

included in the image. Scanning the laser beam across the specimen generates a sharp two-dimensional image of the plane of focus. A series of optical sections at different depths allows a three-dimensional image to be constructed. An intact insect embryo stained with a fluorescent probe for actin protein, by conventional fluorescence microscopy generates a blurry image (Fig. 18.15) due to the presence of fluorescent structures above and below the plane of focus.

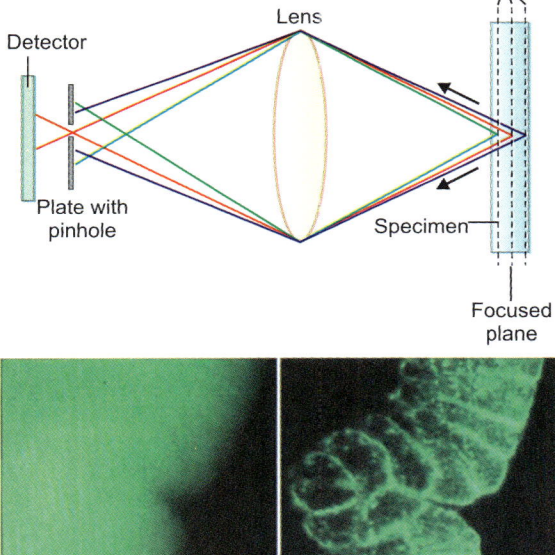

Fig. 18.15: Principle of confocal microscopy. While a very small spot of light originating from one plane of the section crosses the pinhole and reaches the detector, rays originating from other planes are blocked by the blind. Thus, only one very thin plane of the specimen is focused at a time. Two photographs are shown: (A) Blurred image seen, with conventional fluorescence microscopy, due to the presence of fluorescent structures above and below the plane of focus, (B) Confocal microscopy provides a crisp optical section of the cells in an embryo

Confocal Microscopy

The confocal microscopy provides a crisp optical section of the cells in the embryo. It allows precise focusing of light source (laser) on a very thin plane of cell or tissue. Only light from the focused plane reaches the detector, all others are blocked.

STEREOSCOPIC (WIDE-FIELD) MICROSCOPE

The **stereo** or **stereoscopic** or **dissecting micro-scope** (Fig. 18.16) is an optical microscope variant designed for low magnification observation of a sample, typically using light reflected from the surface of an object rather than transmitted through it. A dissection microscope uses low power magnification. Things appear three dimensional with a dissection microscope. The instrument uses two separate optical paths with two objectives and eyepieces to provide slightly different viewing angles to the left and right eyes. This arrangement produces a three-dimensional visualization of the sample being examined. Stereomicroscopy overlaps macrophotography for recording and examining solid samples with complex surface topography, where a three-dimensional view is needed for analyzing the detail.

The stereo microscope is often used to study the surfaces of solid specimens or to carry out close work such as dissection, microsurgery, watch-making, circuit board manufacture or inspection, and fracture surfaces as in fractography and forensic engineering. They are thus widely used in manufacturing industry for manufacture, inspection and quality control. Stereo microscopes are essential tools in entomology.

The stereo microscope should not be confused with a compound microscope equipped with double eyepieces and a bino-viewer. In such a microscope, both eyes see the same image, with the two eyepieces

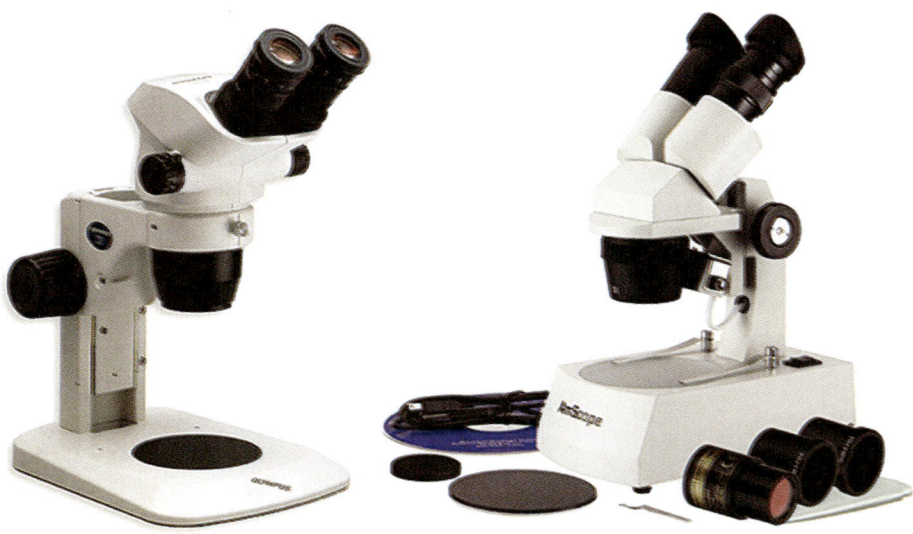

Fig. 18.16: Digital binocular stereo (or dissecting) microscope [WF10X and WF20X Eyepieces, 20X/40X/80X Magnification, and 2X and 4X Objectives].

serving to provide greater viewing comfort. However, the image in such a microscope is no different from that obtained with a single monocular eyepiece.

The wide-field binocular microscopes are constructed with long working distance, high eye-point, and instantly variable magnification with *parfocal* objectives. They are used for the examination of gross specimens under comparatively low magnifications. These microscopes are provided with eye-pieces which are adjustable for interpupillary distance and have positive rack and pinion focusing adjustment. The prism system noticeably increases the perception of depth in the image.

Differences Between Stereoscopic to Normal Optical Microscopes

Unlike a compound light microscope, illumination in a stereo microscope most often uses reflected illumination rather than transmitted (diascopic) illumination, that is, light reflected from the surface of an object rather than light transmitted through an object.

Use of reflected light from the object allows examination of specimens that would be too thick or otherwise opaque for compound microscopy. Some stereo microscopes are also capable of transmitted light illumination as well, typically by having a bulb or mirror beneath a transparent stage underneath the object, though unlike a compound microscope, transmitted illumination is not focused through a condenser in most systems. Stereo-scopes with specially-equipped illuminators can be used for dark-field microscopy, using either reflected or transmitted light.

Great working distance and *depth of field* are important qualities for this type of microscope. The large working distance at low magni-fication is useful in examining large solid objects such as fracture surfaces, especially using fiber-optic illumination. Such samples can also be manipulated easily so as to determine the points of interest. Both qualities are inversely correlated with resolution: The higher the resolution (i.e. the greater the dis-tance at which two adjacent points can be distinguished as separate), the smaller the depth of field and working distance. Some stereo microscopes can deliver a useful magnification up to 100X, comparable to a 10X objective and 10X eyepiece in a normal compound microscope, although the magnification is often much lower. This is around one-tenth the useful resolution of a normal compound optical microscope.

There are two major types of magnification systems in stereo microscopes. One type is fixed magnification in which primary magnification is achieved by a paired set of objective lenses with a set degree of magnification. The other is zoom or pancreatic magnification, which are capable of a continuously variable degree of magnification across a set range. Zoom systems can achieve further magnification through the use of auxiliary objectives that increase total magnification by a set factor. Also, total magnification in both fixed and zoom systems can be varied by changing eyepieces.

ULTRAVIOLET MICROSCOPE

Ultraviolet (UV) microscopes have two main purposes. The first is to utilize the shorter wavelength of ultraviolet electromagnetic energy to improve the image resolution beyond that of the diffraction limit of standard optical microscopes. This technique is used for non-destructive inspection of devices with very small features such as those found in modern semiconductors. The second application for UV microscopes is contrast enhancement where the response of individual samples is enhanced, relative to their surrounding, due to the interaction of light with the molecules within the sample itself. One example is in the growth of protein crystals. Protein crystals are formed in salt solutions. As salt and protein crystals are both formed in the growth process, and both are commonly transparent to the human eye, they cannot be differentiated with a standard optical microscope. As the tryptophan of protein absorbs light at 280 nm, imaging with a UV microscope with 280 nm bandpass filters makes it simple

to differentiate between the two types of crystals. The protein crystals appear dark while the salt crystals are transparent.

An ultraviolet microscope has quartz lenses and slides and uses ultraviolet radiation as the illumination. The use of shorter wavelengths than the visible range enables the instrument to resolve smaller objects and to provide greater magnification than the normal optical microscope. The final image is either photographed or made visible on a fluorescent screen by means of an image converter. The ultraviolet microscope is an instrument in which an ultraviolet light (UV) source is used. The absorption of UV light by the molecules in the specimen is recorded photographically, since the specimen cannot be visualized directly through an eyepiece and it is injurious to the eye. This type of microscope is utilized in detecting nucleic acids, particularly the purine and pyrimidine bases of the nucleotide (Fig. 18.17). It is also helpful in detection of certain amino acid containing proteins.

UV Microscopy

- Light source: Ultraviolet light instead of white light
- UV light wavelength = 180–400 nm
- White light wavelength = 400–700 nm
- Allows visibility of smaller microorganisms (smaller wavelength ⟶ smaller resolution power)
- Allows observation of substances absorbed by microorganisms (become fluorescent under UV light)
- UV radiations—not visible ⟶ images impressed on photographic film (image converter tube)/captured by phototube and projected on screen.

EAR MICROSCOPE

The instrument, a monocular type microscope, consists essentially of an arrangement to illuminate the interior of the ear and a microscopic system to view the magnified image (Fig. 18.18).

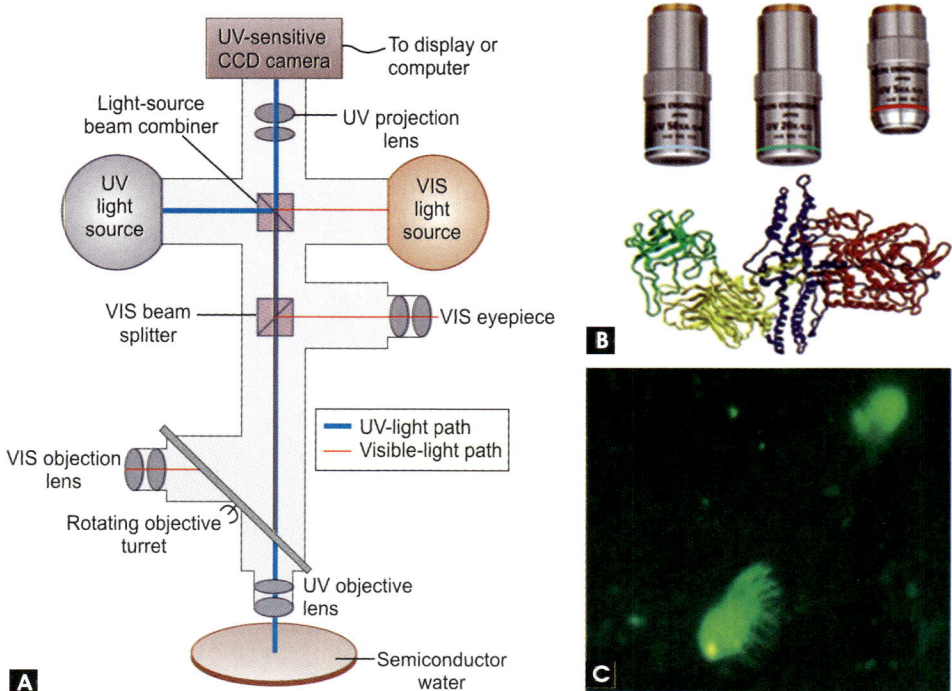

Fig. 18.17: (A) The ultraviolet (UV) microscope actually contains two entirely different optical systems, one for UV and the other for visible light, built into the same microscope (B) and (C) using UV objectives for UV microscopy to identify and locate growing protein crystals.

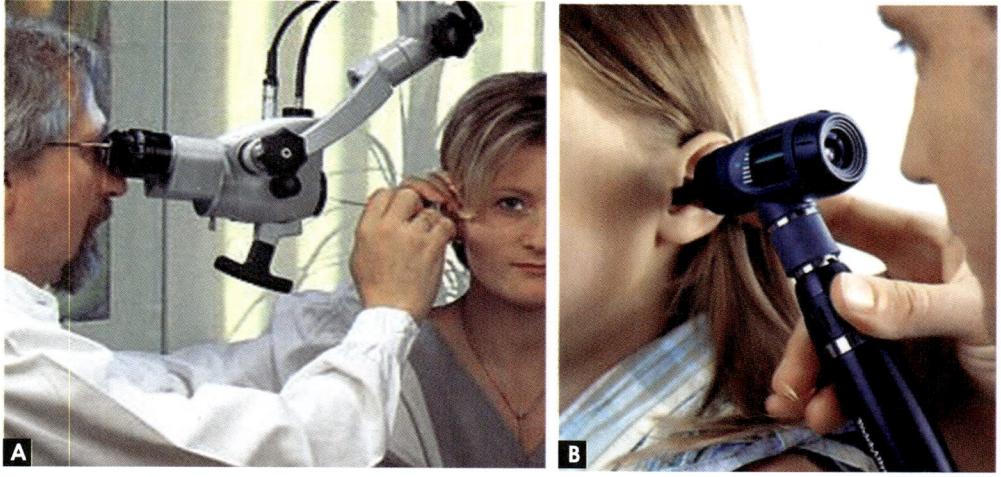

Fig. 18.18: (A) Zeiss OPMI Pico ENT Microscope; (B) An ear microscope.

The illuminating arrangement consists of a light source (6V, 12W standard lamp) with stepped control for intensity variations. The semispherical surface posterior to the semicircular filament of the lamp is silvered to provide a complete circular filament as an effective source of light. The radiations of the lamp, after passing through condenser, are reflected by plane and concave mirrors to form an intense circular light patch. The whole, arrangement is located in the main body of the instrument.

An objective Porro's prism assembly and Ramsden's eyepiece (with iris diaphragm) form the viewing system of the instrument, providing an overall magnification of 6.3X. The object field-size, which can be varied from 2 to 8 mm for all positions of the eyepiece focusing movements, helps exclude the reflections from the speculum walls, thereby increasing the contrast of the image.

The instrument is being used in medical institutes and hospitals and is required by most of the medical practitioners. It is useful for examining the inside of the ear for swellings, discharge, vascular tumors and defects of the tympanic membrane. All the materials and components used in the fabrication of the instrument are available indigenously.

COMPARISON MICROSCOPE

The comparison microscope is an instrument used to compare microscopic items side by side. Although the human eye can be very good at looking minute differences in color and morphology, the brain has a more difficult time remembering and processing these subtle differences. The problem is overcome to great extent when the images from two microscopes are viewed side by side in a single field of view. Human hair comparison particularly the final stages of an examination, are conducted almost exclusively under a comparison microscope. The comparison microscope consists of two independent objective lenses joined together by an optical bridge (Fig. 18.19) to a common eyepiece lens.

Sometimes, an instruction desires to demonstrate the microscope image simul-

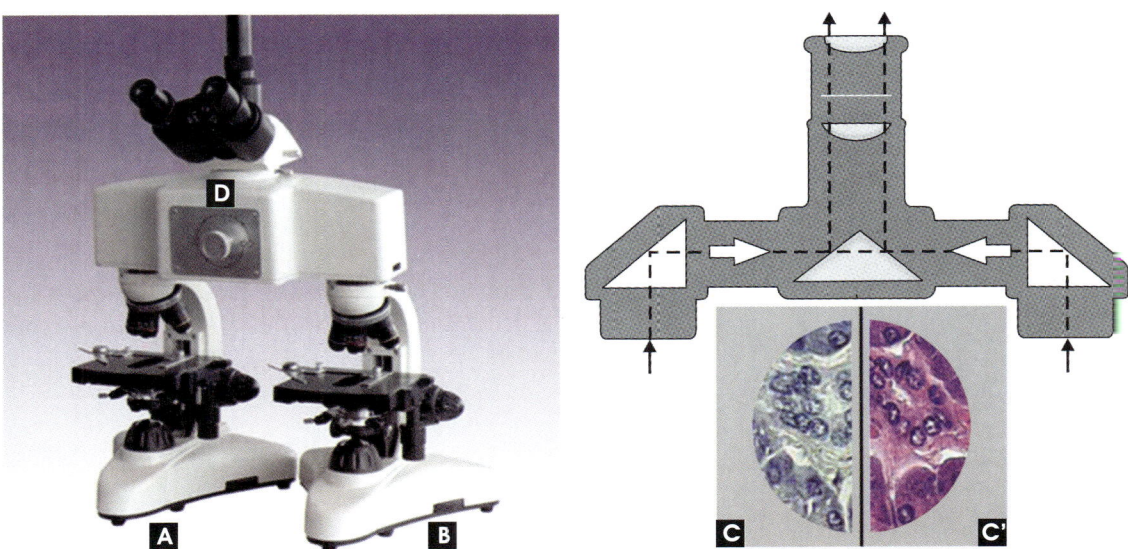

Fig. 18.19: Unlike any other microscope, the comparison microscope looks at two different sections (here the sections stained by different stains) at the same tiem. As its name implies, this microscope is used to compare sections. (A, B) Two different microscopes, (C, C') Sections to compare for difference in the staining results of two different staining procedures, and (D) Comparison eyepiece (optical bridge).

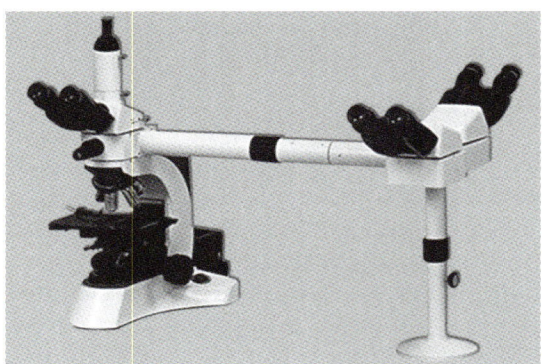

Fig. 18.20: Multi-viewers teaching microscope for 3 persons/demonstration.

taneously to three viewers (Fig. 18.20) for explaining some features.

ELECTRON MICROSCOPE

It must be realized that cell samples for the electron microscope require painstaking preparation. The basic steps of tissue preparation—such as *fixation, dehydration, embedding* in a solid wax or resin, *sectioned* into thin slices, and stained before it is examined. For electron microscopy, similar procedures are required, but have to be much thinner and there is absolutely no possibility of looking at living, wet cells. When the very thin sections are cut, stained, and placed in the transmission electron microscope, much of the part of cell components becomes sharply resolved into distinct organelles—separate, recognizable substructures that are only hazily defined under the light microscope. A delicate membrane, about 5 nm thick, is visible enclosing the cell.

An **electron microscope** is a microscope that uses a beam of accelerated electrons as a source of illumination. As the wavelength of an electron can be up to 100,000 times shorter than that of visible light photons, electron microscopes have a higher resolving power than light microscopes and can reveal the structure of smaller objects. Electron microscopes are used to investigate the ultrastructure of a wide range of biological and inorganic specimens including microorganisms, cells, large molecules, and biopsy samples. Industrially, electron microscopes are often used for quality control and failure analysis. Modern electron microscopes produce electron micrographs using specialized digital cameras to capture the image. The organelles of the cell became known after the electron microscope was invented.

Principle of Electron Microscopy

When electrons move they behave somewhat like light waves and have properties of refraction, diffraction and interference. The wavelength of electrons is inversely proportional to their velocity. The particular wavelength used in an electron microscope is one-twentieth of an Angstrom unit (Å), i.e. about 100,000 times shorter than that of the ordinary light. Thus, high resolution and great magnification are possible. Theoretically, if conditions were identical in the optical and electron microscopes, resolution down to one-fortieth Angstrom would be possible. However, the NA of an electron microscope 'lens' is very small (the diameter of aperture is only a few microns) and does not approach the width of that of an optical microscope. In practice, the best resolution that can be obtained is 3 to 5Å (0.3 to 0.5 nm). Thus the resolution of an electron microscope is approximately a hundred times better than that of the light microscope.

In the 1920s it was discovered that accelerated electrons behave in vacuum just like light. Electrons are negatively charged tiny particles of the atom. An electron is nearly 2000 times smaller and lighter than the smallest atom. The electrons travel in straight lines and have extremely short wavelength (0.05Å) which is about 100,000 times smaller than that of light.

The electrons having very high speed and extremely short wavelength (0.05Å) are capable of quite high magnification and greater resolution than that obtainable from even the best models of light microscopes available. However, electrons possess a very poor penetrating power; and very thin tissue sections (obtained by cutting through special ultrotomes using glass or diamond knives) are required.

When a thin beam of electrons, called the *primary* or *incident electrons* impinges upon a specimen (Fig. 18.21), different types of electrons and electromagnetic waves are emitted as a result of *elastic* (with no loss of initial energy) and *inelastic* (with some loss of initial energy) scattering of primary electrons by various types of atoms present in the specimen being viewed.

A number of things happen when a beam of electrons strikes the surface of any specimen:

1. Some of the electrons are *absorbed*; these cause what is called *amplitude contrast* in the image.

2. Other electrons are *scattered* over small angles, depending on the composition of the specimen; these cause what is called the *phase contrast* in the image.

3. In crystalline specimens, the electrons are scattered in very distinct directions as function of the crystal structure; these cause what is called *diffraction contrast* in the image.

4. Some of the impinging electrons are *reflected*, these are called *backscattered electrons*.

5. The impinging electrons can cause the specimen itself to *emit electrons*; these are called *secondary electrons*.

6. The impinging electrons cause the specimen to *emit photons* (or light); these are called *cathode luminescence*.

7. The impinging electrons cause the specimen to *emit X-rays* whose energy and wavelength are related to the specimen's elemental composition.

8. The impinging electrons are absorbed because of interaction with the specimen. This causes loss of some energy that can be detected by Energy Loss Spectrometer.

Scattering of incident primary electrons results in the production of *Transmitted Electrons*—which penetrate the specimen and are utilized for transmission electron microscopy; *Backscattered* (reflected) and *Secondary Electrons*—which do not penetrate through the specimen, and are deflected back at varying angles. These are collected by detectors and passed through a scan amplifier to reach the cathode ray tube (CRT) of a scanning electron microscope.

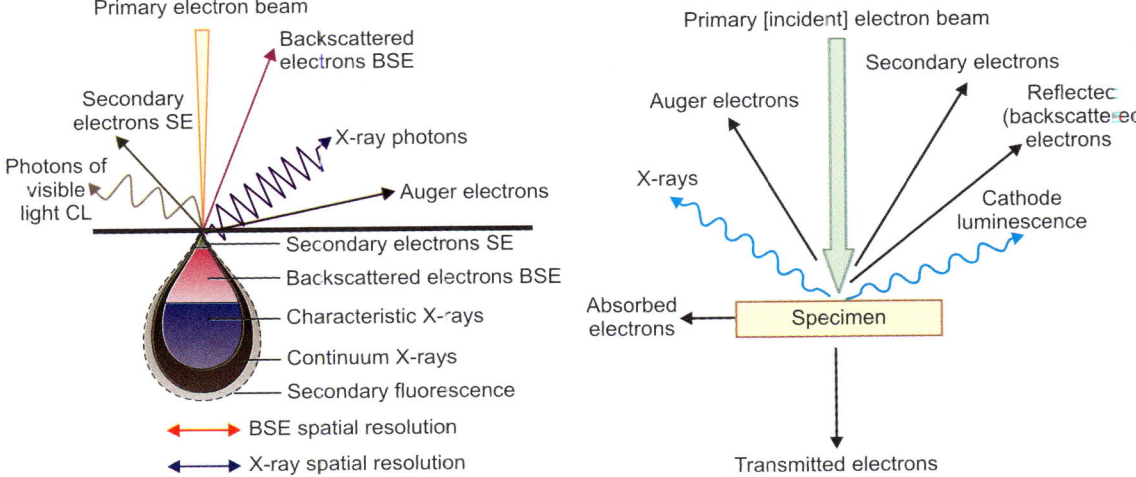

Fig. 18.21: Interaction between the primary (incident) electron beam and the coated surface of a biological (or metallurgical) sample.

Electron microscopy (EM) is a technique for obtaining high resolution images of biological and non-biological specimens. It is used in biomedical research to investigate the detailed structure of tissues, cells, organelles and macromolecular complexes. The high resolution of EM images results from the use of electrons (which have very short wavelengths) as the source of illuminating radiation. Electron microscopy is used in conjunction with a variety of ancillary techniques (e.g. thin sectioning, immuno-labeling, negative staining) to answer specific questions. EM images provide key information on the structural basis of cell function and of cell disease. The electron microscope consists of: (i) A source of supplying a **beam of electron** of uniform velocity, (ii) a **condenser lens** for concentrating the electron on the specimen, (iii) a **specimen stage** for displacing the specimen which transmits the electron beam, (iv) **objective lens**, (v) a **projector lens** and (vi) a **fluorescent screen** on which final image is observed. For permanent record of the image, the fluorescent screen is replaced by photographic film (Figs 18.22 and 18.23).

Types of Electron Microscopes

There are *four* types of electron microscopes: Transmission (TEM), reflection (REM), scanning (SEM), and scanning and transmission combined (STEM).

A. Transmission Electron Microscope (TEM)

The transmission electron microscope (TEM) was the first type of electron microscope to be

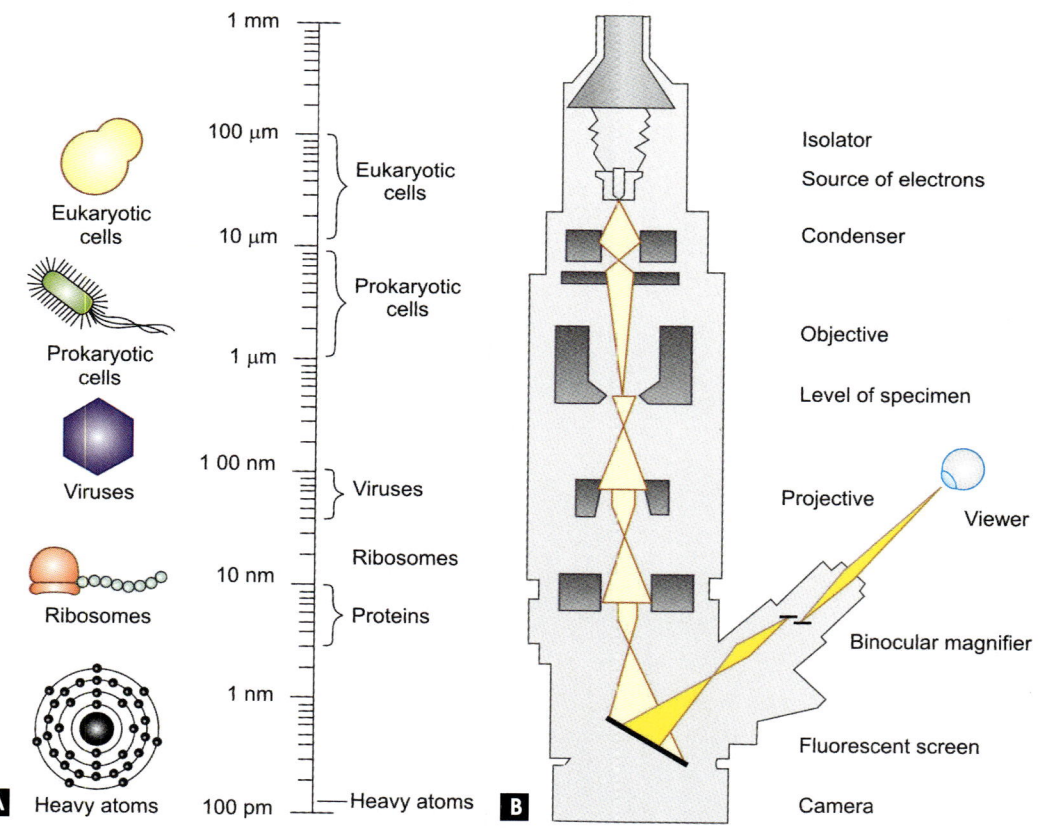

Fig. 18.22: Principle of electron microscopy: (A) Resolution limits of electron microscope, and (B) basic components of an electron microscope.

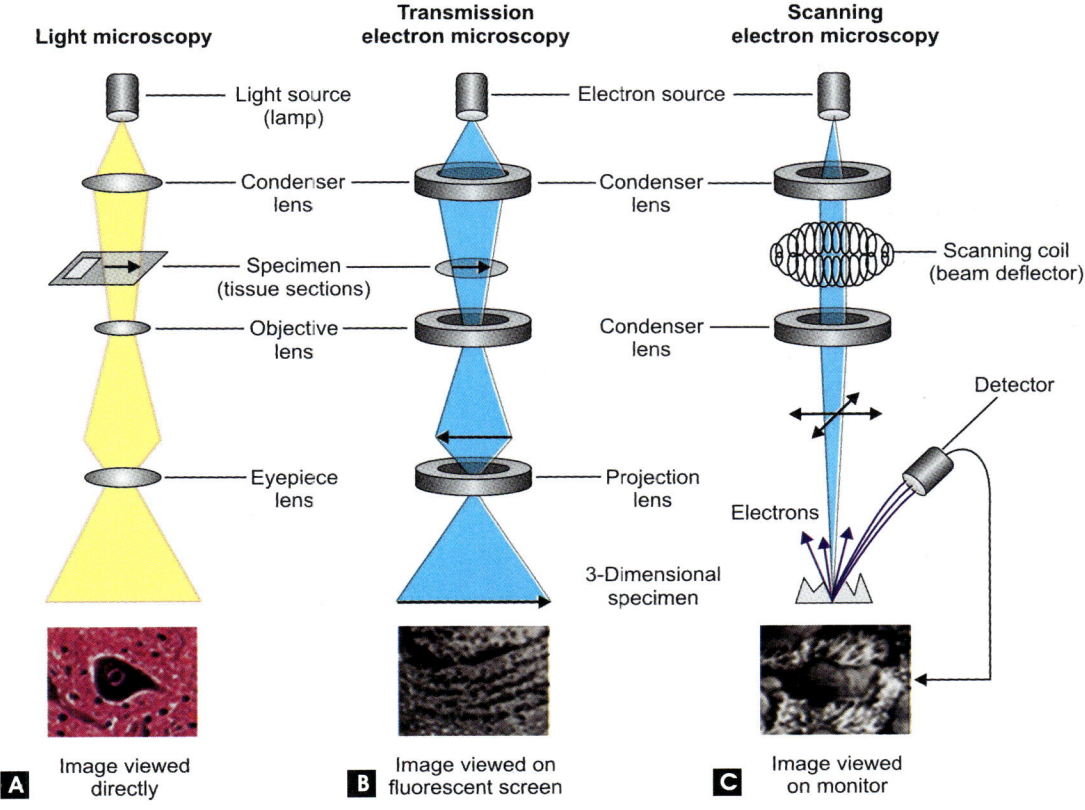

Fig. 18.23: Comparison between (A) the light microscopy, (B) transmission, and (C) scanning electron microscopy.

developed and is patterned exactly on the light transmission microscope except that a focused beam of electrons is used instead of light to "see through" the specimen. The first TEM was built by **Max Knoll** and **Ernst Ruska** in Germany in 1931 with resolution greater than that of light in 1933 and the first commercial TEM in 1939. The TEM (Fig. 18.24) does not use light, but rather a beam of electrons. It utilizes a stream of high speed beam of **accelerated electrons** as a source of illumination and **electromagnetic lenses** to control the electron beam and focus it to form an image (Fig. 18.23). The electrons are deflected by an electromagnetic field in the same way as a beam of light is reflected when it crosses a glass lens. Thus, it can achieve better than 50 pm

resolution and magnifications of up to about 10,00,000 times.

A TEM is used to view thin specimens (tissue sections, molecules, etc) through which electrons can pass generating a projection image. On the other hand, a light microscope (LM) is limited by diffraction to about 200 nm resolution and useful magnifications below 2000X. A TEM is analogous in many ways to the conventional (compound) light microscope as it is used, among other things, to image the interior of cells (in thin sections prepared with the help of an ultratome), the structure of protein molecules (contrasted by metal shadowing), the organization of molecules in viruses and cytoskeletal filaments (prepared by the negative staining technique), and the

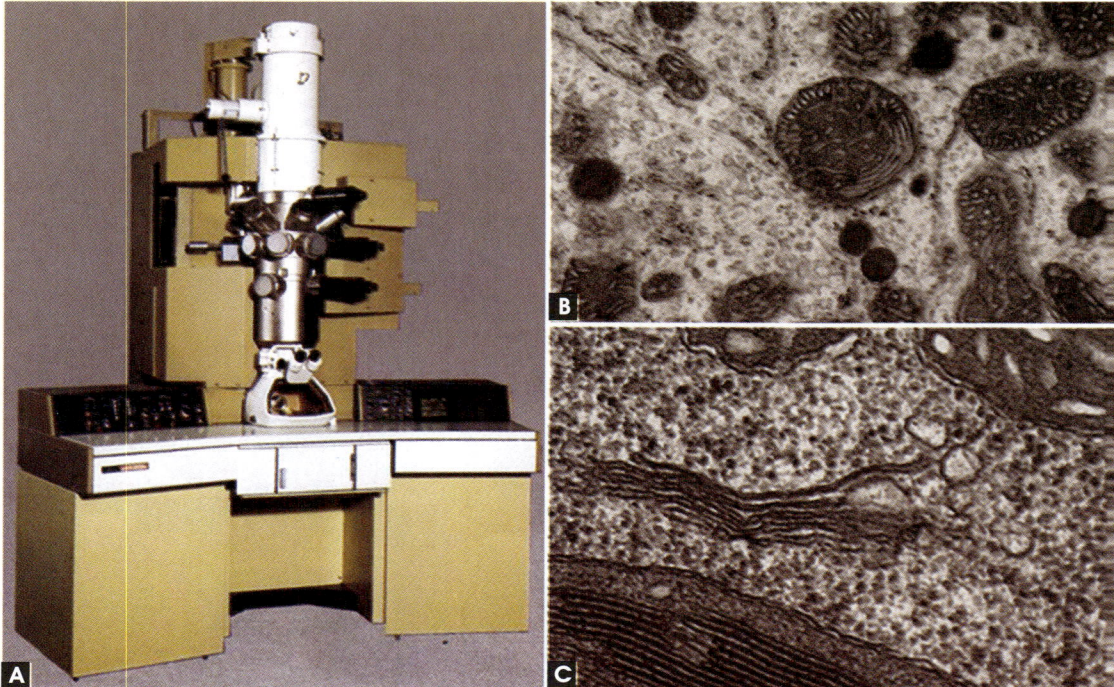

Fig. 18.24: (A) A transmission electron microscope with electromagnetic lenses and high speed beam of electrons, (B) showing structural details—the mitochondria and endoplasmic reticulum, and (C) Golgi apparatus and other finer structures in an animal cell.

arrangement of protein molecules in cell membranes (by freeze-fracture).

The thin sections are stained with a heavy metal (gold or palladium) to make certain part dense, and inserted in the vacuum chamber of the microscope. A 10,000 volt electron beam is focused on the section and manipulated prepared from the image may be enlarged with enough resolution to achieve a total magnification of over 20 million times.

B. Reflection Electron Microscope (REM)

In the **reflection electron microscope** (REM) as in the TEM, an electron beam is incident on a surface but instead of using the transmission (TEM) or secondary electrons (SEM), the reflected beam of *elastically scattered electrons* is detected for focusing the electron beam into a narrow spot which is scanned over the sample in a raster. The raster is the rectangular

pattern of image capture and reconstruction in television.

This technique of reflection electron microscopy is typically coupled with *reflection high energy electron diffraction* (RHEED) and *reflection high-energy loss spectroscopy (RHELS)*. Another variation is spin-polarized low-energy electron microscopy (SPLEEM), which is used for looking at the microstructure of *magnetic domains*. Like TEM, reflection electron microscopy uses a beam of electrons to develop a picture of the target (Fig. 18.25). The technique reads the reflected beam of electrons to form visual representation.

C. Scanning Electron Microscope (SEM)

A scanning electron microscope (SEM) is a type of electron microscope that produces images of a sample by scanning it with a focused beam of high-energy electrons. SEMs

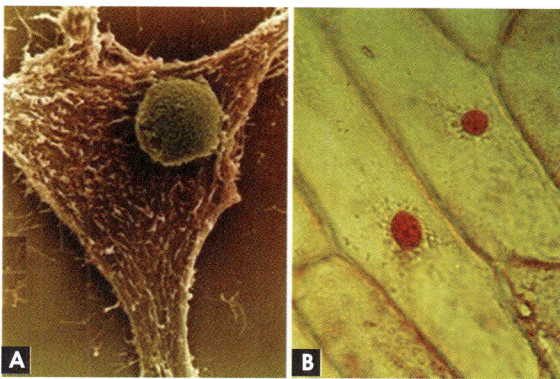

A **B**

Fig. 18.25: Reflection electron microscope (REM) images of (A) a nerve cell, and (B) image of onion cells—100X magnification. The technique involves electron beams incident on a surface, but instead of using the TEM or secondary electrons (SEM), the reflected beam is detected

are one of the most versatile and widely used tools of modern science as they allow the study of both morphology and composition of biological and physical materials to study the surfaces of the cell and organisms. The image formed by this microscope has a remarkable three-dimensional appearance. Typically magnification of scanning electron microscope is around 20,000 times.

The scanning electron microscope (Fig. 18.26) differs basically from a TEM in receiving only the reflected (backscattered) and secondary electrons which are produced as a result of interaction between the gold-coated specimen surface and the primary incident beam of electrons falling on it. The secondary electrons possess very low energy and are

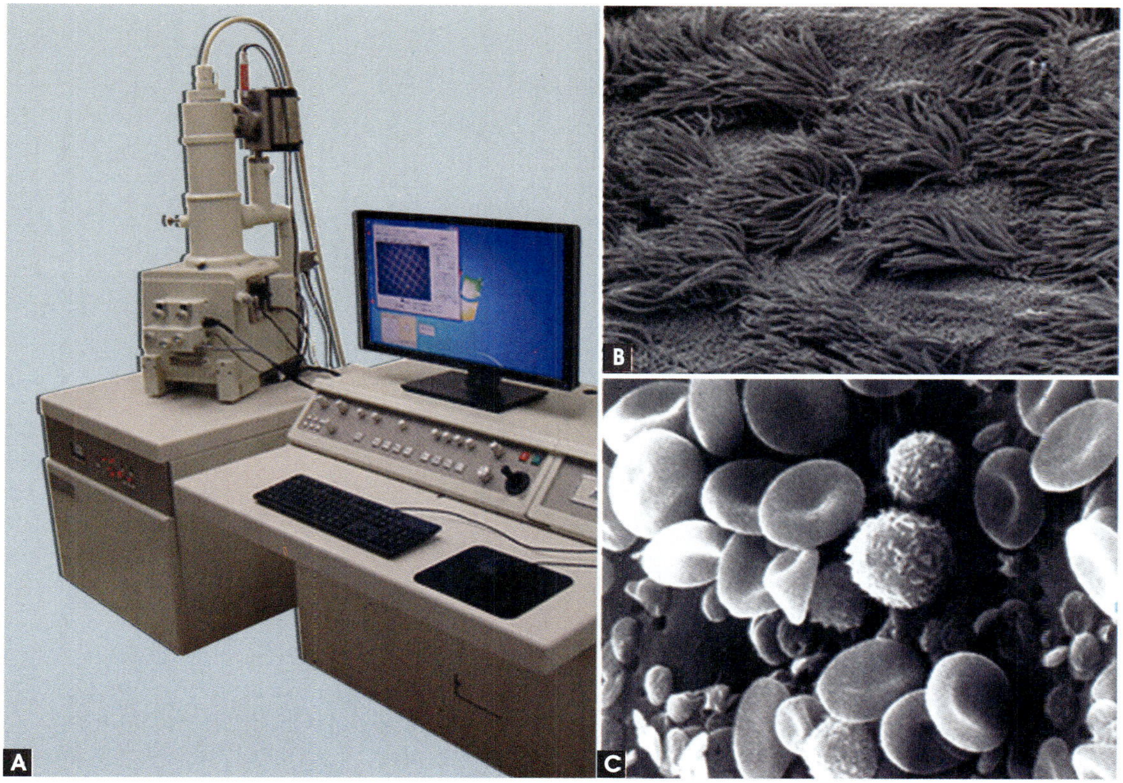

A **B** **C**

Fig. 18.26: (A) Scanning electron microscope (SEM) is a special kind of microscope, (B) SEM image shows ependyma lining the lumen of ventricles inside brain; and (C) SEM image shows circulating human red blood cells. Note that any SEM image is always a gray scale image. Colors can be artificially added later.

considered to be emitted from surface layers within 100Å in depth. The yield of secondary electrons depends on the angle between the direction of primary electrons and the specimen surface. Collected by a detector device and passed through a scan-amplifier, these electrons reach to a cathode ray tube (CRT) acting like a television screen. The surface contours of a specimen are visualized as 3-D image showing remarkable details with great depth of focus. Because of its great depth of focus, a scanning electron microscope is the EM analog of a stereo light microscope. It provides detailed images of the surfaces of cells and whole organisms that are not possible by TEM. The SEM is quite useful for the study of a sufficiently large surface area. A SEM potentially has a diagnostic application and has been regarded by many as a *pretty-picture-machine*.

A scanning electron microscope (SEM) is a type of electron microscope that produces images of a sample by scanning it with a focused beam of high-energy electrons. SEMs are one of the most versatile and widely used tools of modern science as they allow the study of both morphology and composition of biological and physical materials to study the surfaces of the cell and organisms. The image formed by this microscope has a remarkable three-dimensional appearance. Typically magnification of scanning electron microscope is around 20,000 times.

SEM Sample Preparation

Conventional scanning electron microscopy depends on the emission of secondary electrons from the surface of a specimen. The sample preparation for scanning electron microscopy is relatively easier and quicker than for transmission electron microscopy. In the SEM, stabilizing the specimen—is typically done with fixatives (fixation can be achieved, for example, by perfusion and microinjection, immersions, or with vapours using various fixatives including aldehydes, osmium tetroxide, tannic acid, or thiocarbohydrazide. The specimen, which has been coated with a very thin film of heavy metal (gold/palladium) is scanned by a beam of electrons brought to a focus on the specimen by the electromagnetic cells that, in electron microscopes, act as lenses. The quantity of electrons scattered or emitted as the beam bombards each successive point on the surface of the specimen is measured by the detector, and is used to control the intensity of successive points in an image built up on a video screen. The SEM creates striking images of three-dimensional samples with great depth of focus and can resolve details down to somewhere between 3 and 20 nm, depending on the maker of SEM. On the contrary, TEM reveals fine details of internal structure of a cell (Fig. 18.27). The electron beam is generally scanned in a raster scan pattern, and the beam's position is combined with the detected signal to produce an image. SEM can achieve resolution better than 1 nanometer. Specimens can be observed in high vacuum, in low vacuum, in wet conditions (in environmental SEM), and at a wide range of cryogenic or elevated temperatures.

D. Scanning and Transmission Combined Microscope (STEM)

The scanning transmission electron micro-scope (STEM) can be used for accurate and reproducible mass measurements. A conventional **scanning transmission electron microscope (STEM)** is a type of transmission electron microscope (TEM) equipped with additional scanning coils, detectors and necessary circuitry, which allows it to switch between operating as a STEM, or a CTEM; however, dedicated STEMs (Fig. 18.28) are also manufactured.

The basic principle of image formation is fundamentally different from static beam TEM. The small spot size is formed on the

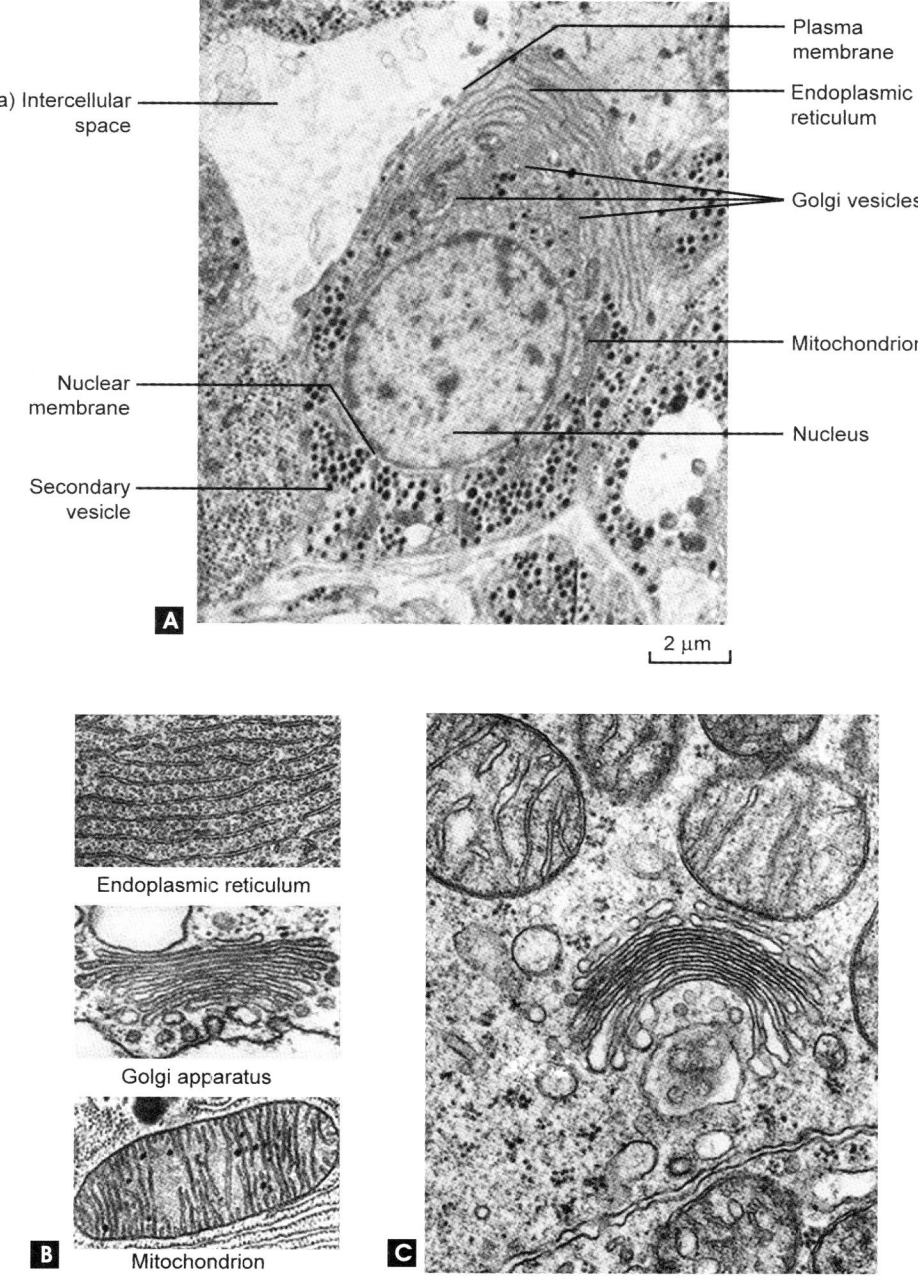

Fig. 18.27: Transmission electron micrographs reveal fine details of internal structure of an animal cel . On the extreme left is shown a complete cell at relatively lower magnification.

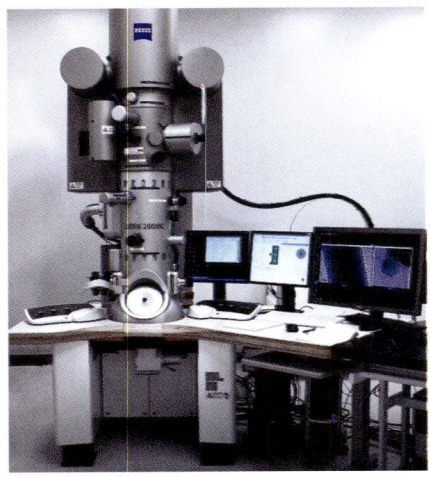
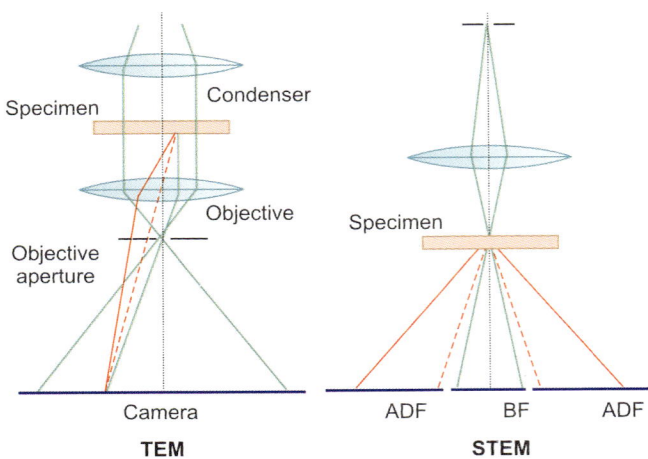

Fig. 18.28: A scanning transmission electron microscopy (STEM) is shown on the extreme left. The optical pathway in a conventional transmission electron microscope (TEM) and a scanning transmission combined electron microscope (STEM) are shown on the right.

sample surface with the condenser lenses. This probe is scanned on the sample surface; the signal is detected by an electron detector, amplified and synchronously displayed (Fig. 18.29). High resolution scanning transmission electron microscopes require exceptionally stable room environments. In order to obtain atomic resolution images in STEM, the level of vibration, temperature fluctuations, electromagnetic waves, and acoustic waves must be limited in the room housing the microscope.

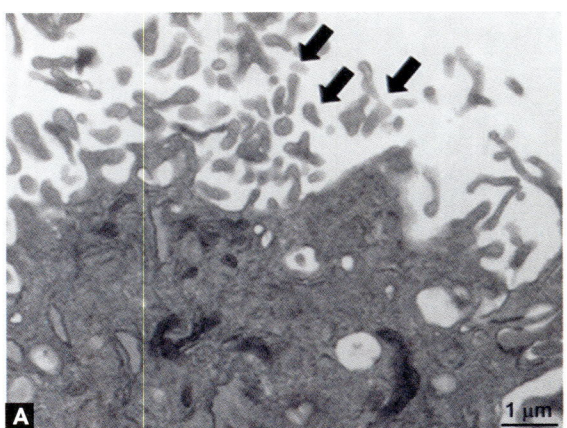
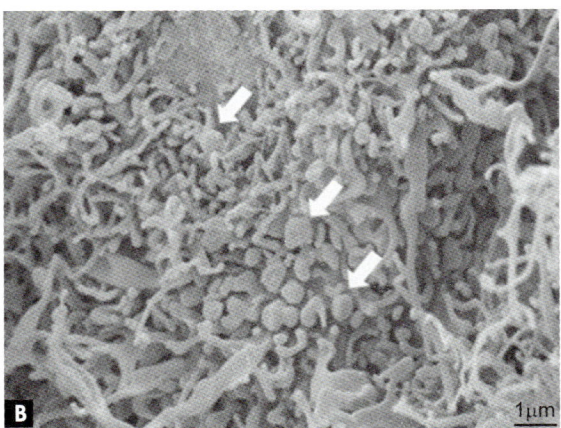

Fig. 18.29: (A) Transmission electron microscopy image shows release of micro-vesicles (black arrows) in a mesenchymal stem cell. (B) The released vesicles (white arrows) on the free surface of the cell are shown.

SUMMARY

Dark-ground Microscopy

Dark-field microscopy is one of the simplest and cheapest contrast enhancing techniques well suited for uses involving live and unstained biological samples, such as a smear from a tissue culture or individual, waterborne, single-celled organisms. To achieve a dark-field image place a dark-field filter "patch stop" into the filter holder of the condenser to prevent light to directly enter the objective. Hence, the objects appear very bright (almost entirely free of artefacts) against a dark background. The images appear spectacular and visually impressive. Even the procedure allows for visualization of objects that are *below* the resolution limit of a microscope. However, the shape of the objects thus seen is not recognizable. Limitation of dark-field microscopy is the low light levels meaning that the sample must be very strongly illuminated. Damage to the sample is thus inevitable. Also, dark-field microscopy is very sensitive to dirt and dust located in the light path; the interpretation of dark-field images, therefore, requires a greater care than in bright-field microscopy because low light level often makes some images invisible.

Phase-contrast Microscopy

This procedure is useful for examining living cells, because the specimen does not need to be stained. It is a contrast-enhancing optical technique that can be utilized to produce high-contrast images of transparent specimens. The phase contrast microscope has the same resolving power as the ordinary light microscope but it permits visualization of different parts of the cell due to differences in their refractive index (Refractive index is defined as the ratio of the velocity of light in a vacuum to its velocity in a transmitting

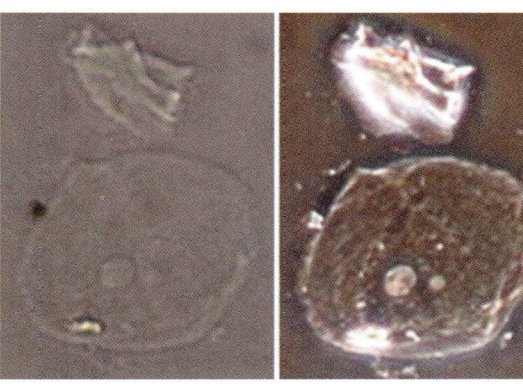

Phase contrast image on right side

medium). A phase contrast microscope is useful for examining living cells, because the specimen does not need to be stained. The phase-contrast microscope introduces contrast into unstained tissues and living transparent cells created by purely optical means, and not by staining differentially.

Because light is transmitted through a structure at a velocity inversely proportional to the refractive index of the structure, light waves emerging from structures with different refractive index will be out of phase with one another. The phase contrast microscope is able to convert these differences in phase to differences in light intensity, producing an image with good contrast. The phase-contrast microscope utilizes interference between two beams of light. Maximum destructive interference will result if, by optical means, the phase difference between the two sets of waves is increased from $\frac{1}{4}\lambda$ to $\frac{1}{2}\lambda$. The phase change is brought by interposing a phase plate annulus (a film of glass of sufficient thickness to alter the phase of green light by $\frac{1}{4}$ wavelengths) at the back focal plane of the objective lens of the phase contrast microscope and other optical plate is placed within the condenser lens system.

Fluorescence Microscopy

The absorption and subsequent re-radiation of light by organic and inorganic specimens is typically the result of well-established physical phenomena described as being either **fluorescence** or **phosphorescence**. The emission of light through the fluorescence process is nearly simultaneous with the absorption of the excitation light due to a relatively short time delay between photon absorption and emission, ranging usually less than a microsecond in

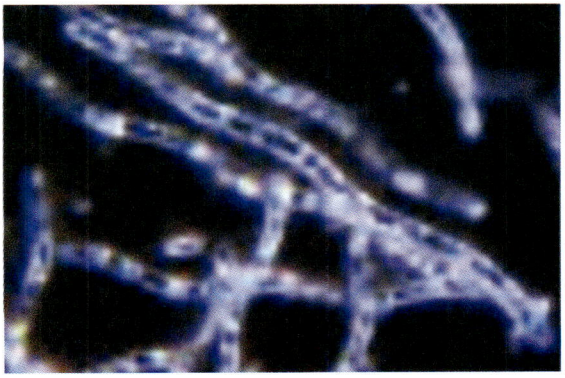

Dark-field image of a bacteria Spirillum

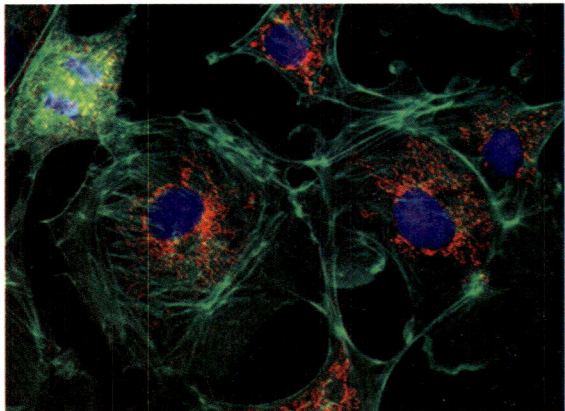

Fluorescent microscope image of a cell showing mitochondria (red), actin filaments (green) and nuclei (blue).

duration. When emission persists longer after the excitation light has been extinguished, the phenomenon is referred to as phosphorescence.

A **fluorescence microscope** is an optical microscope that uses fluorescence and phosphorescence instead of, or in addition to, *reflection* and *absorption* to study properties of organic or *inorganic* substances. Fluorescence microscopy uses a much higher intensity light to illuminate the sample. This light excites fluorescence in the sample, which then emits light of a longer wavelength. The image produced is based on the second light source or the emission wavelength of the fluorescent species—rather than from the light originally used to illuminate, and excite the sample. It must be realized that fluorophores lose their ability to fluoresce as they are illuminated in a process called **photobleaching.** Photobleaching occurs as the fluorescent molecules accumulate chemical damage from the electrons excited during fluorescence. Photobleaching can severely limit the time over which a sample can be observed by fluorescent microscopy.

For other special types of microscopes, reader is advised to read the text under appropriate type.

19

Different Devices and Methods used for Micrometry

WHAT IS MICROMETRY?

The term *micrometry* means measurement of microscopic objects. The standard unit of measurement of structures (objects) as accepted in a histological study is called a *micron* (μ) or a *micrometer* (μm). A micron is one-thousandth part of a millimeter or approximately twenty-five thousandth part of an inch. Other terms used in the past for measurement of histological specimens were millimicron (mμ, one-thousandth of a micron), and Angström unit (Å, one-tenth of a millimicron). Presently, these units of measurements in microscopy are being replaced by units as given below:

Old terms	New terms
Micron (μ)	Micrometer (μm)
Millimicron (mμ)	Nanometer (nm)
Angstrom (Å)	0.1 nm

Out of the following four methods used for micrometry any one can be practiced according to the types of microscope and accessories available.

MEASUREMENTS WITH MECHANICAL STAGE

A light microscope which is provided with a mechanical stage and a glass disc in the eyepiece with a diamond-cut line across its centre may be used for micrometry. A histological slide is placed over the mechanical stage and the desired area of the section brought into focus. The field also shows the diamond-cut line. By manipulating the fine screw of the stage the edge of the object to be measured is brought in such a position that one end of the line is exactly over the edge. The reading of the calibrations on the verniers of the stage screws is noted. The stage is then moved slowly so as to bring the other end of the object at the same end of the line. The reading is noted again. Subtracting one reading from the other gives the measurement value of the object. This method is fairly accurate and largely used for quick work.

MEASUREMENTS WITH OPTICAL DEVICES

The different devices used for measurements of the images visualized at various magnifications in a light microscope are as follows.

Micrometers

There are two types of micrometers—*stage micrometer* and *eyepiece* or *ocular micrometer*. Measurement of an object may be done either by using both the types of micrometers or by using the stage micrometer along with another instrument called camera lucida. Before describing actual procedures for the micrometry with these devices, a brief description of both kinds of micrometers and camera lucida will be given.

Stage Micrometer

A stage micrometer consists of 3" × 1" glass slide with rulings engraved at one corner. The English and metric rulings are both available. These can either be obtained on separate slides or mounted together on the same slide. A scratch 1 mm in length is divided into 10 large divisions, each of which is further subdivided into 10 small divisions (Fig. 19.1). Thus each small division is equal to one-hundredth of a millimeter (engraved on the slide) or 10 microns. If a stage micrometer has both types of rulings mounted under the same cover glass, there are chances to get confused as to which scale happens to be in the field seen. Therefore, separate slides each having only one type of ruling should be preferred. If, however, only one slide is to be purchased, certainly the metric ruling should be selected.

Eyepiece (or Ocular) Micrometer

An eyepiece micrometer is a small circular disc of glass, with divisions ruled and figured on its under surface (Fig. 19.2) which is mounted on the diaphragm of a positive type of eyepiece. Some eyepiece micrometers have only rulings and no figures. An ocular micrometer can be easily placed on the diaphragm and the lines can be brought into focus by unscrewing the eye lens of an eyepiece.

Stage and Ocular Micrometers

The stage micrometer is put on the stage of the microscope and its rulings are focused

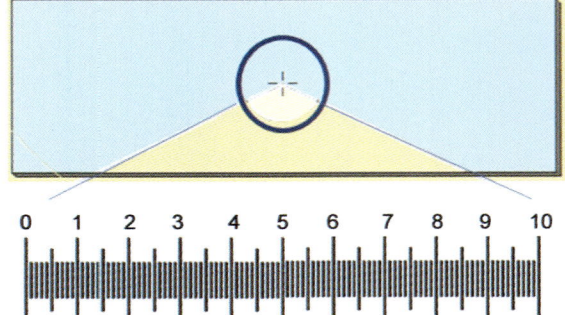

Fig. 19.1: Stage micrometers are slides with markings of a unknown dimension that, when viewed under the microscope can be used to calibrate eyepiece (or ocular) micrometers. A stage-micrometer slide showing a 0.1 mm scratch in the center within the circle. Note the magnified view of the scale as shown in the figure.

under low power of magnification. They may, however, be focused by the appropriate objective according to the size of the object to be measured. The ocular micrometer is now placed on the diaphragm of the eyepiece of the microscope. Generally the lines over the ocular micrometer are visible sharp. If not so, they may be brought into focus by a little up-screwing the eye lens of the eyepiece. In some microscopes, the eyepieces are provided with sliding eye lenses.

This facilitates exact focusing of the rulings of the ocular micrometer (Fig. 19.3) more conveniently. In microscopes with no sliding lenses, the rulings can be distinctly visualized (if they are not sharply focused as such) by slight alteration in the position of the diaphragm of the eyepiece in use. If the ocular micrometer has rulings with figures, it should be so placed on the diaphragm of the eyepiece, that the figures are seen erect. By looking into the microscope the rulings of both the micrometers will be seen.

Calibrating a *Microscope:* To properly *calibrate* your reticle with a stage micrometer, align the zero line (beginning) of the stage micrometer with the zero line (beginning) of the reticle. Now, carefully scan over until you see the lines appearing line up again.

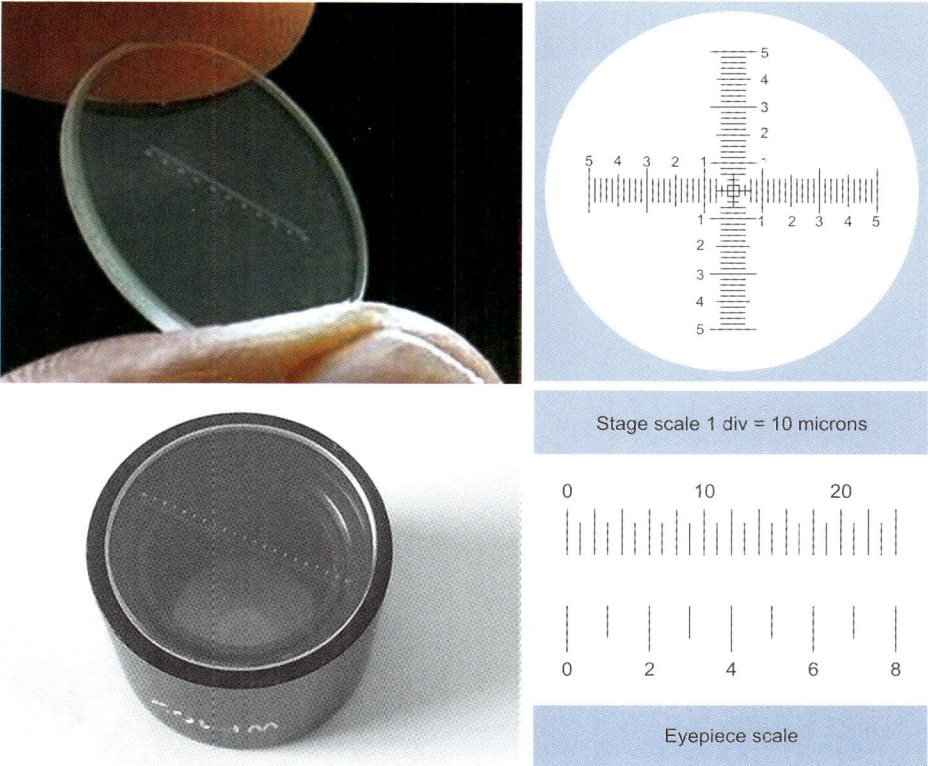

Stage scale 1 div = 10 microns

Eyepiece scale

Fig. 19.2: An ocular (or eyepiece) micrometer is a glass disk that fits in a microscope eyepiece that has a ruled scale, which is used to measure the size of magnified objects. The physical length of the marks on the scale depends on the degree of magnification.

Fig. 19.3: On the right side is shown an ocular (eyepiece) micrometer. When you look into your eyepiece lens, the markings will always be the same but the size of the image superimposed under them will get larger with more magnification.

For taking the measurements, count the number of intervals of the stage micrometer rulings which overlap or correspond to any convenient number of intervals of the ocular micrometer (Fig. 19.4). This helps in determining the *millimeter value* of a single small division of the ocular micrometer. It is advisable to make the millimeter value an integral number for making the further calculations easier. Once the ocular micrometer divisions are valued it becomes too simple to measure the specimen. Just remove the micrometer and place a slide, with sections to be measured, on the stage of the microscope. Next focus the section. The numbers of divisions of the ocular micrometer which just cover the object are counted. Since the millimeter value of divisions of ocular micrometer is already known, the size of the object can be calculated. The millimeter value depends upon two factors: (i) the objective used, and (ii) tube length. In case of the stage and eyepiece scales do not coincide with each other—it is advisable to increase or diminish the draw-tube length or to move the adjustable collar of the objective. If the draw-tube is so adjusted that one division of the stage micrometer equals 10 divisions of the eyepiece scale, then each division of the latter corresponds to one micron.

The only real difficulty faced in this method of micrometry is difficulty of bringing the object in exact alignment with the divisions in eyepiece micrometer. This is truer when using a high-power objective.

Filar Micrometer

This is a special type of eyepiece for more exact measurements. The elaborate but somewhat expensive screw-micrometer consists of a graduated scale mounted on the diaphragm of a positive type of eyepiece, which can be

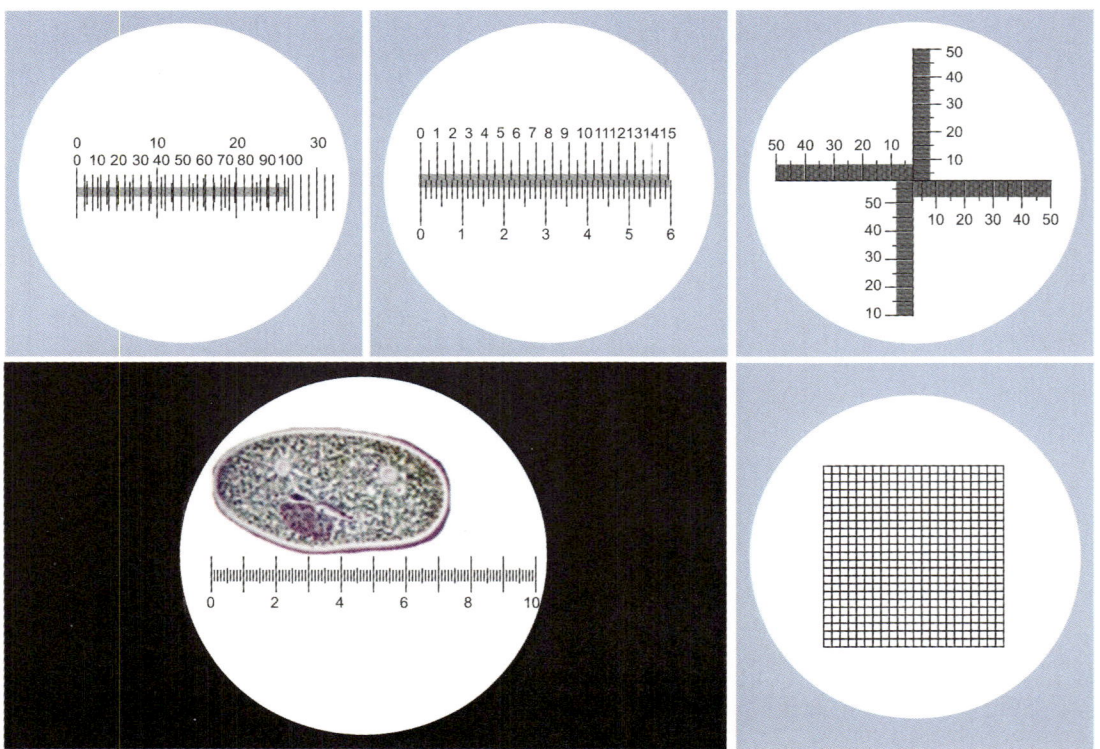

Fig. 19.4: Scales (for calibration) and grid in ocular micrometers of different make.

focused by means of the movable eye lens (Fig. 19.5). It employs a vertical fine hair line made to traverse a fixed scale situated on the eyepiece diaphragm. This type of eyepiece is of the Ramsden, or compensating type and is fitted with an accurate screw contained in a drum which is divided into hundred parts, each of which represents a displacement of 0.01 mm of the hair line. One-tenth of this interval (equal to 0.001 mm–1 µm) can easily be estimated. A fine line running through the center of the field, parallel to the axis of the screw, serves as a guide in orienting the object with reference to the direction of motion of the movable wire. Every second interval of the scale is numbered. The fixed scale is not intended for direct measurement but is constructed so that with each complete revolution of the drum the hair line traverses one division. Calibration is made by comparison with a stage micrometer in the way described for micrometry with the eyepiece micrometer. The screw micrometer eyepiece is more precise and accurate than the eyepiece micrometer and measurements to one-hundredth of the fixed scale divisions are possible.

For highly accurate measurements a **filar micrometer** (Fig. 19.5) replaces the conventional eyepieces listed in Table 19.1, and contains several improvements over conventional reticles. In the filar micrometer, a reticle with a **measuring scale** (there are many variations in scale types) and a **very fine wire** is brought into focus with the specimen (Fig. 19.5B). The wire is mounted so that it can be slowly moved across the viewfield by the calibrated thumb-screw located on the side of the micrometer (Fig. 19.5A). One complete turn of the thumb-screw (divided into 100 equal divisions) equals the distance between two adjacent reticle marks. By slowly moving the wire from one position on the specimen image to another and taking note of the changes in thumb screw numbers, the microscopist has a much more accurate measurement of distance. Filar micrometers (and other simple reticles) must be calibrated with a stage micrometer for each objective with which it will be used.

Special Kinds of Eyepieces

Eyepieces can be adapted for measurement purposes by adding a small circular disk-shaped glass reticle (sometimes referred to as a **graticule** or **reticule**) at the plane of the field diaphragm of the eyepiece. Reticles

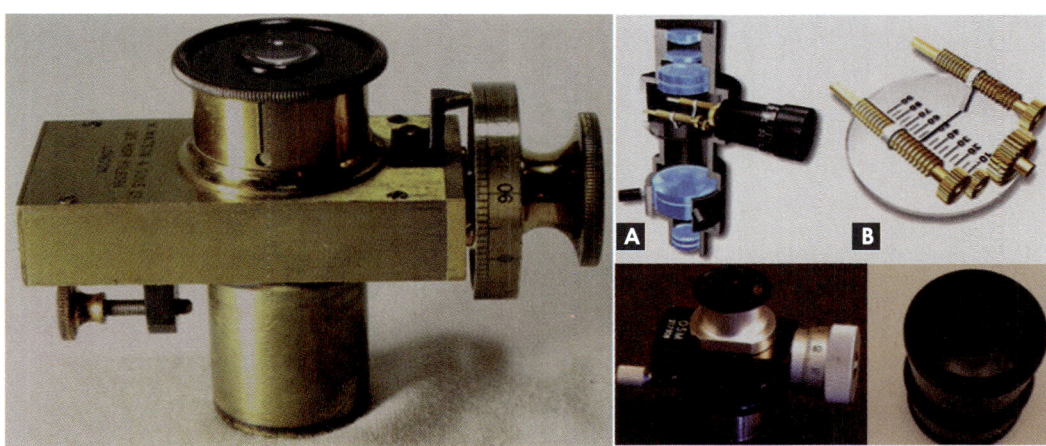

Fig. 19.5: On the left side of the figure is shown Watson 'screw micrometer eyepiece'. Special filar micrometers are used in research where highly accurate measurements are required: (A) Thumb-screw located on the side of the instrument; (B) a reticle with measuring scale and a very fine wire is brought into focus with the specimen (variations in scale types) and a very fine wire is brought into focus with the specimen.

usually have markings, such as a measuring rule or grid, etched onto the surface. Because the reticle lies in the same plane as the field diaphragm, it appears in sharp focus superimposed over the image of the specimen. Eyepieces using reticles must contain a focusing mechanism (usually a helical screw or slider) that allows the image of the reticle to be brought into focus. Several typical reticles are illustrated in Table 19.1.

In addition, there are many other variations of eyepiece reticles, and the reader should consult the many manufacturers of microscopes and optical accessories to determine the types and availability of these useful measuring devices.

Two main types of eyepieces are commonly used:

1. Periplan eyepiece

More advanced eyepiece designs resulted in the *periplan* eyepiece with seven lens elements that are cemented into a single doublet, a single triplet, and two individual doublet lenses (Fig. 19.6). Design improvements in periplan eyepieces lead to better correction for residual lateral chromatic aberration, increased flatness of field (*see* Chapter 15), and a general overall better performance when used with higher power objectives.

Modern microscopes feature vastly improved *plan*-corrected objectives in which the primary image has much less curvature of field than older objectives. In addition, most microscopes now feature much wider body tubes that have greatly increased the size of

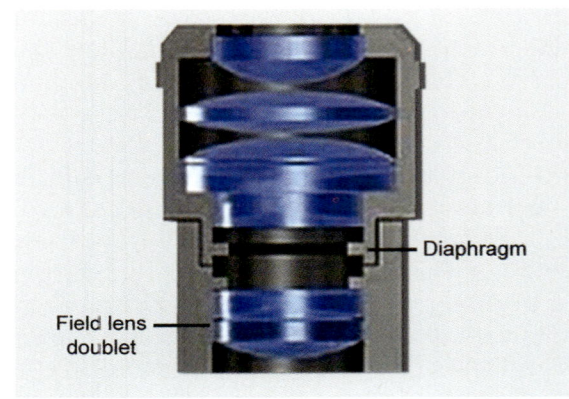

Fig. 19.6: Periplan eyepiece with seven lens elements (*see* text).

Table 19.1	Different common types of eyepiece reticles and their usage

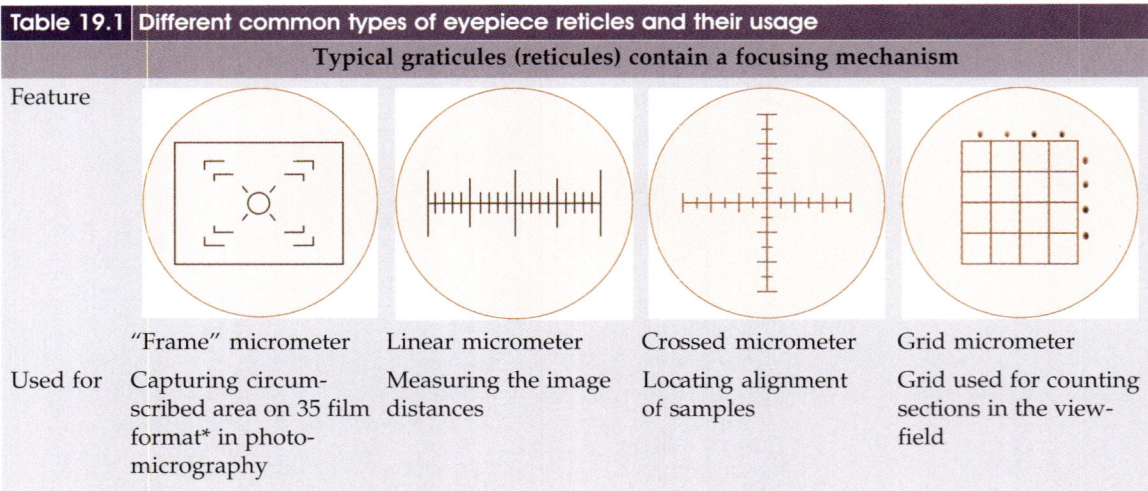

Feature				
	"Frame" micrometer	Linear micrometer	Crossed micrometer	Grid micrometer
Used for	Capturing circumscribed area on 35 film format* in photomicrography	Measuring the image distances	Locating alignment of samples	Grid used for counting sections in the viewfield

- *Besides a 35 mm film format, other film formats (120 mm and 4 × 5 inches) are delineated by sets of "corners" within the larger 35 mm rectangle.
- In the center of the reticle is a series of circles surrounded by four sets of parallel lines arranged in an "X" pattern. These lines are used to focus the reticle and image to be parfocal with the film plane in a camera back attached to the microscope.

intermediate images. To address these new features, manufacturers now produce wide-eyefield eyepieces that increase the viewable area of the specimen by as much as 40 percent. Because the strategies of eyepiece-objective correction techniques vary from manufacturer to manufacturer, it is very important (as stated above) to use only eyepieces recommended by a specific manufacturer for use with their objectives.

2. Aberration-free 10X eyepiece with diopter adjustment

These eyepieces (Fig. 19.7) have a movable "pointer" located within the eyepiece and positioned so that it appears as a silhouette in the image plane. This pointer is useful when indicating certain features of a specimen, especially when a microscopist is teaching students about specific features. Most eyepiece pointers can be rotated in a 360° angle around the specimen and more advanced versions can translate across the viewfield.

Stage Micrometer and Camera Lucida

Stage Micrometer

The stage micrometer is put on the stage of the microscope. Its rulings are focused under the low power of magnification. Some of the micrometer rulings are drawn on a piece of paper with the help of a camera lucida. Then the specimen to be measured is substituted, for the stage micrometer, and the desired field is also drawn on the paper. It is important to note that both the rulings and the specimen, are drawn under the same set-up, i.e. with the same objective and eyepiece in use. If the lines of the stage micrometer, which are 1/100 mm apart, appear 1 mm apart when drawn on a paper; it is at once known that the magnification in use is 100 times. In this method of micrometry, the measurements should be taken about the center of the microscope field. The measurements should not be made towards the edge of the field, especially when using a high-power objective. This is important because owing to curvature of the field, the edges of the field appear more highly magnified than the center.

Camera Lucida

The camera lucida is an optical device used for drawing, with hand, the structural outlines (or details) of any given tissue sample as viewed through a light microscope. It may be used also for measurements of microscopic

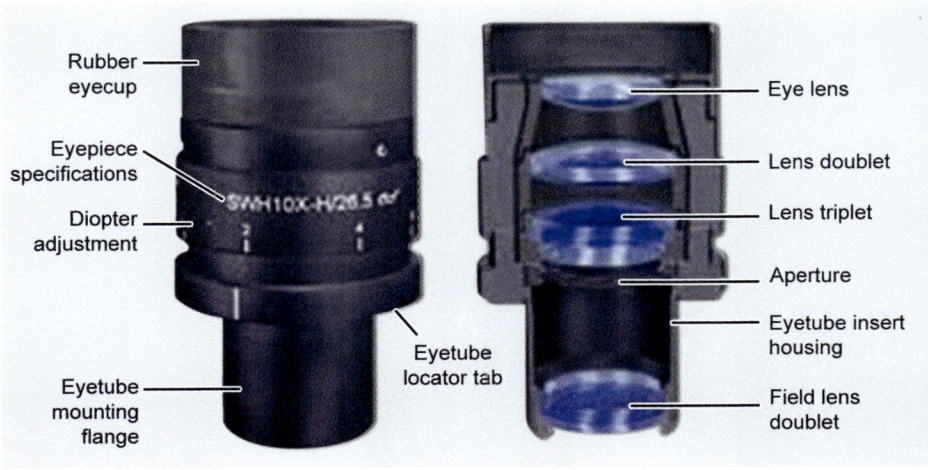

Fig. 19.7: Aberration-free 10X eyepiece with diopter adjustment.

objects. The device can be fitted to the draw-tube of a monocular microscope in such a way that a prism mounted on a suitable carrier (Fig. 19.8) fits over the eyepiece of the microscope. The observer views the section from the top of the prism. The entire field can be seen from above the prism and, by means of two carefully graded series of neutral tint filters, the light can be adjusted so that the specimen and the drawing pencil's tip, held over a plain paper placed over some hard base kept at the level of the microscope stage, are illuminated with equal intensity. Two adjustment screws provide centering over the eyepiece. The prism housing is hinged to swing away on a hori-zontal axis when desired. Both the bar for supporting mirror of camera lucida (generally 105 × 70 mm) and the fitting for fixing angle or inclination are graduated. If one desires to measure any component of a specimen, the section is replaced with a stage micrometer without disturbing the setting. The rulings drawn at the same magnification are then utilized for making the measurements.

Least Count and How to Find it

A prominent concept to understand and measure readings properly (of a measuring instrument) is least count. Least count is a very

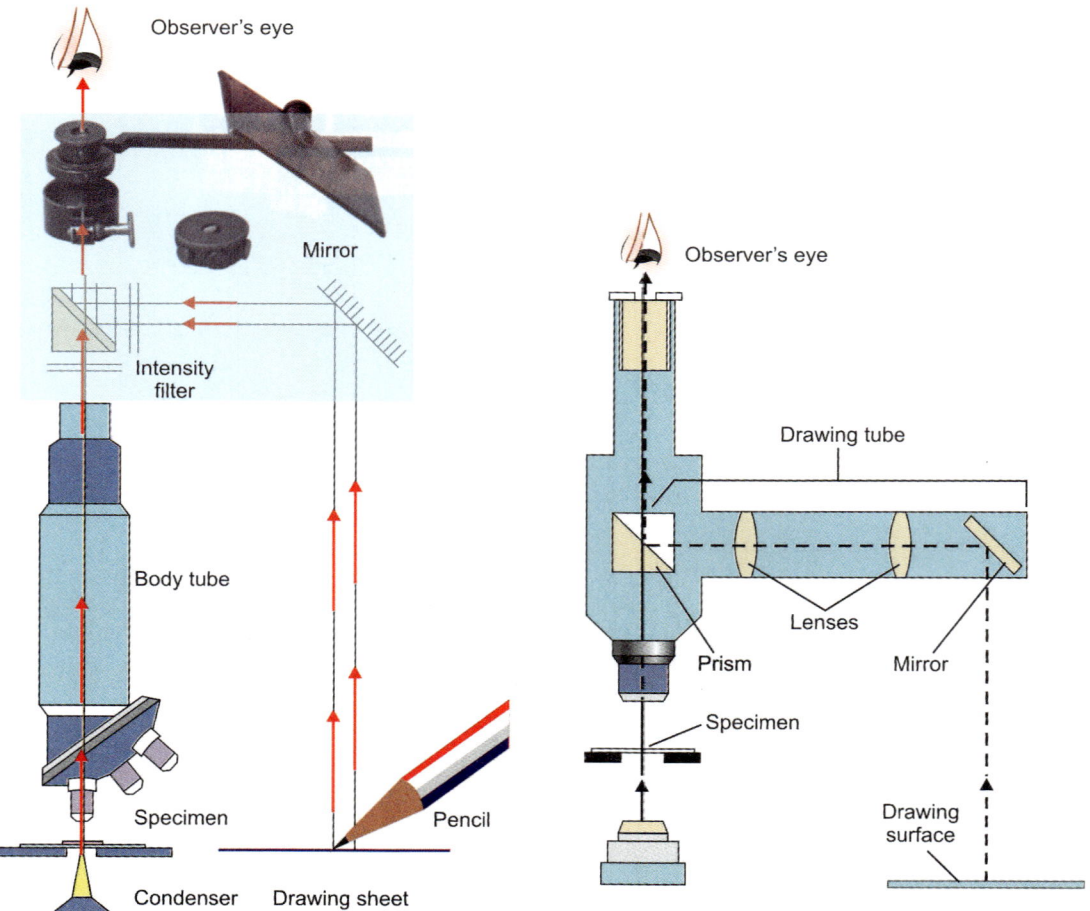

Fig. 19.8: Principle of a camera lucida showing the optical path for drawing a microscope image on a paper.

important concept, introduced to properly measure the reading from a measuring tool. Do not worry if you are not able to understand the concept completely, go through the examples (ruler, vernier calipers and travelling microscope) to nicely understand.

Least count tells you the minimum reading or value that can be measured with a measuring tool or device. Generally, simply multiplying least count with the number of divisions (like in ruler) or fraction of divisions (like in vernier calipers), we get our answer in the units specified. For example, 21 divisions in a ruler would mean 2.1 cm or 21 mm. Least count of a ruler is 0.1 cm or 1 mm.

I have seen many students (and junior faculty as well) confused about how to find least count. The method used by them might be slightly different, tough to remember, so even if they understand it once, the next time they try to do the same, they forget about it, which is not good. The way I like to make it understandable is related to the definition of least count itself. Remember, least count gives you the minimum value that can be measured by the instrument/device/tool. So considering that, least count will be: You can take any number of divisions for finding least count, but those have to be the smallest ones. Let us take "n" small divisions.

$$\text{Least count} = \frac{\text{Value measured in n divisions}}{n}$$

So, overall, least count is based on the concept of unitary method.

MEASUREMENTS WITH MICROMETER RULER

The *micrometer rule* helps to provide an idea about the actual size of any object. The vertical scale of micrometer represents the magnification produced by the microscope. The apparent size of an object is first determined with the help of a stage micrometer. If the apparent size of the object at any given magni-

fication be marked off on the appropriate horizontal line at the points of intersection with the diverging lines, its actual size in microns will be indicated by the top horizontal scale (Fig. 19.9). For example, if at magnification × 100, an object appears to be 1 mm long, its actual size would be 100 × 10 × 0.001, that is 10 microns. For magnifications ten times those given, the divisions will represent values one-tenth of those indicated. For magnifications one-tenth of those given, the values represent ten times those indicated in the figure.

Vernier Calipers

Vernier calipers is similar to ruler but a little more complex and can give more specific results. The accuracy of vernier calipers (Fig. 19.10) is much more than ruler, as the former can give results up to two decimal places (in case of centimeters), whereas the ruler can only give up to 1.

Now let us talk about what is the least count of vernier calipers and how to find it.

The main scale here is the same as the ruler. But the vernier scale has been designed to be another ruler! (or half-ruler to be precise) Instead of simply 10 smallest divisions in the vernier scale, there are a total of 50 smallest ones. Therefore, summon the least count formula:

$$LC = \frac{\text{Value measured in n divisions}}{n}$$

Remember, vernier scale gives the smallest divisions "n", not the main scale. So imagine the smaller 0.1 cm of the main scale being divided into 50 more divisions, such that 0.1 cm gap makes 50 smallest divisions, 0.2 cm makes 100, 0.3 cm gap makes 150, So:

$$\text{Least count} = \frac{\text{Value measured in 50 divisions (smallest)}}{50} = \frac{0.1 \text{ cm}}{50}$$

which is 0.002 cm or 0.02 mm, the least count of the vernier calipers is shown.

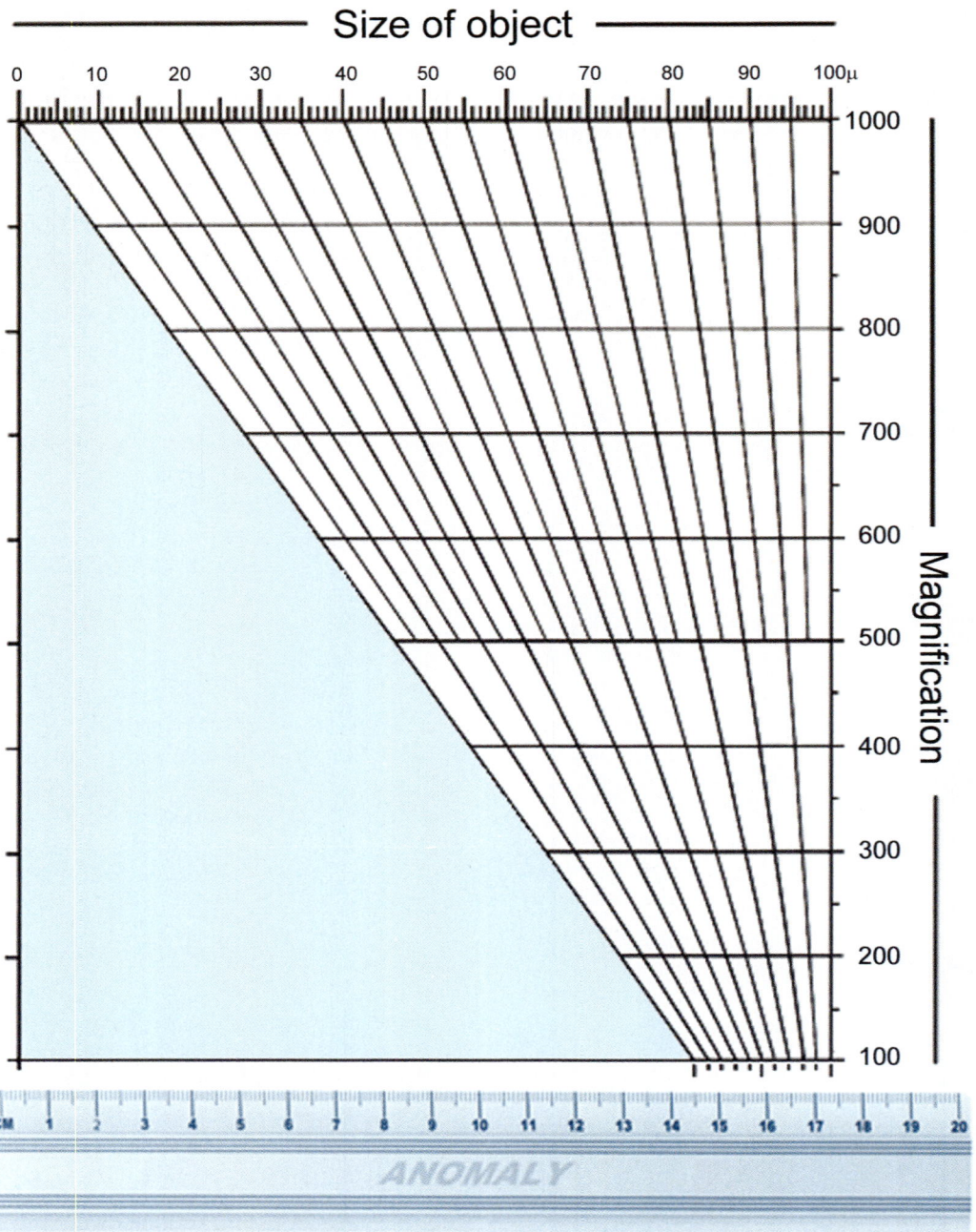

Fig. 19.9: The micrometric ruler It is actually a stage **micrometer**, which contains a small metallized millimeter **ruler** that is subdivided into increments of 10 and 100 **micrometers**.

MEASUREMENTS ON PHOTOMICROGRAPHS

This method of micrometry is undoubtedly the most accurate out of all the methods. Since only straight structures are possible to be measured by using an ocular micrometer, the curved objects must be either drawn by a camera

on the print (photomicrograph) is measured (Fig. 19.11) and its exact value calculated by applying the divider distance to the photomicrograph of the micrometer.

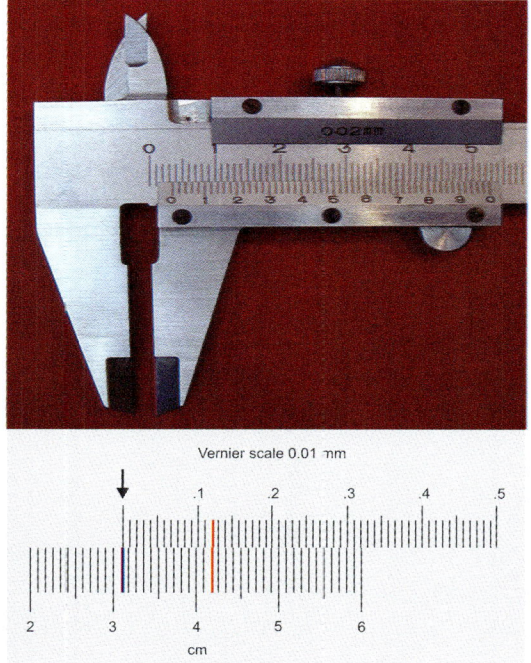

Fig. 19.10: A micrometer vernier.

lucida or reproduced by photomicrography. The object desired to be measured is photomicrographed under suitable magnification. Without disturbing the camera or the microscope setting, a photomicrograph of the stage micrometer is also taken at exactly the same magnification. By means of a pair of fine dividers the length of the desired object

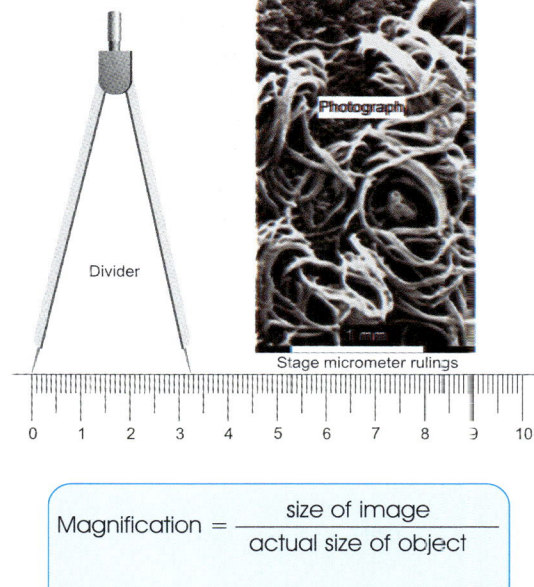

$$\text{Magnification} = \frac{\text{size of image}}{\text{actual size of object}}$$

$$\text{Actual size of object} = \frac{\text{size of image}}{\text{magnification}}$$

Fig. 19.11: A divider may be used for measurement of any structure in a photomicrograph if the rulings of a stage-micrometer are also photographed at the same magnification.

What is Micrometry?

The term *micrometry* means measurement of microscopic objects. The standard unit in a histological study is called a *micron* (μ) or a *micrometer* (μm).

Measurements with Mechanical Stage

This is fairly accurate and largely used method for quick work. A stained slide is placed over the mechanical stage and the desired area of the section focused. The field also shows the markings of diamond-cut line. By manipulating the fine screw of the stage the edge of the object to be measured is brought in such a position that one end of the line is exactly over the edge. The reading of the calibrations on the vernier of the stage screws is noted. The stage is then moved slowly so as to bring the other end of the object at the same end of the line. The reading is noted again. Difference between the two readings gives the measurement of the object.

Measurements with Different Optical Devices

• *Stage micrometer:* Consists of 3" × 1" glass slide with British and English rulings engraved at one corner. A scratch 1 mm in length is divided into 10 large divisions, each of which is further subdivided into 10 small divisions

• *Eyepiece (or ocular) micrometer:* Consists of a small circular disc of glass, with divisions ruled and figured on its under surface which is mounted on the diaphragm of a positive type of eyepiece.

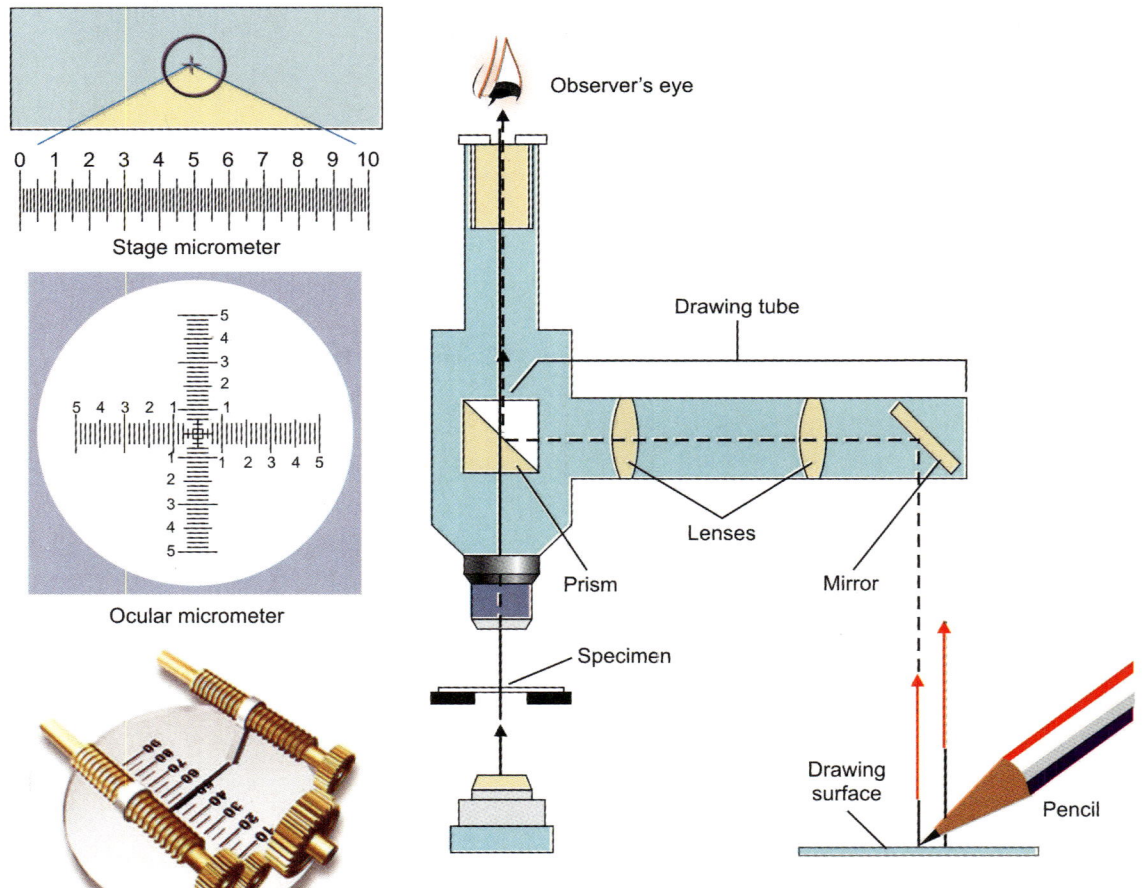

Stage micrometer

Ocular micrometer

Filar micrometer

Observer's eye

Drawing tube

Lenses

Prism

Mirror

Specimen

Drawing surface

Pencil

Principle of camera lucida

Optical devices used for micrometry

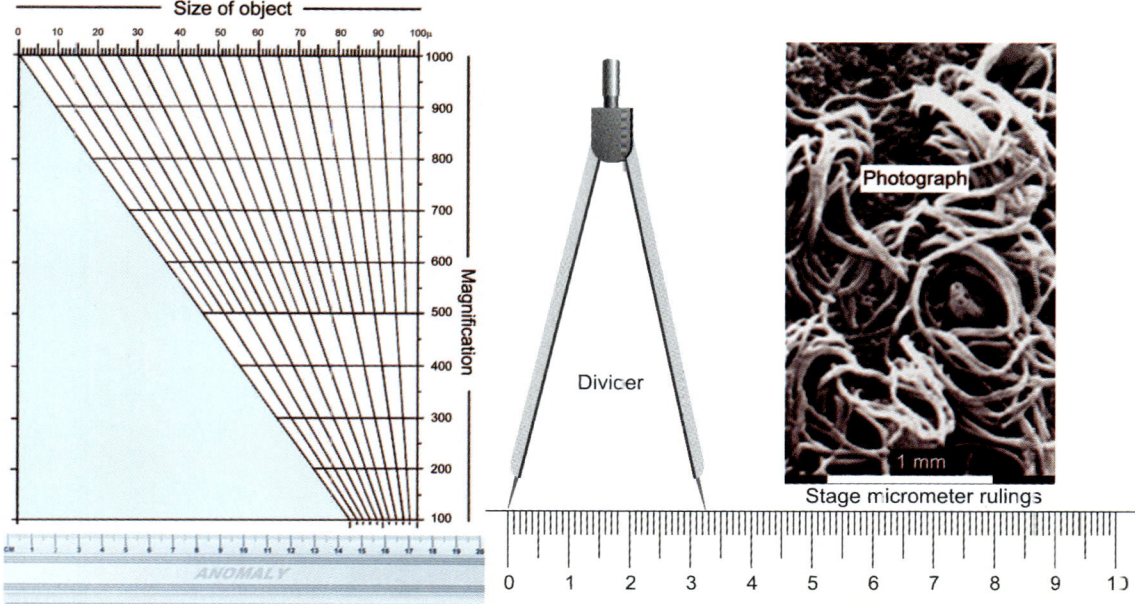

Size of object

Magnification

Photograph

Divider

1 mm

Stage micrometer rulings

0 1 2 3 4 5 6 7 8 9 10

ANOMALY

Principle of camera lucida

Measurements taken on photomicorgraphs

- *Stage and ocular micrometers:* Both are used together. The rulings of both micrometers need to be properly **calibrated**. For doing this, align the zero line (beginning) of the stage micrometer with the zero line (beginning) of the reticle.

- *Filar micrometer:* This elaborate but somewhat expensive screw-micrometer used for more exact measurements consists of a graduated scale mounted on the diaphragm of a positive type of eyepiece

- *Stage micrometer and camera lucida.*

Measurements with Micrometer Ruler

The vertical scale of micrometer represents the magnification produced by the microscope. The apparent size of an object is first determined with the help of a stage micrometer. If the apparent size of the object at any given magnification be marked off on the appropriate horizontal line at the points of intersection with the diverging lines, its actual size in microns will be indicated by the top horizontal scale.

Measurements on Photomicrographs

This method of micrometry is undoubtedly the most accurate out of all the methods. Since only straight structures are possible to be measured by using an ocular micrometer, the curved objects must be either drawn by a camera lucida or reproduced by photomicrography. A photomicrograph of the stage micrometer is also taken at exactly the same magnification. By means of a pair of fine dividers the length of the desired object on the print (photomicrograph) is measured—its exact value calculated by applying the divider distance to the photo.

Annexure
Basic Microscopy and Micrometry

Common Questions with Answers during viva voce of Practical Examination

Questions	Answers
What are modern day light microscopes known as?	• Compound microscopes
Why are they referred to as such?	• Because they use more than just a single lens
What is the light source within a compound microscope?	• An electric bulb with a tungsten filament
How is light gathered into a focused beam within a compound microscope?	• By a condenser lens
Where does the light beam originate?	• At the bottom of the microscope
What does light passing through the specimen enter?	• One of the objective lenses
Where do the objective lenses sit?	• On a movable turret located just above the specimen
How many objective lenses are usually available on a single turret?	• Four
Generally, by what factor do the lenses magnify a specimen by?	1. ×4 times 2. ×10 times 3. ×40 times 4. ×100 times
The image from the objective lens is gathered and further magnified by?	• The ocular lens of the eyepiece
What does this lens usually magnify by a factor of?	• 10
For total magnifications of?	• 40, 100, 400, and 1000
What is focusing of the image performed by?	• The use of knurled knobs that move the objective lenses up or down above the specimen
The quality of an image depends not only on the capability of a lens to magnify but also on its...?	• Resolution
What is the resolution of a lens?	• Its ability to show that two distinct objects are separated by a distance
What is the theoretical limit of resolution of a light microscope?	• 0.25 nano meters
What is this restriction determined by?	• The wavelength of visible light

Photomicrography

20

General Principles of Photography and Photomicrography

PHOTOGRAPHY

The term *photography* is used for making a picture formed by chemical reaction of light on a sensitive film. A micrograph (or correctly called a **photomicrograph**) is a photograph or digital image taken through a microscope or similar device to show a magnified image of an item. This is opposed to a macrographic image, which is at a scale that is visible to the naked eye. The following points regarding photography should be considered if excellent results are desired.

Water Supply

The water supply in a particular area is dependent upon the nature of soil, and always contains even after purification, at the water works, small quantities of minerals and gaseous impurities. Of these, we are more concerned here with lime, magnesia, iron, carbon dioxide, and oxygen. These impurities should be as far as possible removed; since lime and magnesia cause turbidity, and oxygen prematurely oxidizes the developer. This object is partially achieved, in the case of tap water, by boiling which precipitates some of the mineral salts and almost completely drives off the dissolved oxygen in water. The use of *filters* is nowadays common in laboratories. The use of distilled water is but rarely called for, and in such exceptional cases, this fact is specially mentioned in the formulae.

Where the water supply is very hard, due to a high content of calcium salts, it is advisable, in order to prevent precipitation, to adopt the following modified procedure in filling a seventy liter tank with solutions prepared according to the following schedule:

- **Solution I**—the developing agent is dissolved in 20 liters of water at about 40°C (100° F).
- **Solution II**—remaining chemicals are likewise dissolved in 20 liters of water at about 40°C (100°F).

When completely dissolved, solution I is first poured into the tank. Solution II added to it and the quantity then made up to 70 liters with cold water. Any precipitate should be filtered out by means of an ordinary funnel with filter paper or cotton wool. A separate filter should be kept for each solution. The filter should be thoroughly cleaned each time after use.

Purity of Chemicals Used

Only pure chemicals should be used—not the commercial, or 'technical' quality. The use of impure chemicals jeopardizes the work and is false economy. So far available, the fine crystalline or powder forms of chemicals are to be preferred, on account of their more ready solubility.

Accessory Apparatus

The best containers to keep photographic chemicals are wide mouth glass bottles, undamaged enamel jugs or buckets, vessels of stainless steel or plastic. Iron, copper, aluminium, zinc, or brass containers must not on any account be used. Separate vessels and stirring rods should be kept for the different solutions and these should be distinctly marked.

Order in which Chemicals should be Dissolved

The sequence in which the chemical ingredients of any solution are to be dissolved must be followed very strictly. Before adding each new chemical the previous one should be completely dissolved. Generally the sequence is: Sodium sulphite, metol, sodium or potassium carbonate, caustic soda or potash, and potassium bromide. This sequence is modified in the case of metol developers. Since metol does not readily dissolve in sulphite solution, it must be dissolved before the sulphite, after first dissolving a few grams of sulphite to prevent oxidation of the metol.

Making-Up of Volume

The total volume of solutions in the formulae has been adjusted to one liter, so that the percentage content of any particular substance in any of the solutions can be directly compared. The chemicals are dissolved in about 3/4th of the final volume and when all are dissolved; water is added to make up to the total volume.

Temperature of Solution

For safety, the temperature of the solution should not be allowed to rise above 30° C to 40° C (85° F to 100° F) because at about 50° C (120° F) there is a danger of oxidation of the developing agent by atmospheric oxygen. When solution is complete, the final solution should be cooled to the room temperature of 18° C to 20° C (65° F to 68° F).

Consistent Activity of Developers

This is insured by making developer at least one day before it is required and keeping it ready for use in a fairly large glass vessel, such as a flask, which can be made air-tight with a rubber ring. In this way the oxygen already dissolved in the solution is used up, and the activity of the developer brought to a constant value. Any precipitate that may form will sink to the bottom of the vessel and can be filtered off before use.

Absolute Cleanliness

The need for absolute cleanliness which exists in all chemical work need not be too greatly stressed where photographic processes are involved. All utensils such as dishes, measures, stirring rods, funnels, and thermometers, etc. must be kept scrupulously clean by cleaning with soap or any other detergent using brushes of different kinds (Fig. 20.1).

A mere rinse is very rarely sufficient, and every trace of residue should be cleaned with the use of a suitable oxidizing agent. The following solution is recommended.

Water	1000 ml
Potassium bichromate	50 g
Sulphuric acid (conc.)	100 ml

Note

Always add the sulphuric acid drop-by-drop to the aqueous solution; **never the reverse.**

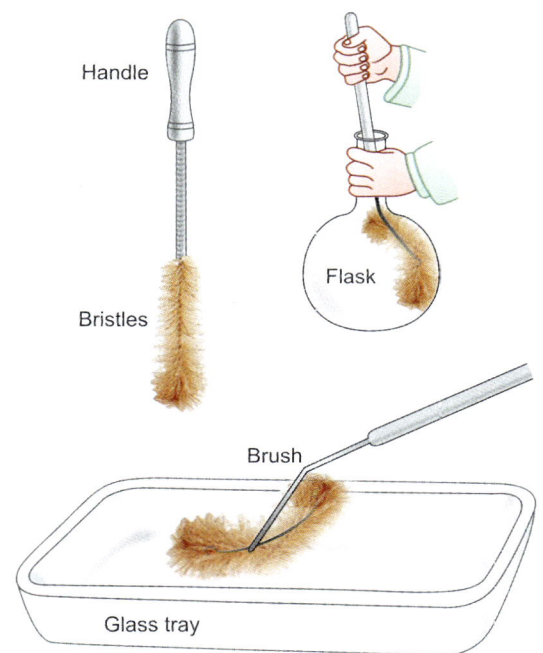

Fig. 20.1: Types of brushes used for cleansing of glassware in a photography laboratory.

The utensils should first be rinsed well, and then thoroughly washed out with this solution, all traces of which are finally removed with a five per cent solution of bisulphite dye, followed by thorough washing with water.

PHOTOMICROGRAPH AND PHOTOMICRO-GRAPHY

Micrographs are widely used in all fields of microscopy. The word *photomicrograph* is derived from two words: (i) *photograph* which is used to designate a picture produced through the instrumentality of light; and (ii) *micrograph*, which means an enlarged reproduction of a minute object made either by drawing with free hand or produced through photographic process. The neuropathologist Solomon C Fuller designed and created the first photomicrograph in 1900.

Two more terms are used frequently. The *first* term *photomacrograph* means a photographic image of a relatively large object magnified only a few times but not over 10 diameters. The *second* term is *microphotograph* often used wrongly for photomicrograph. The second term is quite different from photomicrograph, and means a minute photograph which must be examined with a microscope to reveal its details. On the contrary, a photomicrograph can be seen without any aid with unaided eyes.

The term *photomicrography* means the art of reproducing pictures of microscopial images generally of a tissue section. The process can be attributed to advances both in microscopy and in photography. Nowadays photomicrography is largely used as tool in teaching and research. In teaching, the reproduction of microscopic images, in the form of *transparencies* suitable for projection, facilitates demonstration to a large group of students. In research, there is no better method than photomicrography for communicating results in scientific journals. Moreover, photomicrographs are of immense value for comparative studies; placed side by side these offer a far easier means of noting similarities and differences that can be obtained by examining the sections (or preparations) separately under the microscope. Also, the photomicrographs mounted for display provide a medium for exhibiting the result of microscopical observations to large audiences.

Types of Photographs

Photomicrograph

A **light photomicrograph** (commonly called a **micrograph**) is a photograph prepared using an optical microscope through *photomicroscopy*. At a basic level, photomicroscopy may be performed simply by hooking up a regular camera to a microscope, thereby enabling the user to take photographs at reasonably high magnification. Roman Vishniac was a pioneer in the field of photomicroscopy, specializing in the photography of living creatures in full motion. He also made major developments in light-interruption photography and **color photomicroscopy**.

Electron Micrograph

An **electron micrograph** is a micrograph prepared using an electron microscope. However, the term *electron micrograph* is not used in electron microscopy. Common designation is a **micrograph**.

Digital Micrograph

Digital micrograph is a digital picture obtained either directly with a microscope or by scanning of a photomicrograph. Digital micrographs are commonly obtained using a USB microscope attached directly to a home computer or laptop.

Magnification and Micron Bar

Micrographs usually have micron bars, or magnification ratios, or both. Magnification is a ratio between size of object on a picture and its real size. Unfortunately, magnification is somewhat a misleading parameter. It depends on a final size of a printed picture, and therefore varies with variation in picture size. Editors of journals and magazines routinely resize a figure to fit the page, making any magnification number provided in the figure legend incorrect. A *scale bar*, or *micron bar*, is a bar of known length displayed on a picture. The bar can be used for measurements on a picture. When a picture is resized, a bar is also resized. If a picture has a bar, the magnification can be easily calculated. Ideally, all pictures destined for publication/presentation should be supplied with a scale bar; the magnification ratio is optional. All but one (limestone) of the micrographs presented on this page do not have a micron bar; supplied magnification ratios are likely incorrect, as they were not calculated for pictures at the present size.

The following terms in relation to photomicrography need attention:

Field and Projected Image Diameters

The *field diameter* is a term used for the actual diameter of the illuminated circle seen when looking into a microscope through the top lens of a microscope eyepiece. The field itself designates whatever is included within the illuminated circle. The diameter of field of view is generally expressed as:

$$D = \frac{\text{Diameter of eyepiece diaphragm (in mm)}}{\text{Magnification of objective}}$$

The field of view number is normally 18 for most of the eyepieces except in wide field eyepieces when this number may be up to 30.

The *projected image diameter* is the diameter of the circular image of the field as it is projected on the focusing screen by the camera used for photomicrography of a given specimen sample.

Camera Extension

The *camera extension* is the distance from the eyepiece to the focusing screen. It really means the distance between the *eyepoint* (just in front of the eyepiece) and the plane of the photographic film (on which image of a section is being formed).

Resolution and Magnification

Resolution

The *resolution* of a microscope can be defined as the ability to separate minute details that are very close together. This is usually measured as the *minimum distance* between two minute circular objects of high contrast at which they can still be separated. The minimum distance is called *limit of resolution*, which is inversely related to the instrument's maximal capacity to resolve—the resolving power. The smaller the limit of resolution, the higher the resolving power.

The *resolving power* depends primarily on the numerical aperture (NA) of the objective (and to some extent on the microscope condenser also); and is regarded as the minimum distance between two points, whose images can be seen as distinct from each other. The approximate resolving power of microscope can be expressed by formula:

$$R = \frac{NA}{0.61 \times \lambda}$$

Where,

R = Resolving power of microscope (in nm)
λ = Wavelength of light used (in nm)
NA = Numerical aperture of objective

For increasing the resolving power, NA of lenses should be maximum; and the wavelength of light be minimum.

Magnification

The *magnification* of a microscope is a function of both the objective lens and the eyepiece. It means the number of times larger (in linear dimension) an image is over the object. The total magnification is obtained by multiplying magnification of objective with that of the eyepiece (Table 20.1).

Table 20.1	Total magnification is a product of magnifications achieved by objectives and eyepieces		
	Objective	Eyepiece (ocular)	Total magnification
Low power	4X	10X	40X
Medium power	10X	10X	100X
High power	40X	10X	400X

The expression for denoting the magnification is:

$$\text{Magnification} = \frac{\text{Image size}}{\text{Object size}}$$

The magnification also bears important relationship of the optical tube length (OTL) of a microscope with the focal lengths of its objective and eyepiece. It may be denoted by the equation:

$$\text{Magnification} = \frac{OTL}{f \text{ objective}} \times \frac{200}{f \text{ eyepiece}}$$

Where,

OTL = Optical tube length (the distance between the back focal plane of an objective and focal plane of eyepiece. Contrary to the mechanical tube length, the optical tube length varies for each objective and can be calculated by multiplying the magnification of objective and its focal length).
f objective = Focal length of objective (in mm)
f eyepiece = Focal length of eyepiece (in mm

Note

The optical tube lengths for the common objective magnifications are depicted below.

Magnification of objectives	Focal length of objective		Optical tube length
× 5	25 mm	1"	125 mm
× 10	16 mm	2/3"	160 mm
× 40	4 mm	1/6"	160 mm
× 100	2 mm	1/12"	200 mm

Since in photomicrography, a camera is attached to the microscope system for obtaining an image projected on a sensitive film, the final magnification of a microscope depends upon three factors (Figs 20.2 and 20.3).

i. Camera bellow's length, i.e. the distance between the photographic plate and the top lens of the eyepiece.

ii. Magnification achieved by the objective.

iii. Magnification achieved by the eyepiece.

Depth of Focus (Field)

If an object, only a few feet from the camera, is focused, everything in front and back is out of focus. The sum of distance on both sides of plane of sharp focused image is called the *depth of focus*. Lenses with large apertures possess correspondingly less depth of focus. When greater depth of focus is desired (i.e. everything in front and back of the main object is required to be in focus), one has to do

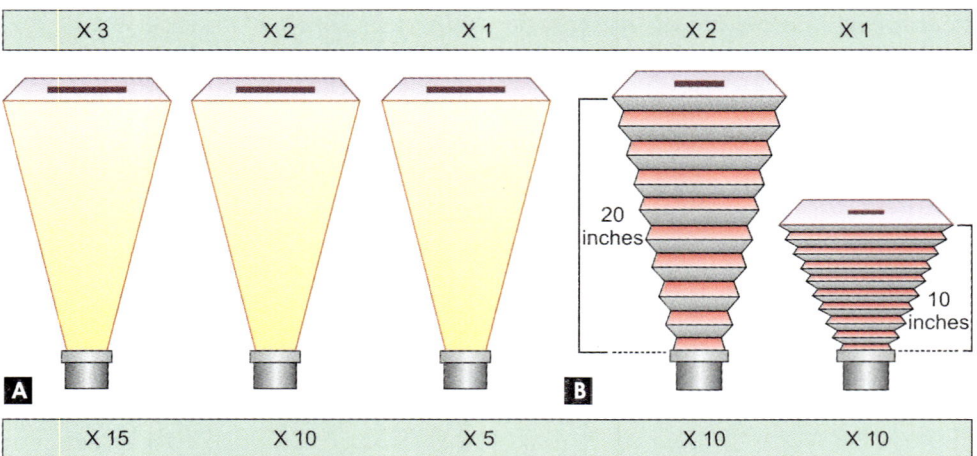

Fig. 20.2: Diagrammatic representation of factors affecting magnification of a photomicrograph: (A) Change of eyepieces of different magnification; and (B) varying bellow extension length in photomicrograph equipment. For the effect of variation of the objective power on the magnification, refer to Fig. 20.3.

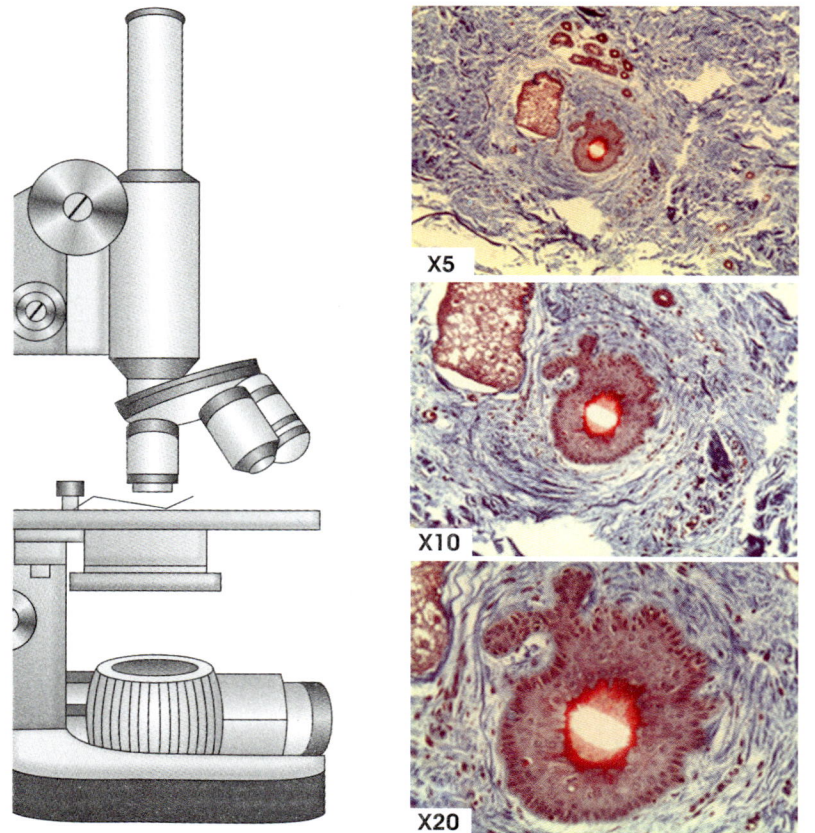

Fig. 20.3: Relationship between objective lenses of a microscope on the magnification (here a portion of skin is photomicrographed with different objective lenses).

stopping down—by closing the diaphragm. In other words, one has to work with a lower aperture. In photomicrography, the optimum depth of focus can be best secured by high-power eyepieces and long-bellow camera extensions in such a way that the thickness in the object space (section viewed) is in acceptably sharp focus at any one time. The depth of field varies very much according to the numerical aperture (NA) of the objective lens.

An objective with a lower NA has a greater *working distance* (between the top of the covers lip and the front lens of objective).

The working distance goes on decreasing with higher powers of objectives (Fig. 20.4). The refractive index of the mountant and the wavelength of the light also affect the depth of field.

Different magnifications of objectives along with related parameters are shown below:

Objective magnification	Focal length (in mm)	NA	Field diameter (in mm)	Working distance	Depth of focus (µm)
X5	25	0.15	3.2	23	20
X10	16	0.28	1.6	07	10
X40	04	0.75	0.4	01	02
X100	02	1.30	0.15	0.25	01

 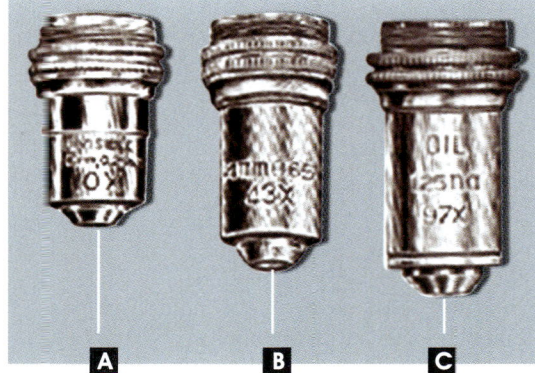

Fig. 20.4: Working distances of objective lenses of different power. Notice that this distance is indirectly proportional to the power of the lens from the plane of section under examination (represented with a brown line): (A) Maximum with low power (5X), (B) medium with high power (40X), and (C) minimum with oil-immersion (100X) objective. The working distance in all cases has been exaggerated from actual for clarification.

SUMMARY

Photography

The term *photography* is used for making a picture formed by chemical reaction of light on a sensitive film. A *photomicrograph* or a *digital image* is taken through a microscope; opposed to a *macrographic image*, which is at a scale that is visible to the naked eye.

The following points regarding photography need attention for getting excellent results.

- Water supply—impurities in the ordinary 'water-supply' such as lime, magnesia, iron, carbon dioxide, and oxygen should be removed as far as possible because:
  - i. Lime and magnesia cause turbidity
  - ii. Oxygen oxidizes the developer prematurely
- Purity of chemicals used—instead of commercial, or 'technical' quality, only pure chemicals should be used.

- Accessory apparatus—photographic chemicals (mainly solutions, etc.) **must** be kept in wide mouth glass bottles, undamaged enamel jugs or buckets, and stainless steel or plastic containers—**never** in containers made up of brass, iron, aluminum, and zinc metals.
- Temperature of solutions—should not be allowed to rise above 30 to 40° C (85 to 100° F) because at about 50° C (120° F) there is a danger of oxidation of the 'developing solutions' by atmospheric oxygen.
- Consistent activity of developers—best insured by making developer at least one day before it is required and keeping it ready for use in a fairly large and air-tight glass vessel. This minimizes the oxygen already dissolved in the solution, and the activity of the developer brought to constant. Also, any precipitate that may form will sink to the bottom of the vessel and can be filtered off before use.

Photomicrography

The term photomicrography means a procedure to take a *photomicrograph*—a word derived from two words: (i) *Photograph* which is used to designate a picture produced through the instrumentality of light; and (ii) *micrograph*, which means an enlarged reproduction of a minute object. The following terms are related to photomicrography:

- **Field and projected image diameters**
  i. *Field diameter* is the actual diameter of the illuminated circle seen when looking into a microscope through the top lens of an eyepiece.
  ii. *Projected image diameter* is the diameter of the circular image as projected on the focusing screen.
- **Camera extension**—*camera extension* is the distance from the eyepiece to the focusing screen

- **Resolution** and **magnification**

Resolution

The term *resolution* of a microscope is its ability to separate minute details that are very close together. This is usually measured as the *minimum distance* between two minutes circular objects of high contrast at which they can still be separated. The minimum distance is called *limit of resolution*, which is inversely related to the instrument's maximal capacity to resolve—the **resolving power**.

The smaller the limit of resolution, the higher the resolving power.

The *resolving power* depends primarily on the following factors (also refer to Chapter 14 for factors influencing resolution of a light microscope):

Numerical aperture (NA) mainly of the objective and to some extent also of the microscope condenser. It is regarded as the minimum distance between the two points, whose images can be seen as distinct from each other. The approximate resolving power of microscope can be expressed by formula:

$$R = \frac{NA}{0.61 \times \lambda}$$

Where,

R = Resolving power (in nm)
λ = Wavelength (in nm) of light
NA = Numerical aperture of objective

For increasing the resolving power, NA should be maximum; and wavelength minimum. The wavelength of light, usually measured in meters or Ångströms (1 Å = 10–10 m); the frequency at which light waves pass by is measured in units of per seconds (1/s).

Magnification

The term *magnification* of microscope means the number of times larger (in linear dimension) an image is over the object. It is a function of both the objective lens and the eyepiece.

The expression for denoting the magnification is:

$$Magnification = \frac{Image\ size}{Object\ size}$$

Since in photomicrography, a camera is attached to the microscope system for obtaining an image projected on a sensitive film, the final magnification of a microscope depends upon three factors:

1. Camera bellow's length
2. Magnification by objective
3. Magnification by eyepiece

- **Depth of focus (field)**

The sum of distance on both sides of plane of sharp focused image is called the *depth of focus*. Lenses with large apertures possess correspondingly less depth of focus. When greater depth of focus is desired, one has to do *stopping down*—by closing the diaphragm.

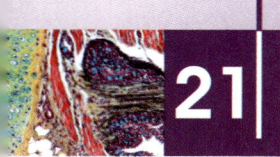

21

Photomicrography: Introduction, Methods and Techniques

WHAT IS PHOTOMICROGRAPHY?

Photomicrography is photography through a compound microscope. The range of total magnification is usually from 10X to 1000X. Of course, total magnification will depend on the magnification of the optical setup, on the size of the image sensor and on the size of the paper print (or on the size of the image displayed on a computer screen). A very important aspect of photomicrography is the illumination, which includes the type of light source, collector lenses, condenser, and correction filters.

Sometimes the term "microphotography" is mistakenly taken in this context. Micro-photography is the production of a very small photograph. Two important applications of microphotography is the creation of computer chips and was the creation of microdots to hide information. Written secret messages were embedded inside a dot the size of speck of dirt. Microdots were invented in the 1920s by Emanuel Goldberg. A photograph made through a microscope. Also called

microphotograph. To photograph (an object) through a microscope

There are many different setups used for photomicrography. As already mentioned, the main focus of this text is about the combination of a consumer grade digital camera and a compound microscope. But the principles apply also for digital cameras, which do not have a non-removable lens.

The secret of good photomicrography is nine-tenths good microscopy and one-tenth good photography. The technique of photo-micrography consists of two distinct parts.

Operation of Photomicrographic Apparatus

The aim is aligning of the microscope in position; then aligning the light source, either to project on the center of the microscope mirror, or in case of illuminators designed to project the light vertically into the condenser, to ensure that they are in line. The primary requisite of this step is the securing of correct optical alignment of all parts from the center of light source to the center of the camera ground glass.

Purely Photographic Work

This consists of mainly dark room processes, and the general routine conditions which need to be satisfied for the photographic work (like freedom from vibrations in the room and freedom from the dust and from chemical fumes specially acids) should be obtained. Proximity of the dark room to the room where photomicrographic setting of microscope, etc. is done must be preferred.

Brief Historical Account

Historically, **Thomas Wedgwood** (1771–1805) was the first to propose the concept of photomicrography by stating that the image of small objects, produced by means of the *solar microscope*, may be copied without difficulty on a prepared paper. This will probably be a useful application of the method; that it may be employed successfully. However, it is necessary that the paper be placed at but a small distance from the lens.

A solar microscope forms the image of an object that has been illuminated by sunlight and observed through the objective lens of a microscope. This image is projected onto a white screen (e.g. a wall) or recorded through a photographic device. It is believed that **William Henry Fox Talbot** (1800–1877) created the first photomicrograph using a solar microscope.

Since 1834, Talbot used small cameras to record images on photographic paper. His all photomicrographs were at magnifications below 20X. An image of the transverse section of a stem recorded in 1839 clearly showed the low magnification he was working with (image available on the Internet).

Worth mentioning is the work by **Joseph Bancroft Reade** (1801–1870) who produced a series of satisfactory photomicrographs in 1837. It is possible that **Reade** was one of the first persons to successfully record a "fixed" photomicrograph using a solar microscope.

In 1893, **August Köhler** (1866–1948) published an important article about "improved illumination technique".

Already in 1904, Köhler had used short wavelengths to observe fluorescence. This work led four years later to the introduction of the first fluorescence microscope and opened new possibilities for creating photomicrographs of fluorescent signals. Before the invention of the fluorescence microscope, ultraviolet radiation was used to increase the resolution but not to intentionally excite fluorophores. The earlier photomicrographs with long wavelength radiation appear to have been made by Köhler around 1912–13.

Optical Systems in Photomicrography

For photomicrography with light microscope, two types of optical systems are available:
  i. With finite tube length (in earlier makes)
 ii. With infinity-corrected tube length (newer makes)

The types of optical system can be looked upon by the **makings on the objectives** (Table 21.1).

Table 21.1	Interpretation of the kind of optical system by taking microscope markings
Marking on objective	**Interpretation of the optical system**
160/0.17	• Finite optical system • No 'tube-lens' between the objective and the intermediate image (*vide infra*)
Mechanical tube length of 160 mm (160 mm, 170 mm and 210 mm were common tube)	• Infinity-corrected optical system provides an internal tube lens (*vide infra*); and the intermediate image can be considered "**finished**" for all practical purposes. No addition/correction need to be made to improve the image quality.

The compound microscope objectives, with both the finite and infinity-corrected tube lengths, form a *real*, circular **intermediate image** of the section located in the "space" just below (approximately 10 mm) the edge of the eyepiece tube.

There is however one very crucial difference. With infinity optical systems there is an additional lens inside the microscope body between the objective and the intermediate image. It is most commonly known as the tube lens.

SUITABLE MATERIAL TO BE PHOTOGRAPHED

Mostly the histological sections from any biological sample are prepared without any thought to future photomicrography. The sections, obtained and stained by routine procedures, sometimes do not fulfill the 'ideal' conditions required for photomicrography to get a high quality photomicrograph. Such sections, therefore, are only of a little use when best results are expected. Hence, the following points deserve special attention for better results.

i. Sections should be properly *flattened* on slides. They must be of *uniform thickness* and optimal *staining intensity*.

ii. When low-power photomicrography is planned, the thickness of sections should range between 12 and 15 μm. However, with the use of high-power or oil-immersion objectives, thinner sections of 5 μm or less thickness are required.

iii. The common practice of keeping a section of average thickness for a longer time in a stain should always be discouraged, as it is by no means critical. Instead, the desired staining intensity should be regulated by the thickness of sections.

iv. Care should be taken to use a *minimum amount of mountant* under the cover slip over the stained section. Any excess of the mounting medium unnecessarily increases the distance between the section and the objective lens, thereby decreasing the working distance of high-power and oil-immersion objectives.

v. The drying of slides with sections should be *slow* preferably at room temperature. Unless some urgency is required, haste in this process should always be avoided.

CAMERAS USED FOR PHOTOMICROGRAPHY

Cameras used for photomicrography fall into *two* broad groups: (i) **Cameras with a fixed attached lens** form a *virtual image* from the microscope, such as the formed above a regular viewing eyepiece; (ii) **Cameras with no attached lens** have a direct access to the film or sensor. They form a *real image* that can either be "projected" or can be placed directly onto the film/sensor. Cameras in this group have variable film/sensor sizes. Some small "C-mount" lens-less *digital cameras* have sensors as small as 3×4 mm, while film commonly used may be up to 4×5 inches in dimension.

A **virtual image** is one that cannot be projected. It requires a lens (like the lens in your eye) to form, and then project, a real image onto the light sensing material—your retina, film, or a camera sensor. The place we encounter a virtual image in a microscope is directly above the viewing eyepiece(s). The eyepiece creates a virtual image made of parallel rays and formed at infinity. The additional lens of the eye, or a camera, is needed to produce a real image from this virtual image.

A **real image** is one that can be placed directly upon a piece of film, or a camera sensor. If we place a screen at the appropriate distance in front of the lens of the projector, the image that previously existed in "space" is now placed upon the screen surface and can be seen. We are dealing with a real image. In the microscope, a real image of the subject is formed by the objective.

Cameras Taking 35 mm Films

These days the 35 mm format will be, by far, the most common film size, with frame dimensions of 24 × 36 mm. A single lens reflex camera (SLR) can be used either with, or without an attached lens.

Small attached cameras taking 35 mm film (Figs 21.1 and 21.2) have several advantages but also certain disadvantages and limitations.

Advantages

1. Economy of operation—cost per exposure is extremely low.
2. Space required for the entire equipment is small.
3. The camera can be used for photographing of living micro-organisms and others according to one's choice.
4. Roll-film cameras are both convenient and rapid when a large series of photomicrographs in sequence, or under identical conditions, is required.
5. By far the most inexpensive for those wishing to take photomicrographs in color (Fig. 21.2).
6. The cameras are ideal for monochrome or color transparencies.

Fig. 21.1: Nikon dark box m-35 S photomicrographic microscope camera body and adapters used in photomicrography.

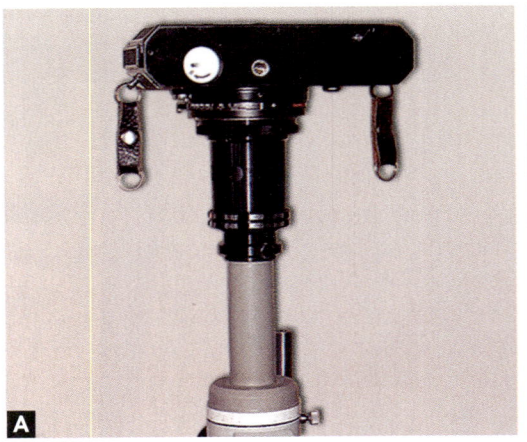
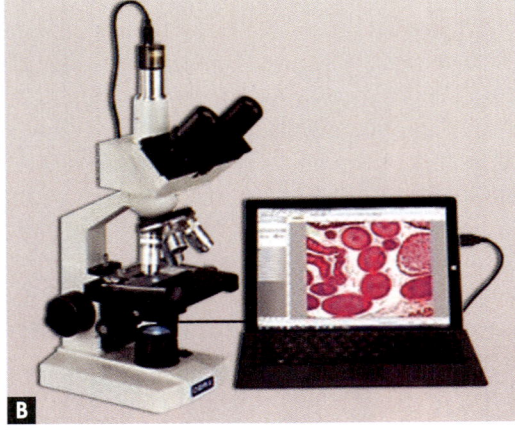

Fig. 21.2: (A) Nikon 35 mm camera body fitted with adapter at top of a microscope after removing the eyepiece, (B) OMAX 40X–2500X LED Digital Lab Trinocular Compound Microscope with 5 MP camera.

Disadvantages

1. It is not practical to take, develop, and study a single negative, as can be done when plates or cut-films are used. Preparation of a single negative is possible by wasting some film and then reloading the unused portion of the film.

2. Because of the small size of the negative produced, an enlarger is required, the cost of which should be included in the total outlay required, for making the final prints.

3. Prints made by enlargement have a consequent loss of definition.

4. A small picture also necessitates rapid exposures. Fast exposures are ideal under ordinary photographic conditions, but in photomicrography such short exposures are not preferable and hence must be avoided.

5. Various exposures on a single roll cannot receive individual treatment as to the type of developer (soft or contrast) and time of development.

6. The type of emulsion and film characteristics cannot be changed from one picture to the next. The different preparations recorded on any given length of film each suffer from the restrictions imposed by the particular emulsion.

7. There is extreme lack of flexibility in the amount of magnification obtainable. This is due to the fixed projection distances. It can be compensated for, to some degree, by variation in the amount of enlargement used for the final print—but this may result in excessive enlargement.

8. Miniature cameras are unsuited for low-power macrographs.

9. From the psychological point of view, there is one final objection to the use of miniature cameras in photo micrographic work. This lies in the very cheapness of an exposure and the ease with which exposures can be made.

The result is that instead of trying to make each exposure as nearly perfect as it can be, one is tempted to take a dozen shots of each object under slightly varying conditions, in the hope that one, at least, out of the lot, "ought to be good".

Larger Attached Cameras

The large-size photo micrographic cameras are for use in either vertical or horizontal position, with any standard microscope and light source. When used in vertical position, the camera can be swung aside—by means of the inclination joint at the base—to permit direct observations through the eyepiece or for gross photography. The base is of heavy metal, size 10 × 13 inches; with platform and clamps for the microscope; and graduated optical bed 24 inches long carrying the camera parts. The camera model shown in Fig. 21.3 is supplied with double plate holder for 5 × 7 inch plates, with reducing kits for 3¼ × 4¼ and 4 × 5 inch plates. A shutter facilitates making exposures. The camera back accommodates the ground glass focusing screen with clear center, interchangeable with the plate holder. Some of the larger cameras (designated as Type K) are for use in vertical position only (Fig. 21.3).

Fig. 21.3: A vertical outfit for taking a photomicrograph. On the left is shown a camera with large bellow and a negative-carrier on the top; on the right a set up for attachment of a 35 mm camera is shown for comparison.

This type of camera has a projection distance of 250 mm, corresponding to the far-point distance of the human eye, so that it provides the same magnification photographically as the microscope gives when used visually. It consists of a tapered metal box about 9 inches long. The camera back is of standard form and takes 3¼ × 4¼ inch double plate holder, or film pack adaptor. Mounted at the lower aperture of the camera is an automatic shutter with speeds of 1/25, 1/50, and 1/100, bulb and time. A ground glass viewing screen is provided for focusing the image.

Advantages

1. They are more flexible.
2. Sheet films can be used for color work.
3. They are of advantage for low-power macrographs, for which the miniature cameras are unsuited.

PRELIMINARY TESTING OF EQUIPMENT

The very first essential in photomicrography is to have as perfect a microscopic image as possible, because the sensitive photographic plate merely receives what is presented to it, and unlike the eye has no power of accommodation for different depths of focus. A high quality photomicrograph is one that reproduces the microscopic images accurately and clearly. For having a good photomicrograph it is necessary to have (i) suitable material, (ii) efficient equipment, (iii) knowledge of the equipment, (iv) knowledge of the subject matter.

These enable the correct field to be selected for photography and important features to be portrayed to advantage. The presentation of a photomicrograph which includes the method for display and use of annotations, is equally important because although nothing can be done to transform a poor photomicrograph into a good one, the converse is unfortunately true. The preparation of material for microscopy and its subsequent photography are so allied that the closest cooperation is called for between the histology technician preparing the material, the photographer, and the researcher requiring the photomicrograph.

For standard first class routine photomicrography, one should depend only on well-designed equipment of stable construction. Any photomicrographic set-up essentially consists of three basic components: (i) *A source of illumination*, (ii) *a microscope*, and (iii) *a camera*. These components may be purchased as separate units; microscope with integral light source and a camera; or light source, microscope, and camera assembled in the separate unit. Ideally two camera bodies should be available, one taking 35 mm film and the other taking cut-films or photographic plates. Expense is not the one item to consider in the selection of the photomicrographic apparatus; neither is the size of photomicrograph which can be taken from a particular set-up; both advantages and disadvantages may be noticed in every commercial model, large and small, simple and complex. If the researcher is familiar with his or her own special problems, an intelligent choice of the photomicrographic equipment can then be made. The following points about each of the three basic components should be kept in mind.

The setting up of any photo micrographic apparatus is always required except in models where not only the source of illumination and camera but also the microscope is an integral part of the system of apparatus. The aims are: (i) The alignment of the microscope with the optical axis of the camera, (ii) its permanent fixation in position, followed by (iii) the alignment of the light source, either to project on the center of the mirror, or in case of illuminators designed to project the light vertically into the condenser to ensure that these are in the line. The primary requisite is to secure the correct optical alignment of all parts from the center of light source to the center of the camera ground glass.

Euscope (*eu-* good + G. *skopeô*, to view) is an instrument for showing on a screen an enlarged image from a microscope. It is a device for use with any standard laboratory microscope, preferably with substage condenser; suitable for examination of microscopic fields in enlarged dimensions projected on a white opaque screen. Made from a natural and comfortable position with both eyes a euscope eliminates a large part of eyestrain and fatigue that usually accompany continued work with a microscope (Fig. 21.4). For photo micrographic use, a photographic or paper holder, permitting the use of 4 × 5 inch negatives, replaces the white opaque screen at the end of euscope. A felt covered shield, placed in the eyepiece end of the viewing box after the image is focused, keeps out all foreign rays.

Where large areas are to be photomicrographed, at low magnification and without an eyepiece, *micro tessar lenses* (Fig. 21.5) are used. These lenses cover an angular field of approximately 55°. The micro tessar lenses are furnished in a regular photographic lens mount for use on the front board or shutter of the camera. These lenses are swifter and quieter in auto focus operations compared to the micro motors. Often the focus acquisition in video is better on these lenses and hence

Fig. 21.5: Micro tessar lenses. A Carl Zeiss Jena Tessar 50 mm f2.8 standard lens + caps, Hood & Filter–M42 fit is shown in the lower part.

they are preferred over the normal lenses having a micro motor driven apparatus.

Photomicrography with Horizontal Set-Up

The center of the camera ground glass constitutes one end of the optic axis. This is naturally fixed end. The camera ground glass should be placed at the distal end of adjustable movement, i.e. the bellows fully extended (Fig. 21.6). Next the lamp should be set up at the opposite end. At this stage, if the lamp be lighted and its condenser focused along the optic axis, it should project a circle of light symmetrically around the diagonal cross made at the center of the ground glass. Now, one is ready to insert the microscope into the system. The microscope is bent over into its horizontal position and the mirror removed. Objective, eyepiece, and substage condenser are omitted, but the iris diaphragm of the substage is allowed to remain. In case a pinhole eyepiece cap is not available, a substitute is of opaque cardboard with a pinhole in the exact center should be fitted into the eyepiece tube. Now the substage diaphragm is closed to a pinhole. The position of the microscope of its sole-plate must now be adjusted until as light from the light source, passing through its condenser,

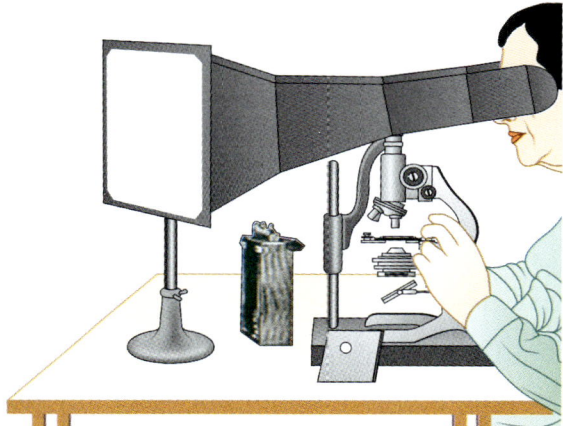

Fig. 21.4: A euscope minimizes eyestrain during prolonged working with a microscope.

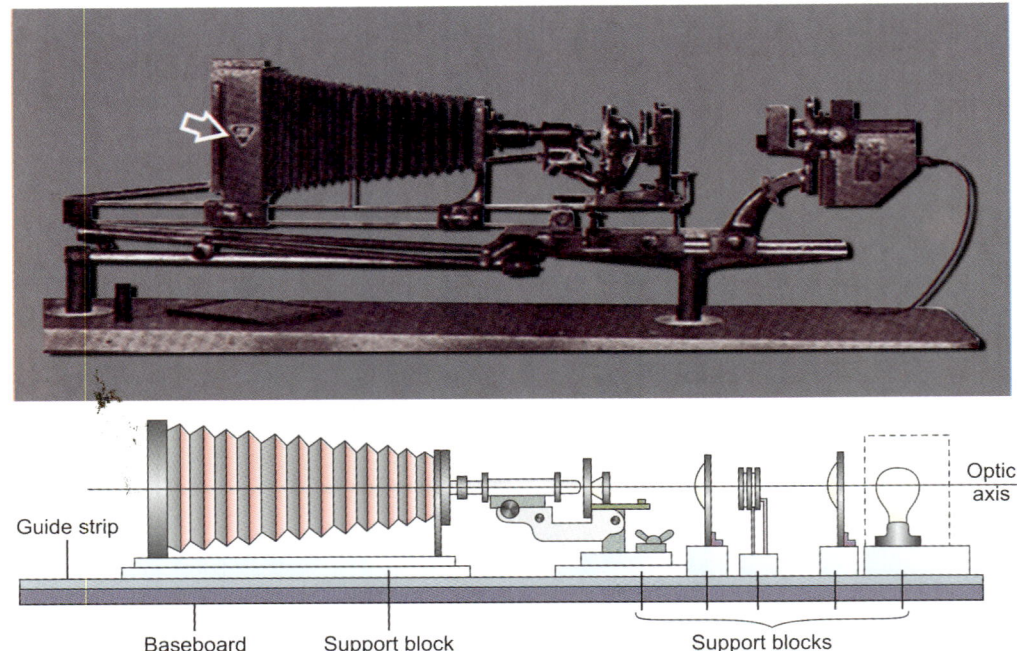

Guide strip

Baseboard   Support block   Support blocks   Optic axis

Fig. 21.6: A horizontal set-up suitable for photomicrography using photographic sheet-films in cassettes (arrow).

then through the two pinhole diaphragms of the microscope, registers a small disc of light on the cross lines of the focusing screen. Not until this is accomplished is the alignment perfect. When this condition is met, it can be assumed that the quality of the optical parts of the microscope is such that the final alignment is assured, except for the minute adjustment which must be made to align each objective with the condenser.

Photomicrography with Vertical Set-Up

In setting up of a vertical apparatus alignment does not differ in theory from that of the horizontal outfit. The complication which makes the alignment of the vertical apparatus somewhat harder is that there are two optic axes at an angle to each other. The point common to both is the center of the microscope mirror. The angle between two axes may be a right angle or an acute angle, differing with various makes of the equipment. In some substandard models of microscopes, there is a

poor designing of the trunnion pivots of the mirror so that the pivots do not lie in the optical axis of the microscope (Fig. 17.1D). Thus the reflected light cannot be directed at 90° in the path of the optical axis of the microscope due to the poor designing. In standard equipment, however, the pivots lie at the intersection of light beam to the microscope with the optical axis. A poor condition results when the reflected beam of light cannot lie in the optical axis. However, the best condition with such models of microscopes is achieved when the trunnion pivots lie in the axis of microscope not at the point of intersection of axes but lie below the axis of light beam, to locate the silvered surface of the mirror at the intersection of the two axes. An ideal condition in the operation of mirror with the microscope in vertical alignment is shown in Fig. 17.1A where trunnion pivots lie in the plane of the mirror surface at the intersection point. The front surface of the mirror makes an angle of 45° with the optical axis of the microscope which

Fig. 21.7: Vertical attachments shown on the left and right sides are fitted in a triocular light microscope (in the center) after removal of eyepiece. The light reaches along the path shown by the red arrow.

makes a right angle with the light beam to the microscope axis. For a vertical outfit (Fig. 21.7) an opaque but light colored card diaphragm is to be put over the mirror, the hole being ¼" to ¾" in diameter. The light passes through the two pinhole diaphragms of the microscope, to the center of the ground glass of focusing screen.

SECURING OPTICAL ALIGNMENT

First make sure that a small diagonal cross is marked in the center of the ground glass. The optical alignments are ordinarily of two types depending upon whether the apparatus is of the *vertical* or *horizontal* type. Correct alignment is much easier to secure with a horizontal camera and microscope than with the vertical set up. Therefore, an understanding of the steps in alignment of a horizontal set up will be described first and that of a vertical outfit later.

After securing primary optical alignment, the next step is to assemble a low-power (10X) objective, a low-power eyepiece, and the substage condenser on the microscope. The camera is then moved out of the way so that visual work may be done. A well-stained

section is placed on the stage and the objective focused upon it. Those with microscope experience will have an advantage here, but let us suppose that the student is just starting and knows nothing about the operation of the microscope.

With the object in place and the condenser racked up until it nearly touches the slide, the diaphragm being opened wide, and the object will be brightly illuminated when the lamp is turned on. After adjusting the light, the eye is placed over the eyepiece and objective focused on the object.

We can ask a question here. Are we now ready to move the camera into place, turn the light on to full brilliancy, focus the image on the ground glass, and expose for a picture? Many persons who have been taking photo-micrographs for a long time might answer "yes", or "why not", but the answer is decidedly "no", because we have not yet assured ourselves that the light is critical.

Critical Illumination and Critical Image

An object is said to be *illuminated critically* when it is placed at the apex of a solid axial cone of light (Fig. 21.8), the aperture of which is not

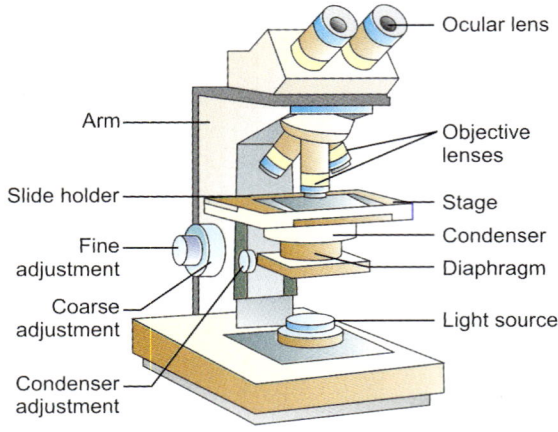

Ocular lens

Arm

Objective lenses

Slide holder

Stage

Fine adjustment

Condenser

Diaphragm

Coarse adjustment

Light source

Condenser adjustment

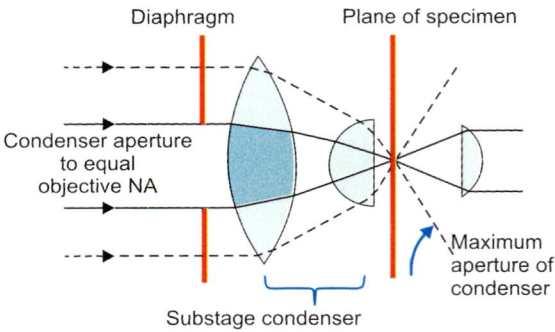

Diaphragm   Plane of specimen

Condenser aperture to equal objective NA

Maximum aperture of condenser

Substage condenser

Fig. 21.8: Principles of critical illumination. Note that the image of the light source is formed in the plane of the object.

less than three quarters of the numerical aperture of the observing objective lens of the microscope. The image of the light source is formed in the plane of the object. The image of an object is *critical* when it is obtained by means of an objective of fine quality which has been placed in correct adjustment for that object by its screw collar, or by the alteration of the tube length, after the illumination is set *critical*.

Setting-up Critical Illumination

Whenever perfect results in the microscope are desired, or in other words, whenever we want to get best out of the instrument, it is necessary to set up critical illumination. The following

steps must be carefully practiced to fulfill the undernoted criteria.

   i. Illuminating rays should be symmetrically disposed about the optical axis of the microscope and they should coincide in area with the back lens of the objective.

  ii. The diameter of this area should cover the entire field being visualized (photographed).

 iii. Illumination should possess uniform intensity throughout.

Steps for Setting-up a Photomicrography Camera

1. Half close the field (iris) diaphragm in front of the light source.
2. Place a slide with stained section on the microscope stage.
3. Rack the substage condenser up until it nearly touches the slide.
4. Open the substage diaphragm to its full extent.
5. Focus the section with the objective of choice.
6. Close the substage diaphragm.
7. Move the lamp-condenser back and forth until a sharp image of the light source is formed on the substage diaphragm.
8. Open the substage diaphragm to its full extent.
9. Check that the section is still in sharp focus.
10. Move the substage condenser down until the circle of light seen illuminated by the field diaphragm is in focus.
11. Center the illuminated circle with the help of centering screws on the substage condenser.
12. Adjust the field diaphragm so that the circle of light is just larger than the area to be photographed.
13. Remove the eyepiece from the system and look down the microscope at the back lens of objective.

14. Close the substage diaphragm until its diameter (circle of illumination which is clearly seen with the unaided eye) is seen to coincide with the diameter of the back lens of the objective. The movements of the diaphragm can be distinctly appreciated as the circle of illumination increases and decreases.

15. Replace the eyepiece in its position.

16. The light is now critical and critical illumination has been achieved.

Note

i. Steps 6 to 9 (both inclusive) may be omitted if the microscope possesses an integral illumination system;

ii. Although moving the condenser farther away from the slide (i.e. the condenser is lowered) will be found to increase the circle of illumination; and this method is often followed by those ignorant of the correct operation of the microscope, it should never be resorted to, as it immediately destroys the critical set-up we have taken so much pains to produce.

After the critical illumination has been attained replacing the eyepiece will reveal the best possible resolution of the object. The entire field of view is probably not entirely covered, even with the field diaphragm opened to its full extent.

A very important thing to be noted is that the critical illumination remains critical only with that objective with which section has been focused while attaining the critical set-up. As soon as any other objective is swung into the system, the set-up of light remains, by no means, critical since the ordinary nose-pieces are not centering. Hence, as such, the set-up is not suitable for photomicrography. The lack of centrality does result from non-alignment of the objective and substage condenser. After centering has been accomplished with one objective if another objective is swung into the system, only by rare chance the second objective may be found to have the circle of illumination lying in the exact center of its field. Hence the steps are to be repeated for each objective used.

The original 35 mm transparency slide copier ... the sort that projected its imagery onto a 36 mm × 24 mm film plane in an SLR works just as well with something like a Nikon D700 FX format camera since the optical arrangement suits the full framed digital camera as did the older film SLR. For the DX format digital camera with a sensor about half its size the old copier does not work since it projects an image which is far too large.

SUMMARY

What is Photomicrography

Photomicrography is photography through a compound microscope. Sometimes, the term "microphotography" is mistakenly used in this context. **William Henry Fox Talbot** (1800–1877) created the first photomicrograph using a solar microscope. **Joseph Bancroft Reade** (1801–1870) produced a series of satisfactory photomicrographs in 1837. In 1893, **August Köhler** (1866–1948) published an article about "improved illumination technique".

Two types of optical systems are available: (i) with finite tube length (in earlier makes), (ii) with infinity-corrected tube length (newer makes)

Cameras used in Photomicrography

Two types of cameras may be used for taking a photomicrograph: (i) Cameras taking 35 mm films, and (ii) larger attached cameras

Cameras taking 35 mm film

These cameras have several advantages but disadvantages as well.

Advantages

- Extremely low cost per exposure
- Convenient and rapid when a large series of photomicrographs in sequence
- Photographing of living micro-organisms is possible
- Ideal for monochrome or colour transparencies

Disadvantages

- Not practical to take, develop, and study a single negative
- Due to small size of the negative produced, an enlarger is required
- Short exposures are not preferable and hence must be avoided

- Type of emulsion and film characteristics cannot be changed from one picture to the next.
- Due to the fixed projection distance extreme lack of flexibility results in the amount of magnification obtainable.

Larger Attached Cameras

Large size cameras are for use in either vertical or horizontal position (vide infra).

Suitable Material to be Photographed

- Sections should be properly *flattened* on slides; of *uniform thickness*; and optimal *staining*
- Thickness of sections should range between 12 and 15 μm for low-power; and 5 μm or less for high-power and oil-immersion objectives.
- *Minimum amount of mountant* over the stained section should be put under the cover slip.
- Sections should have been dried *slow* (preferably at room temperature).

Preliminary Testing of Equipment

- Horizontal set-up apparatus (*see* details on page 311)
- Vertical set-up apparatus

The camera back accommodates the ground glass focusing screen with clear center, interchangeable with the plate holder.

Securing Optical Alignment

Critical Illumination and Critical Image

An object is said to be *illuminated critically* when it is placed at the apex of a solid axial cone of light (Fig. 21.8), the aperture of which is not less than three quarters of the numerical aperture of the observing objective lens of the microscope.

The image is *critical* when it is obtained by means of an objective of fine quality which has been placed in correct adjustment for that object by its screw collar, or by the alteration of the tube length, after the illumination is set *critical*.

After the critical illumination has been attained replacing the eyepiece will reveal the best possible resolution of the section.

Set-up for Photomicrography Camera

See step-by-step procedure in the text (to avoid repetition).

22

Photographic Emulsions and Processing Solutions

The sensitive photographic black and white emulsions consist of minute *grains* (ranging from 0.1 to approximately 2 microns) of silver halide suspended in gelatin. **Photographic emulsion** is a *colloid* used in photography. Although customarily used in photographic literature, photographic emulsion is not a true *emulsion*, but a *suspension* of solid particles (silver halide) in a fluid (gelatin in solution). Most commonly, in silver-gelatin photography, it consists of silver halide crystals dispersed in gelatin. The emulsion is usually coated onto a substrate of glass, films, paper, or fabric. The suspension is supported on glass, film (Fig. 22.1), or paper. The halide used in making the negative materials is silver bromide, which contains small amounts of iodide. The halide employed for positive materials may consist of silver bromide, silver chloride or a mixture of the two. The proportions of these different salts may differ.

SPEED AND GRANULARITY

The speed or granularity of photographic emulsion refers to produce a given density.

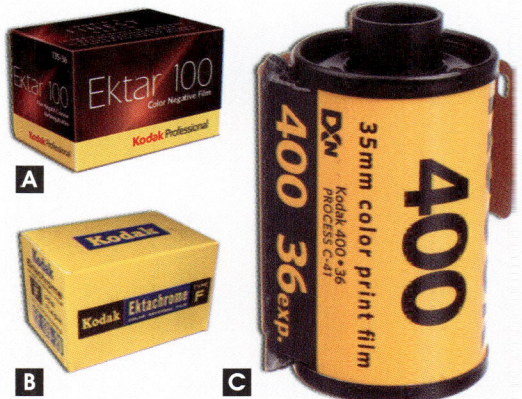

Fig. 22.1: Color photographic films: (A) Slow speed high-contrast 35 mm roll film, 36 exposure Kodak Professional Ektar 100 film; (B) Color transparency positive film; and (C) 35 mm color print film of 400 ASA.

Thus, an emulsion requiring more exposure to reach a given density will be slower in speed than an emulsion requiring less exposure for the same degree of blackening. Speeds for emulsions designed primarily for use with visible light are expressed as number on arbitrary but defined standard scales notably

DIN	ASA	DIN	ASA	DIN	ASA
Table 22.1	**Speeds of the photographic films**				
9	6	20	80	31	1000
10	8	21	100	32	1300
11	10	22	125	33	1600
12	12	23	160	34	2000
13	16	24	200	35	2600
14	20	25	250	36	3200
15	25	26	320	37	4000
16	32	27	400	38	5200
17	40	28	500	39	6400
18	50	29	650	40	8000
19	64	30	800	41	10400

the ASA (American Standards Association) and BS (British Standard) scales which coincide, and the DIN (Deutsche Industie Norme) scale. These scales are not relevant for electron exposures and they give no guide to the speed of the emulsion when exposed to electrons. The following chart (Table 22.1) may be useful for noting equivalent speeds of film emulsions available in the market.

The grain size of an emulsion is controlled by the alkalinity and temperature of the solution, amount of sulphite in it, and, of course, details of operating techniques in the process of development.

An ideal developer would produce a negative (or print) in which the density of each portion of the image would be in exact proportion to the intensity of the light producing that portion. It would also produce a silver deposit where the exposure had been very short; and would give a fine grain. It is important to learn that a *contrast* subject requires over-exposure and under-development. A *flat* subject requires under-exposure and over-development. If one follows these broad rules, he can be certain of a high average of results. But it is equally important to remember that development of a photographic emulsion cannot reveal an image where exposure (a combination of the light intensity and the length of time it is permitted to act on

the emulsion) is below a certain minimum. Of course, the minimum varying with the emulsion is characteristic of the illumination, and the developer used. However, few, if any, developers may be regarded as ideal ones.

It is to be noted that although halides of other metals may also serve the purpose of photographic emulsion, silver halides are chosen for: (i) their property for being able to be reduced by the action of light alone—although this is not practicable as too long an exposure would be required; and (ii) the reason that the image produced on the sensitive emulsion after being exposed to light remains invisible (latent image) until it is *developed* by the application of developing agent.

PROPER FILM LOADING IN CAMERA

Film must be trimmed to have a longer leader (Fig. 22.2), and exactly two perforations must show at the bottom. If you do not trim your film properly to provide a longer leader tongue, the film will get jammed in the gate and not load. Standard out-of-the-box film will not load properly until trimmed. If you do trim your film, it slides right into the film-socket and loads very easily. To trim it, I count back 13 perforations along the bottom, and cut it out with my Swiss Army knife. Be sure not to cut across a perforation, as the edge might get caught trying to load. Leave two perforations

Fig. 22.2: Photographic film Loading. Notice that the film must be trimmed to have approximately 10 cm long leader and exactly two perforations must show at bottom (*see* text).

pulled out of the canister out at the bottom as shown in Fig. 22.2, and you are good to go.

Beginners who do not follow these directions go through all sorts of dangerous antics, like sticking their fingers through the shutter, to try to load the film improperly. The reason one needs to make the leader longer is because the film has to slide past the film gate, with the pressure plate in place. The back does not swing open.

When you trim your film properly, there is no bottom edge that has to jump over the other film rail, which is what hangs-up people who do not read directions. When properly trimmed film is loaded, only after it is in place do you advance the full width of the film over the gate, so all works as intended. Now that you know this, you will be fine as you read the printed manual to load the film.

PRINCIPLES OF DEVELOPING AND FIXATION

The *developing* of photographic emulsions is a chemical process (Tables 22.2 to 22.9) in which the silver halide grains, which have been exposed to light, are reduced to metallic silver. A photographic emulsion which has been exposed to light produces a *latent image* of metallic silver from grains of silver halide present in emulsion prior to exposure. Application of a developing agent intensifies the *latent image*. This results in the blackening of the latent image, which now becomes a visible image. For the process of development—the chemical solution used is called a *developer*.

A relatively smaller proportion of 'metol' as opposed to 'hydroquinone' increases the time of development. In other words, the development is said to be slower. On the contrary, a larger quantity of 'metol' reduces the time of development, but there is a greater risk of *fog*. A faster development, (without larger quantity of 'metol') can be achieved by adding to a developer, alkalies (sodium carbonate; sodium or potassium hydroxide and borax). These chemicals are, therefore, called

Table 22.2	Chemicals with their formulas used in a photographic developer	
Ingredients	Examples	Chemical formulas
Reducer	Metol (Monomethyl-para amino phenol sulphate)	$OH.C_6H_4(NH.CH_3).\frac{1}{2} H_2SO_4$
	Hydroquinone	$C_6H_4(OH)_2$
Activator or accelerator	Sodium carbonate or	Na_2CO_3
	Sodium hydroxide or	NaOH
	Potassium hydroxide or	KOH
	Borax (sodium tetraborate)	$Na_2B_4O_7.10H_2O$
Preservative	Sodium sulphite or	$Na_2SO_3.7H_2O$
	Potassium metabisulphite	$K_2S_2O_4$
Restrainer	Potassium bromide	KBr

Table 22.3	Ingredients of three kinds of negative developers		
Ingredients	**Agfa-42** normal contrast	**Agfa-40** high contrast	**Agfa-44** soft-working (fine grain)
Metol	0.8 g	1.5 g	1.5 g
Sodium sulphite (anhydrous)	45.0 g	18.0 g	80.0 g
Hydroquinone	1.2 g	2.5 g	3.0 g
Potassium carbonate	–	18.0 g	–
Sodium carbonate (anhydrous)	8.0 g	–	–
Potassium metabisulphite	4.0 g	–	–
Borax	–	–	3.0 g
Potassium bromide	1.0 g	1.0 g	0.5 g
Development time at 18°C	**10–12 minutes**	**4–5 minutes**	**18–20 minutes**

Table 22.4	Ingredients of three kinds of photographic paper developers		
Ingredients	**Agfa-100** standard contrast	**Agfa-105** soft-working	**Agfa-108** high contrast
Metol	1.0 g	3.0 g	5.0 g
Sodium sulphite (anhydrous)	13.0 g	15.0 g	40.0 g
Hydroquinone	3.0 g	–	6.0 g
Sodium carbonate (anhydrous)	26.0 g	–	–
Potassium carbonate	–	15.0 g	40.0 g
Potassium bromide	1.0 g	0.4 g	2.0 g
Development time at 18°C	**1½–2 minutes**	**1½ minutes**	**2 minutes**

Table 22.5	Developer for transparent paper sheets
Ingredients	**Agfa-110**
Caustic potash	26.0 g
Sodium sulphite (anhydrous)	100.0 g
Hydroquinone	60.0 g
Potassium bromide	3.0 g
Development time at 18°C	**1 minute**
Fixing time	**5 minutes** in acid-fixing bath
Washing	**10 minutes** in running tap-water. Avoid prolonged washing

accelerators or *activators*. Since developing agents have an affinity for oxygen, addition of *preservative* prevents absorption of oxygen from atmosphere. Sodium sulphite and potassium metabisulphite either alone or with sodium sulphite are used in a developer as preservatives. With larger amount of sodium sulphite used—the developing solution keeps better. The final ingredient of most developing solutions is called a *restrainer*. The main function of a restrainer is prevention of development of unexposed silver halide grains. Commonly used restrainer is potassium bromide. It so controls the process of reduction that only exposed silver halide grains are reduced, thus the blackening produced is maximum in the region the sensitive emulsion has been exposed most.

Table 22.6	Formulas of different stop baths
Agfa-200 acetic acid stop-bath	
Water	1000 ml
Glacial acetic acid	20 ml
	(may be 30)
Agfa-201 potassium metabisulphite stop-bath	
Water	1000 ml
Potassium metabisulphite	40 g
Agfa-203 stop-bath	
Water	1000 ml
Sodium sulphite	100 g
Glacial acetic acid	20 ml

Table 22.7	Ingredients of fixing baths for photographic paper and films	
Ingredients	**Agfa-300 (for papers)**	**Agfa-301 (for films)**
Sodium thio-sulphate (Hypo)	200 g	250 g
Potassium meta-bisulphite	20 g	15 g
Water up to	1000 ml	1000 ml
Fixation time	**5–10 minutes**	**10–15 minutes**

Table 22.8	Hardening baths for photographic prints/enlargements	
Ingredients	**Agfa-401 (for mild hardering)**	**Agfa-402 (for extreme hardening)**
Water	1000 ml	1000 ml
Alcohol	–	500 ml
Formaldehyde (35%)	120 ml	120 ml
Time for immersion	**5–10 minutes**	**1–10 minutes**

Table 22.9	Hardening baths for photographic films/plates	
Ingredients	**Agfa-406 (for mild hardering)**	**Agfa-410 (for extreme hardening)**
Chrome alum	15 g	–
Sodium sulphite (*anhydrous*)	–	150 g
Potassium metabisulphite	15 g	–
Sodium carbonate	–	20 g
Formaldehyde (35%)	–	20 ml
Water up to	1000 ml	1000 ml
Time for immersion	**3–5 minutes**	**2–3 minutes**

Prevention of unexposed silver halide granules from developing inhibits what is called *chemical fog*. In the end, the developing process is finally 'stopped' by transferring the emulsion into another solution called the fixative, which allows no further development even on exposure to light.

DIFFERENT KINDS OF EMULSION DEVELOPERS

Although formulas of various developers used for the negative and paper emulsions are beyond the scope of this book, some are given here as a reference for those who do not have an easy access to the photography literature.

Developers for Photographic Negative Films and Plates

There are three kinds of negative developers for tank use. For preparing any one of them, the ingredients are dissolved, in the order given, in approximately 750 cc of tap-water, and then the volume of the solution is made up to one liter. Solution is filtered before use preferably in a glass container. The following table may be seen for the required specifications of negative development.

Developers for Photographic Papers

The paper developers commonly used for development of prints or enlargements to obtain pure black tones on Agfa lupex or Brovira papers and for other commercially available process photographic papers are, like the negative developers, for standard-contrast, high contrast, and soft-working results.

Each developer is prepared by dissolving the ingredients, in the given order, in approximately 750 cc of tap-water and the final volume made up to one liter.

For Agfastat transparent photographic paper sheets, a rapid-high-contrast developer (Agfa-110) may be prepared as per details given below in the usual manner.

Stop Baths

The stop-bath solutions, kept between the developer and fixer solutions, are recommended to avoid quick deterioration of fixing-baths from contamination with developers' ingredients. The following types of stop-baths are routinely used.

Fixing Baths

After an emulsion has been developed and rinsed in a stop-bath, the next step is to place it immediately in a fixing-bath. Formulas for making one liter of fixing-bath are as follows.

Note

Matt and semi-matt photographic papers should always be treated in a hardening-fixing-bath, to ensure that they will not stick to the blanket of the drying machine. The following bath for prints is useful.

Agfa-302 Hardening-fixing-bath

Preparation

Mix thoroughly the following to **freshly prepared** 10% solution of potash alum (by dissolving the alum in water by heating and then cooling the solution).

i.	Sodium sulphite (anhydrous)	7.5 g
ii.	Glacial acetic acid	12 ml
iii.	Agfa-300 fixing-bath	1000 ml

Hardening baths

These baths are useful for the protection of the photographic emulsions both for films and plates and for photographic papers. However, the formulae for films and papers are different and are therefore given separately.

NEGATIVE AFTER-TREATMENT

Sometimes after development, a negative emulsion seems to be a little under-exposed or over-exposed. To bring such emulsions close to the normal, negatives require the so-called *after-treatment*. Whereas under-exposed negatives need a treatment with *intensifiers*, if over-exposure is a problem *reducers* are helpful. There processes allow satisfactory prints to be made from negative either too dense or thin and lacking in intensity.

Intensifiers

Intensification builds up a greater metallic deposit, either of silver or some other element, on that already present, to provide greater intensity.

Agfa-601 Mercuric Chloride Intensifier

Solution A

Water	100 cc
Mercuric chloride	2 g

Solution B

Water	100 cc
Ammonia (25% solution)	10 cc

Procedure: The negative is treated in solution 'A' until the silver image is bleached white right through. After thorough washing it is then blackened in solution 'B', in which it should, however, not be left any longer than is necessary. Bleaching can also be effected in a 5 per cent solution of sodium sulphide or in any metol-hydroquinone developer for as long as needed.

Chromium Intensifier

A simple, easily controlled and satisfactory intensifier, this intensifier has met with increasing popularity, owing to ease and certainty of its operation, and the permanency of intensified image. It also permits a considerable degree of intensification.

Solution A

Water	500 cc
Potassium dichromate	25 g

Solution B
Hydrochloric acid (conc.)

Procedure: For use first bleach the negative in a solution containing solutions 'A' and 'B'. Then blacken in a developer. By varying the proportions of solutions 'A' and 'B', the desired degree of intensification may be achieved.

For moderate intensification		For strong intensification	
Solution 'A'	12.5 cc	Solution 'A'	12.5 cc
Solution 'B'	0.3 cc	Solution 'B'	1.5 cc
Water	100.0 cc	Water	100.0 cc

Reducers

Reduction is a chemical process which dissolves out a portion of silver deposit in the emulsion so as to make it less dense.

Farmer's Reducer

This is a cutting reducer used for lessening the density of heavy negatives and at the same time increasing their contrast.

Solution A

Sodium thiosulphate (Hypo)	240 g
Water to make	1000 cc

Solution B

Potassium ferricyanide	19 g
Water to make	1000 cc

Procedure: For use mix 1 part of solution 'B' and 4 parts of solution 'A' in 32 parts of water.

> **Note**
>
> Solution 'A' and 'B' should be stored separately and must be mixed immediately before use.

Flattening reducer

This is recommended for lessening the density and contrast of heavy negatives.

Solution A

Potassium ferricyanide	35 g
Potassium bromide	10 g
Water up to	1000 cc

Solution B Ansco-47 metol-hydroquinone developer

Water (52°)	750 cc
Metol	1.5 g
Sodium sulphite (anhydrous)	45.0 g
Sodium bisulphite	1.0 g
Hydroquinone	3.0 g
Sodium carbonate	0.8 g
Water to make	1000 cc

Procedure: Bleach in solution 'A' and after thorough washing redevelop to desired density and contrast in solution 'B' or any other negative developer, and fix and wash in usual manner. The entire process should be carried in subdued light. The *development time* at 20 degrees is 6 to 8 minutes.

SUMMARY

Speed and Granularity of Emulsions

Photographic emulsion is a *light*-sensitive *colloid* used in film-based photography. Although customarily used in photographic literature, photographic emulsion is not a true *emulsion*, but a *suspension* of solid particles (silver halide) in a fluid (gelatin in solution). Most commonly, in silver-gelatin photography, it consists of silver halide crystals dispersed in gelatin. The emulsion is usually coated onto a substrate of glass, films, paper, or fabric. Photographic black and white emulsions consist of minute *grains* (ranging 0.1–2 μm) of silver halide suspended in gelatin. The speed or granularity of photographic emulsion refers to produce a given density. An emulsion requiring more exposure to reach a given density will be slower in speed than an emulsion requiring less exposure for the same degree of blackening.

Speeds for emulsions designed primarily for use with visible light are expressed as number on arbitrary but defined standard scales notably the **ASA** (American Standards Association) and BS (British Standard) scales which coincide, and the **DIN** (Deutsche Industie Norme) scale. *See* text for ASA/DIN chart.

The grain size of an emulsion is controlled by the alkalinity and temperature of the solution, amount of sulphite in it, and, of course, details of operating techniques in the process of development.

Proper Film Loading in Camera

Any photographic roll film must be trimmed to have a longer leader and must show, at the bottom, with exactly two perforations.

Principles of Developing and Fixation

The *developing* of photographic emulsions is a chemical process in which the silver halide grains, which have been exposed to light, are reduced to metallic silver. A photographic emulsion which has been exposed to light produces a *latent image* (of metallic silver) from grains of silver halide present in emulsion prior to exposure. Application of a developing agent intensifies this *latent image*—resulting in the blackening of the latent image and converting it to a *visible image*. For the process of development—the chemical solution used is called a *developer*. Finally the developing process is 'stopped' by transferring the emulsion into another solution called the *fixative* that allows no further development even after exposure to light.

Different Kinds of Emulsion Developers

For selection of 'developers' for photographic negative films, plates, and photography paper; and different kinds of 'stop baths' and 'fixing baths' together with 'intensifiers' and 'reducers' required for negative after-treatment, the readers are advised to consult text of this chapter.

An Overview of Techniques in Photography

DEVELOPING OF PHOTOGRAPHIC FILMS AND PLATES

After photomicrography development and fixation of films/plates is required for looking the results. Solutions of 'Developer', 'Stop-Bath', and 'Fixer' are needed for this purpose. These should always be prepared with precautions mentioned in Chapter 20 on basic points to be observed in photography. For films development requires a developing tank (Fig. 23.1). It is *loaded* in a darkroom with the

exposed film and carefully closed with its cover. Some manufacturers market developing tanks which may be used for daytime loading of roll films, one end of the film being clipped inside the container. Then outside the dark room it is filled with developer solution. Constant agitation throughout time of development makes proper development.

The developer is replaced with fixer solution. After the required time the fixer is poured back into the container for the next time use. The film is then thoroughly washed in running tap water and dried at room temperature in a dust free room. *Haste in drying of the film should always be discouraged.* Photographic plates and cut-films are developed in the trays filled up with the developer solution. The whole process is carried out in a dark room until plates are fixed. They can be dried by hanging by clips.

MAKING PRINTS AND ENLARGEMENTS

Determination of Time of Exposure

Generally, the prints are made by placing a film-strip or negative in direct contact with the emulsion surface of the suitable grade of

Fig. 23.1: Photographic film developing tank.

photographic paper. Three grades of papers commonly used are: *Soft* (for hard negatives, and producing less contrast), *Normal* (for normal negatives, and produces normal contrast), and *Hard* (for softer negatives, and produces greater contrast). If enlargements are to be made, the film-strip or negative is placed, with its shining emulsion surface towards the light source of the enlarger, in the *negative carrier* of the enlarger (Fig. 23.2).

The desired area of the negative is magnified by increasing the distance between the paper and the negative in the carrier. All the time a safety bulb (usually red light) in the photographic laboratory is the only light in which the dark room procedures are carried out. With experience, one can find out the correct time of exposure for a particular negative. For beginners, a *strip exposure method*, with varying time periods of exposure (Fig. 23.3) are advised. Correct exposure may be judged by trial and error. The time of exposure is affected by several factors. The main ones are described here.

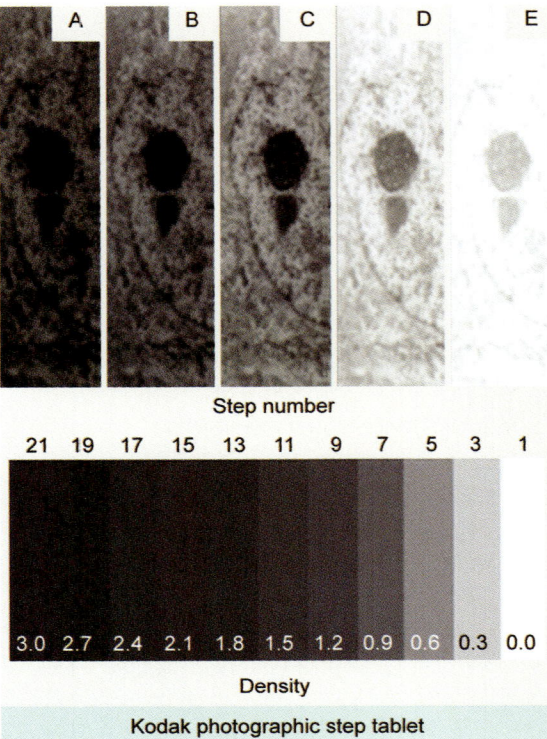

Fig. 23.3: Strip exposure method for determining the correct exposure time: (A) Much over-exposed, (B) Slightly over-exposed, (C) Correct exposure, (D) Slightly under-exposed, and (E) Much under-exposed. Step numbers of Kodak photographic step tablet for deciding the correct exposure time are shown.

FACTORS INFLUENCING TIME OF EXPOSURE

In **photography**, **exposure** is the amount of light per unit area (the image plane luminance **times** the **exposure time**) reaching a **photographic** film or electronic image sensor, as determined by shutter speed, lens aperture and scene luminance.

How to Correct Exposure?

Use a faster shutter speed and **adjust** the aperture accordingly. The rule of thumb to avoiding camera shake while hand holding your camera is to use the reciprocal of the 35 mm equivalent focal length in use. Say you zoom to 125 mm (35 mm equiv.), then use a

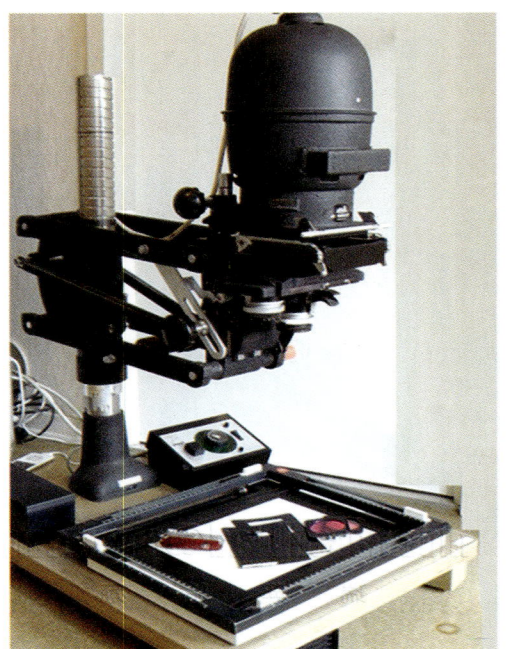

Fig. 23.2: Enlarger equipment for photographic purposes.

shutter speed of 1/125 sec. or faster to avoid camera shake.

The intensity of light and the time during which it acts on a sensitized photographic emulsion are equivalent factors in effecting a definite amount of darkening. It is important to note that with very low intensity of light, no blackening whatever results regardless of the extent of exposure time. The following factors play important role in determining the correct time of exposure:

1. *Intensity of light source*: Film density is a measure of the light-stopping ability of film and is related to the opacity and transmittance of the film. Transmittance is defined as the ratio of light transmitted by the film divided by the total amount of light incident on the film surface. When 50 percent of the incident light is transmitted through the film, the transmittance for that film is equal to 0.5. Opacity is defined as the reciprocal of transmittance, so a film having a transmittance value of 0.5 would have a corresponding opacity of 2.0. Density is defined as the logarithm of opacity, so that a film having opacity of 2.0 will have a density of 0.3 and, as discussed above, will transmit 50 percent of light incident on the film surface. Film density is dependent upon the quantity of metallic silver present in the developed image (Fig. 23.3).

2. The plate or film speed (*see* DIN/ASA chart, *refer* to Table 22.1).

3. Filter factors vary for different types of light sources. The filter factor will vary also with the kind of plate being used. The basic exposure without a filter must be multiplied by the filter factor.

4. *Effect of magnification*: When the image is ten times the object size, the superficial areas of the image and object are to each other as the square of ten is to one, i.e. 100 1. Increasing the magnification to 20X, i.e. twice the previous amount, decreases the light intensity to 1/400 the original value. This is only ¼ as much as obtained with a 10X magnification. *The light intensity varies inversely as the square of the magnification.*

5. *Effect of numerical aperture (NA)*: With increase in NA there is direct increase in the resolution. For an objective with double NA (other conditions being identical), the light intensity or the time of exposure should be raised 4 times; i.e. square of the number of times the NA is increased.

6. *Effect of density in the object*: The more the light is absorbed by the objective, the longer must be the time of exposure. Incorrect exposures obviously in those laboratories where exposure-meters are not available cause errors in photomicrographs (Fig. 23.4). One of the common rules of ordinary photography also applies in photomicrography—that is, *always expose for the shadows and let the highlights take care of themselves.*

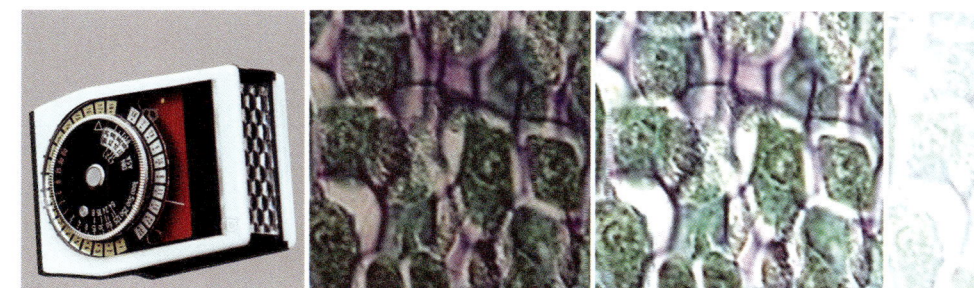

Fig. 23.4: A series of photomicrographs illustrating the most common exposure errors that occur with color transparency film. An exposure-meter is shown on the left.

Comparison of ASA and DIN values of available photographic films

ASA	DIN	Example of films
6	9	Original Kodachrome
8	10	Polaroid PolaBlue
10	11	Kodachrome 8 mm film
12	12	Gevacolor 8 mm reversal film, later Agfa Dia-Direct
16	13	Agfacolor 8 mm reversal film
20	14	Adox CMS 20
25	15	Old Agfacolor, Kodachrome II and (later) Kodachrome 25, Efke 25
32	16	Kodak Panatomic-X
40	17	Kodachrome 40 (movie)
50	18	Fuji RVP (Velvia), Ilford Pan F Plus, Kodak Vision2 50D 5201 (movie), AGFA CT18, Efke 50, Polaroid type 55
64	19	Kodachrome 64, Ektachrome-X, Polaroid type 64T
80	20	Ilford Commercial Ortho, Polaroid type 669
100	21	Kodacolor Gold, Kodak T-Max (TMX), Fujichrome Provia 100F, Efke 100, Fomapan/Arista 100
125	22	Ilford FP4+, Kodak Plus-X Pan, Svema Color 125
160	23	Fujicolor Pro 160C/S, Kodak High-Speed Ektachrome, Kodak Portra 160NC and 160VC
200	24	Fujicolor Superia 200, Agfa Scala 200x, Fomapan/Arista 200, Wittner Chrome 200D/Agfa Aviphot Chrome 200 PE1
250	25	Tasma Foto-250
320	26	Kodak Tri-X Pan Professional (TXP)
400	27	Kodak T-Max (TMY), Kodak Tri-X 400, Ilford HP5+, Fujifilm Superia X-tra 400, Fujichrome Provia 400X, Fomapan/Arista 400
500	28	Kodak Vision3 500T 5219 (movie)
640	29	Polaroid 600
800	30	Fuji Pro 800Z, Fuji Instax
1000	31	Ilford Delta 3200 (see Marketing anomalies below)
1250	32	Kodak Royal-X Panchromatic
1600	33	Fujicolor 1600
2000	34	
2500	35	
3200	36	Konica 3200, Polaroid type 667, Fujifilm FP-3000B
	37	
	38	
6400	39	
12500		No ISO speeds greater than 10000 have been assigned officially as of 2013. Polaroid type 612

Note: Also refer to Table 22.1 for reference

PROCESSING OF REVERSAL MATERIALS

By this process a black-and-white negative is transformed into a positive *transparency*. A negative image is produced in the usual manner and then treated in a bleaching solution. The silver halide remaining is exposed to a white light and then developed to produce a positive image.

Reversal Process Developers

There are a number of photographic developers for the development of photographic emulsions for converting the negative material into transparencies. Some of the formulas are given here:

Ingredients	Codes of reversal developers		
	D-8	D-158	D-198
Water	750.0 ml	750.0 ml	750.0 ml
Elon	—	03.2 g	02.2 g
Sodium sulphite (anhyd.)	90.0 g	50.0 g	72.0 g
Hydroquinone	45.0 g	13.3 g	08.8 g
Sodium hydroxide	37.5 g	—	—
Sodium carbonate (anhyd.)	—	69.0 g	48.0 g
Potassium bromide	30.0 g	00.9 g	04.0 g
Potassium thiocynate	04.0 g	—	02.0 g
Add water to make	1000.0 ml	1000.0 ml	1000.0 ml

Steps of the Reversal Process

1. Fill the developing tank with any one of the above solutions (say D-158) and maintain its temperature at 20° C (68.0° F).
2. Load the film and place it in the tank avoiding air bubbles.
3. Develop for 12 minutes; agitate for 5 seconds every 15 seconds throughout the development.
4. On completion of development, quickly remove the film-spiral and wash in running tap water for 3 minutes.

5. Return the developer to its container and retain it for the second development.
6. Fill the tank with any one of the following solutions, termed *bleaching baths*.
   a. *R-21.A. Bleaching bath*
Water	1000 ml
Potassium bichromate	50 g
Sulphuric acid (conc.)	50 ml

   Dissolve potassium bichromate in cold water. Add sulphuric acid slowly and stir constantly. For use, dilute 1 part of this solution with 9 parts of water.
   b. *Ilford bleaching bath*
      Solution A
Water	500 ml
Sulphuric acid (conc.)	10 ml

      Solution B
Water	500 ml
Sulphuric acid (conc.)	10 ml

   For use, equal parts of solutions A and B should be freshly mixed for each film.
7. Replace the spiral and bleach the film for 5 minutes in the tank. Agitate continuously.
8. Switch on the main lighting of the photographic room.
9. Wash in running tap water for 2 minutes.
10. Clear the film for 2 minutes by placing in any of the following *Clearing solutions*.
    a. *R-21.B. Clearing solution*
Water	1000 ml
Sodium sulphite (anhyd.)	50 g
Sodium hydroxide	1 g

       Solution is used without dilution.
    b. *Clearing solution*
Sodium/Potassium metabisulphite	25 g
Water	1000 ml
11. Wash in running tap-water for 2 minutes.
12. Remove the film from the spiral and expose at 18" from a 100 watt tungsten lamp for 30 to 120 seconds.

13. Return the film to the spiral and redevelop for 6 minutes at 20°C (68°F) in the developer retained from the first development until maximum density is obtained (*See* Step-5).

14. Place in an acid hardening fixing bath for 10 minutes.

15. Wash in running tap-water for 30 minutes.

16. Rinse in water containing a few drops of *wetting agent*, available commercially.

17. Remove the film from the spiral.

18. Dry the film in a dust-free atmosphere or in a drying cabinet.

GLAZING OF PRINTS AND ENLARGEMENTS

Three methods for the glazing of photographic prints are available. The simplest method suitable for glazing a small number of prints, involves the use of chromium plated metal sheets. The following steps may be taken:

1. Clean thoroughly the chromium surface of the glazing plate with methylated spirit. It may then be followed with a cloth moistened with 'glazing solution'.

2. Fill a dish with glazing solution diluted according to the instructions.

3. Remove the prints from the print-washer and place in the glazing solution for 5 minutes. Agitate to remove air-bubbles.

4. Take the prints singly from the glazing solution and lay face down on the glazing sheet. The prints should not overlap.

5. When all prints are in position, place the sheet between Fotonic paper and remove the surplus fluid by placing a print roller (Fig. 23.5) over the paper. Excess pressure will cause the prints to slide out of position.

6. Dry the prints by removing the glazing sheet between the Fotonic paper and stand upright away from draughts at room temperature. The prints will fall from the sheet when dry. Do not attempt to hasten the process by applying heat or using an electric fan.

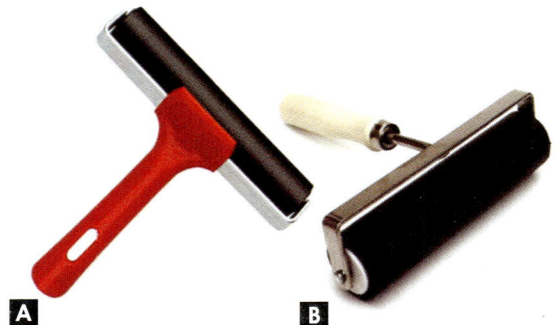

Fig. 23.5: (A) Rubber print roller 6 inch, and (B) Hauer 15 cm print darkroom roller.

COMMON FAULTS IN PHOTOMICROGRAPHIC TECHNIQUES

1. Faults in Exposure of Film

Exposure of film to unwanted chemicals or excess light will also have a deleterious effect on photomicrographs that may be difficult, if not impossible, to correct. Film should not be stored in fume hoods or other areas where it may be subject to attack by volatile liquids and gases such as formaldehyde, acetone, or ether. The effect that any particular chemical may have on a given film emulsion is unpredictable. Film speed may increase or decrease, color balance can be dramatically affected, and contrast will usually be lost. Another laboratory hazard, especially in biochemistry, molecular biology, and medical labs, is inadvertent exposure of film to high-intensity radiation sources. These sources include cobalt-60 isotopes, radium, and X-rays that are commonly used in hospitals and some research laboratories. Baggage scanners at airports also present a radiation hazard. In order to avoid exposure of film to potentially damaging radiation sources, do not store the film in areas where radiation is being used or radioactive chemicals are housed. Protect film (both exposed and unexposed) with either lead or concrete shielding, where appropriate, and place stored film into circulation as soon as possible.

2. Prints During and After Glazing

Some of the common faults in the prints during and after glazing is done, are depicted below.

Faults in prints	Causes
Sticking of prints	1. Dirty surface of sheet 2. Insufficient glazing tempe-rature 3. Damage of chromium surface
Uneven glazing of prints	1. Dirt 2. Irregular contact of the prints 3. Uneven soaking of the prints
Unglazed pits on the prints	1. Dust particles

3. Is Your Film Actually Exposed?

When the film is very dark or totally black, there is a possibility that it has not been exposed. This can be indicative of a serious problem with the film advance mechanism on automatic cameras or a malfunction of the camera shutter system. If the entire roll is unexposed, yet the camera appeared to be functioning properly during photomicrography, check to make certain the camera is receiving light from the microscope. Also examine the take-up spool to make sure it is working properly and the film is correctly attached. Many automatic exposure systems will not allow film to be exposed when insufficient light is being received, but older manual systems do not have this safety feature. In film rolls where only a single or couple of frames are unexposed, check the camera to make sure film is advancing properly.

Color Balance Errors

The test of properly balanced microscope illumination, matched to a film's emulsion design, is to have the clear background of a photomicrograph appear white. Two things need attention of the operator:

1. A bluish background indicates too high a color temperature, while a yellowish background signifies too low a color temperature.

2. Using a film that has not been correctly balanced for the light source color temperature in photomicrography can lead to unusual color shifts, resulting in unnatural hues that do not faithfully reproduce the colors seen in the microscope eyepieces.

There are 3 options for the DX format photographer wishing to copy slides with their camera:

1. Cobble together an assemblage of extension rings, taking lens and makeshift slide holder to perform the task.
2. Modify a traditional slide copier to suit the DX format.
3. Buy a readymade contemporary slide copier that fits onto the camera's lens' filter thread (Fig. 23.7D).

MOUNTING OF TRANSPARENCIES

After a reversal film is dried, the individual exposures can be cut; and these are now ready to be mounted in the *frames* available in the market. Routinely, the paper-mounts are used. The transparency is placed with its glazed surface facing up. Pressing the edges help seal the mount. For better protection, plastic and metallic mounts (Fig. 23.6) with or without thin glass plates to hold the transparencies are also available. The mounted slides in order and labeled. The transparencies may be stored in plastic wallets or plastic boxes. The placing of mounted transparencies in the slide projector is to be done according to the instructions of manufacturers. Generally a red dot is made on the mount at the top of the right hand corner.

COPYING AND DUPLICATING TRANSPARENCIES

A 35 mm slide duplicate is a same-size copy of an existing slide. Kodak E-cube slide

Fig. 23.6: Plastic slides for mounting of 35 mm strips of transparencies' individual frame. Boxes for 35 mm and 5 x 5 slide mounts are shown.

Fig. 23.7: (A) Duplicator for 35 mm slide, (B) for film-strip, and (C) transparency slide viewer.

duplicating film or duplicate 35 mm slides by duplicating equipment (Fig. 23.7). You can make exact same-size, slightly enlarged-size, or slightly reduced-size copies, as well as correct minor color shifts. *High quality Ektachrome duplicates* make the extremely low cost negative–positive slide.

Exact color matching from original film other than Ektachrome may not be possible, but can be attempted for additional charges. Thin plastic Quickpoint 35 mm slide mounts with a white front and a grey rear are standard, and mounts with different rear colors are available at extra cost.

Fig. 23.7 (D and E): Originally designed for slide copying with film cameras. For digital copy use, it is mostly limited to a full-frame 35 mm DSLR.

SUMMARY

Developing of Photographic Films and Plates

An 'exposed' photographic roll film is *loaded* in a *developing tank* inside the darkroom and carefully closed with its cover. The tank may now be brought outside the dark room and subsequently filled with appropriate developer solution. Constant agitation throughout time of development makes proper development. After the stipulated time the developer is replaced with fixer solution. The film is then thoroughly washed in running tap water and dried at room temperature in a dust free room. *Haste in drying of the film should always be discouraged.*

Photographic plates and cut-films are developed in the trays filled up with the developer solution. The

whole process is carried out in a dark room until plates are fixed. They can be dried by hanging by clips.

Making Prints and Enlargements

- Contact prints—of the same size as that of negative are made by placing a film-strip or cut-film negative

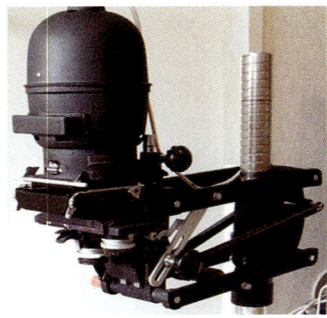

in direct contact with the emulsion surface of suitable grade of photographic paper

• For making enlargements the film-strip or negative is placed, with its *shining emulsion surface towards the light source* of the enlarger, in *negative carrier* of a photographic enlarger.

Time of Exposure

In photography, exposure is the amount of light per unit area (the image plane luminance times the exposure time) reaching a photographic film For beginners, a *strip exposure method*, with varying time periods of exposure is recommended.

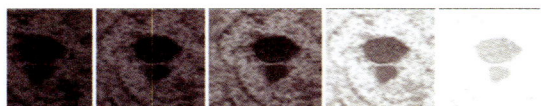

Correct exposure may be judged by trial and error.

The time of exposure is affected by several factors listed below:

• Intensity of light source
• Plate or film speed (ASA/DIN values)
• Filter factors (vary for different types of light sources)
• Magnification

• Numerical aperture (NA)
• Density in the object—incorrect exposures are encountered if exposure-meters are not available

Processing of Reversal Materials

• Reversal process developers
• Steps of the reversal process (*see* detail in the text)

Glazing of Prints and Enlargements

See text.

Common Faults in Photomicrography Techniques

1. Faults in exposure of film
2. Prints during and after glazing

Faults in prints	Causes
Sticking of prints	1. Dirty surface of sheet 2. Insufficient glazing temperature 3. Damage of chromium surface
Uneven glazing of prints	1. Dirt prints 2. Irregular contact of the prints 3. Uneven soaking of the prints
Unglazed pits on the prints	1. Dust particles

3. Is your film actually exposed?

Mounting of Transparencies

• Routinely, the paper-mounts are used. The transparency is placed with its glazed surface facing up.
• A mark made as a red dot at the right upper corner is suitable for keeping the mounted slides in order and labeled.

Copying and Duplicating Transparencies

See text.

24

Digital Photomicrography

DIGITAL IMAGING AND THE MICROSCOPES

Whilst the basic processes have not changed, the universal availability of digital cameras has made the recording of high-quality images available to all microscopists, independent of skill or experience, and the instant viewing of results allows errors to be quickly corrected and a new image captured for all but the most active of subjects. Furthermore, the digital photomicrograph can now be processed by **consumer-level software** to create or analyze images that previously were the preserve of expert professionals, or simply were not possible at all.

Despite these changes in the way photomicrographs are recorded and processed by digital means, there has been surprisingly little published literature on the subject; what has been published is primarily on the Internet and is at best diffuse. It is therefore the intent of this chapter to review the current capabilities and provide practical recommendations for recording digital images with the microscope.

Digital Image Sensor

Whilst it is **not** essential to have an understanding of how digital cameras work, a basic knowledge can be beneficial in making informed decisions regarding selection of a camera, and also in understanding why some post-capture image processing is necessary in photomicrography.

At the heart of digital imaging is the **image sensor**. This consists of an array of light-sensitive receptors, embedded into a microchip containing the wiring and circuitry necessary to record light levels captured from each receptor (Table 24.1). The receptors, termed *pixels* (an abbreviation of picture elements) generally consist of photodiodes embedded in a well. The photodiodes convert photons of light striking the sensor into electrons in a proportional relationship (the more photons striking, the more electrons generated). The charge generated is measured by the microchip circuitry (Fig. 24.1A), converted to a digital signal and processed by in-camera software. The photodiodes are only responsive to the intensity of light, and not its color. Color information is introduced into the digital signal in one of two ways. In conventional digital cameras, a transparent filter mask (termed a Bayer mask, after its inventor) is located immediately in front of the sensor;

Table 24.1	Summary of maximum output image size for various uses of the image		
	Output image size (inches)		
Camera sensor MP	Computer screen 96 ppi	Domestic quality print 200 ppi	Publishing quality print 300 ppi
1.3	13.3 × 10.8	6.4 × 5.2	4.3 × 3.5
2.0	17.0 × 12.8	8.2 × 6.1	5.4 × 4.1
4.0	23.7 × 17.8	11.4 × 8.5	7.6 × 5.7
6.0	32.0 × 21.3	15.4 × 10.2	10.2 × 6.8
8.0	36.5 × 23.3	17.5 × 11.2	11.7 × 7.5
12.0	44.7 × 29.7	21.4 × 14.2	14.3 × 9.5
18.0	54.0 × 36.0	25.9 × 17.3	17.3 × 11.5

Fig. 24.1A: Integrated circuit from a memory microchip showing the memory blocks, the supporting circuitry and the fine silver wires which connect the integrated circuit die to the legs of the packaging.

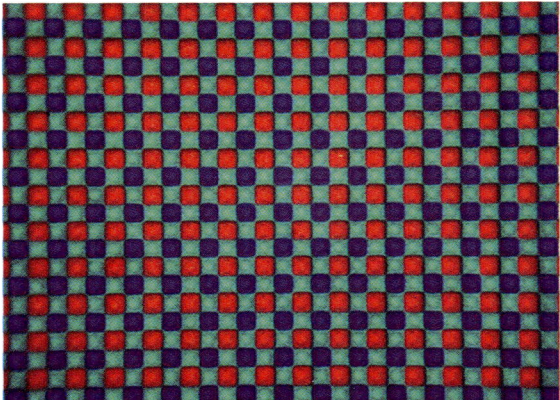

Fig. 24.1B: Bayer filter mask showing red, green and blue filters corresponding to individual pixel sites of a camera sensor.

this mask has a matrix of red, green and blue filters, with one color located above each individual pixel (Fig. 24.1B).

Each pixel therefore records light intensity, and its location under the Bayer mask is used by camera software to determine the color of light at that position in the image. The Bayer mask is generally not an even distribution of red, green and blue filters; green locations are predominant to produce an image that more closely approximates human vision and 'typical' subjects such as landscapes. In dedicated photomicrographic cameras, a color filter wheel is often provided in front of the sensor. Three separate images are recorded with the red, green and blue filters rotated in place in succession; the three images are then combined to produce the final color image. This provides better color fidelity, but cannot be used for motile or dynamic subjects.

Resolution and Image Quality

Camera manufacturers can vary both the number of *pixels* on a sensor, and their individual (physical) size; more pixels can be accommodated into a sensor chip of given size by reducing the dimensions of each pixel. *Resolution*—the ability to see two adjacent points in the image as separate—is determined by both the number of pixels and their size. For the two adjacent points in the specimen to be recorded as individual elements in the image, the microscope must first of all be able to resolve the points. The objective must be of sufficiently high numerical aperture to resolve the structure, and the microscope must be correctly set up; the highest resolution digital camera cannot record information that is not present in the optical image. Secondly, the two adjacent points in the image must fall onto separate pixels, and these pixels must be separated from each other to show a 'gap' between the two structures.

1. Larger pixels have the advantage that they can capture more light before becoming saturated and have a higher signal to **noise ratio**; they are thus more appropriate for recording images of low light intensity where long exposures are necessary such as fluorescence microscopy.

2. Smaller pixels capture less light before becoming saturated and have a higher signal to noise ratio, but provide greater image resolution. Smaller pixels require more software processing to reduce electronic (background) noise in the sensor signal.

Thus the microscopist is presented with a compromise decision; *larger pixels for light sensitivity* or *smaller pixels for resolution*. The correct answer depends on the types of images to be recorded; larger pixels for fluorescent or confocal images or smaller pixels for bright-field, phase and interference contrast images. For the recreational photomicrographer, much of this is academic; the choice of camera will be based on cost, and sensor resolution in terms of the *number* of pixels.

Quality of the image depends on the **output device**—computer screen, projector, or print. The relationship between image resolution and output device is perhaps the most misunderstood aspect of digital photography with the terms 'PPI' and 'DPI' used interchangeably.

• **PPI** stands for 'pixels per inch' and determines the *size* of the image on the output device. The number of pixels used to record the digital image file is fixed by the camera sensor; this cannot be changed. However, these can be displayed or printed at a variable number of pixels (each represented as one colored dot in the image) per inch.

• Lower PPI values will result in a larger output image size but the 'dot' nature of the image will become more apparent as the PPI value is lowered and the final image will have a 'grainy' appearance. For professional printing (such as a book or magazine) 300 PPI is the required quality; printing of photographs on a domestic printer will generally be at 180 to 200 PPI and will be of acceptable quality; graininess will start to become apparent below this value. Computer monitors (PC) generally work up to 96 PPI.

• **DPI** (dots per inch) relates to **dot density** and is a measure of the output device (computer screen or printer) resolution; either the number of dots of ink that can be applied to paper, or can be displayed on screen. The DPI value will affect the quality of the displayed image but not its size; *lower DPI will result in a grainier image*. DPI is usually controlled by the printer software and is usually adjusted under Print Quality Draft, Normal, Fine and Best settings affect the DPI and control image quality, not size. Size is usually set by photo editing software and may be set directly as PPI, or more usually with current software, directly as image size in inches or centimetres.

CHOICE OF DIGITAL CAMERA

Often digital photomicrography is attempted with a pre-existing camera, and almost any digital camera can be utilized to record images from the microscope; several ingenious arrangements to couple less-suitable cameras to microscopes have been published. However, if selecting a new camera there are several important considerations (summarized below). The most fundamental decisions relate to how the camera is to be used, and how the images are to be used.

- If the camera is to be used for general photography as well as photomicrography, a conventional consumer level compact camera or digital single lens reflex (DSLR) is necessary to provide a lens function.

- Alternatively if the camera can be wholly dedicated to the microscope, a specialized microscope camera system may be appropriate. However, it is the balance between sensor *resolution* (number of pixels) and cost that is probably most important. Digital cameras have evolved rapidly in their short existence; a ten-fold increase in sensor resolution of consumer-level cameras has been achieved in as many years whilst prices have dropped in real terms. In the early days of transition from conventional film to digital recording, a number of articles asked 'how many pixels are required to equal (35 mm) film quality?' The relevance of this question is now much less significant because how we use our photographic images has also fundamentally changed. Probably the majority of digital photographs will never be printed but may be displayed on a computer or television screen, or in a digital 'photo-frame'. Transparencies and slide projectors have disappeared and been replaced by the laptop computer and LCD projector. Printing of photographs can now be achieved much more simply at home (or professionally) and the 'darkroom' has been replaced by the 'digital darkroom'. The quality of the displayed image is dependent on the display device (or inkjet printer) as much as the camera sensor. The most important question therefore in purchasing a digital camera (whether for conventional or photomicrographic use) is therefore 'how will I use or display the digital image'?

The next question relates to overall sensor *size* (Fig. 24.2). The physical dimensions of the sensor will have an impact on how much of the field of view of the microscope can be

APS-sized DSLR sensor

Webcam sensor

Full-frame DSLR sensor

Compact camera sensor

Fig. 24.2: Variation in camera sensor sizes (to the same scale) from a 1.3 MP webcam to a 21 MP full-frame DSLR sensor.

projected onto the sensor and therefore recorded. Larger sensors will record more of the visual field of view, assuming no change in projection optics. The situation is however, complicated by compact cameras that have a built-in zoom lens, and several images with smaller fields of view (Fig. 24.3) can now be 'stitched' together by software to make a larger overall image.

Both sensor *resolution* and sensor *size* have a major impact on cost of the camera; the third (and often over-riding question) in selecting a camera is 'how much do I want to spend?'

The cheapest starting-point for digital photomicrography is a dedicated webcam (Fig. 24.4).

These are supplied with optics to replace the microscope eyepiece and simply drop into place in the microscope tube. Connection to a computer via a USB link provides power to the camera, and software included with the camera provides the control. Both still and

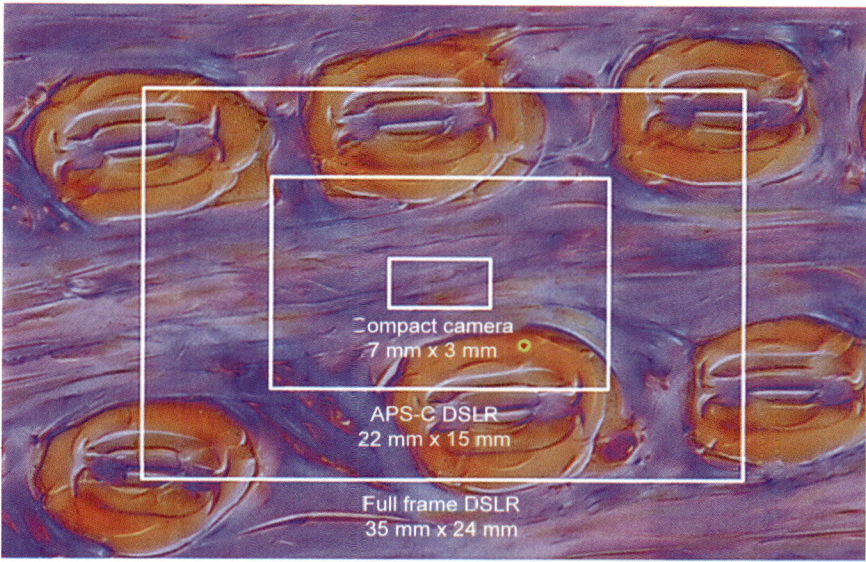

Fig. 24.3: Field of view recorded for different sensor sizes (assuming equivalent relay optics).

Fig. 24.4: Photomicrographic webcam camera fitted to eyepiece tube.

video images can be recorded to the computer. The great advantage of webcams is their low cost and ease of use. Against this, sensor resolution is very low and the camera can only be used in combination with a computer. A typical webcam of 1.3 MP will produce prints of tolerable quality up to 6″ × 4″.

When conventional film cameras were used for photomicrography, the camera would always be used with no lens attached; it was unheard of to use a fixed lens (e.g. rangefinder) camera with the microscope. However, compact digital cameras (with a fixed zoom lens) have been very successfully used for photomicrography. For some years, the Nikon Coolpix 950 (and subsequently 990/995/4500 models) were the most popular compact camera because of their ease of coupling to the microscope using an eyepiece with male screw thread corresponding to the female filter screw thread at the front of the camera lens (Fig. 24.5) and because of the tilting view screen which makes focusing easy.

However, these models have been out of production for several years and are now only available secondhand. Older models are now prone to terminal failure without warning.

A wide range of compact digital camera is now available and some are more suitable for photomicrography than others. In looking for a model suitable for use with a microscope, the following features are important:

Fig. 24.5: Nikon Coolpix camera with Leitz Periplan eyepiece fitted to the lens filter thread.

- A female screw thread on the front of the lens designed to allow filters or lens adapters to be fitted. This thread can be used via an adapter, to connect the camera to the microscope.
- Zooming of the lens is achieved through internal movement, and the front lens element does not protrude in front of the lens barrel beyond the filter screw thread.
- Availability of a remote control to trigger the shutter without physical contact with the shutter button (which would cause vibration to occur), or
- Remote control of the camera functions directly from a computer via a USB link.

For photomicrography, compact digital cameras have many advantages; low cost, high sensor resolution (especially in 'top end' cameras that match the resolution of digital SLRs), and relative ease of use. The main disadvantage comes from the fixed camera lens which is less flexible than a removable-lens DSLR and can introduce optical artefacts (such as 'hot spots' or ghosting) in the recorded image. These can often be eliminated in many instances by trial and error if a range of projection optics is available; there seems to be no way of determining if a particular camera and projection lens is compatible other than trying it. Undoubtedly the best approach is to follow a recommendation on model of compact digital camera from an existing microscope user, either through personal contact or via an Internet forum.

Arguably the best 'all-round' camera for photomicrography is the DSLR. The removable lens allows for easy coupling using a T-mount adapter and photomicrographic systems designed for use with 35 mm film cameras (and now available at *very* low cost second-hand) can be used, providing advantages such as a focusing telescopes (still more accurate than using the camera or computer screen) and vibration-free shutter mechanisms. Most currently-available DSLRs have CMOS sensors with resolution of at least 12 MP; sensor size is usually smaller than 35 mm film and is often based on the APS-C film format of approximately 22×15 mm (the exact dimensions vary between manufacturers). So-called *full-frame* sensor DSLRs are becoming more available (especially those by Canon) and have a sensor the same size as 35 mm film (Fig. 24.6). This provides a wider field of view and can be advantageous if using optical relay systems originally designed for 35 mm film. However, the wider field of view can be prone to vignetting at the edges if sub-optimal relay

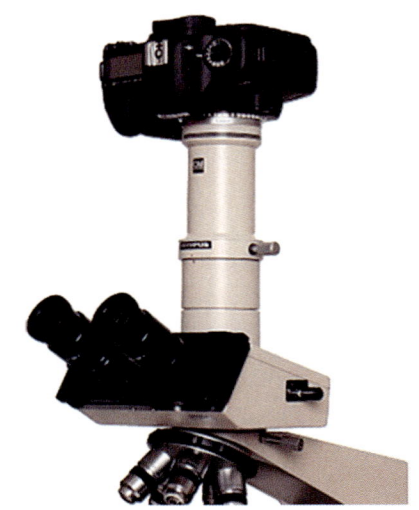

Fig. 24.6: Canon digital SLR with adapter, replacing an Olympus OM 35 mm SLR.

optics are used. The main disadvantage of DSLRs is the vibration that can occur from movement of the camera mirror when the shutter is released. For low magnifications or with long (>3 seconds) exposures this has a little practical impact, but for shorter exposures or higher magnifications the vibration can cause blurring of the recorded image.

In selecting a DSLR for use with the microscope, the following features are desirable:

• Live View: This function is available on a growing number of DSLRs available on the market and allows a real-time image to be displayed on the camera screen, or if coupled to a computer with appropriate software, on the computer screen. Live View makes focusing much more reliable than viewing through the camera eyepiece; most DSLRs do not have interchangeable focus screens designed for finding accurate focus at high magnifications. The 'refresh' rate of Live View is much lower than for video and therefore it is less useful for tracking and focusing rapidly-moving objects under the microscope; in these instances coupling the camera to a relay lens system with focusing telescope is to be preferred.

• Remote control of the camera from a computer via a USB link. Some manufacturers include this software free with the camera but others do not, and the software can be an expensive addition. There are also excellent 'third party' remote control software programs (Fig. 24.7) but it is wise to check compatibility before buying a camera.

• Some cameras (e.g. Canon) have a 'silent' Live View option (usually buried in the 'Special Settings' menus). In conventional Live View mode, the camera mirror flips into the 'up' position and the mechanical shutter opens allowing the image to be 'seen' by the digital sensor and displayed. When the shutter button is triggered, mechanical shutter closes and then re-opens to provide the correct exposure; in some systems the mirror also flips down to allow exposure sensors located in the reflex housing pentaprism to determine the correct exposure. These actions induce some vibration which can cause slight blurring of high-magnification images. Designed to prevent camera shutter noise from scaring wildlife, silent Live View mode utilises an electronic shutter that does not require the

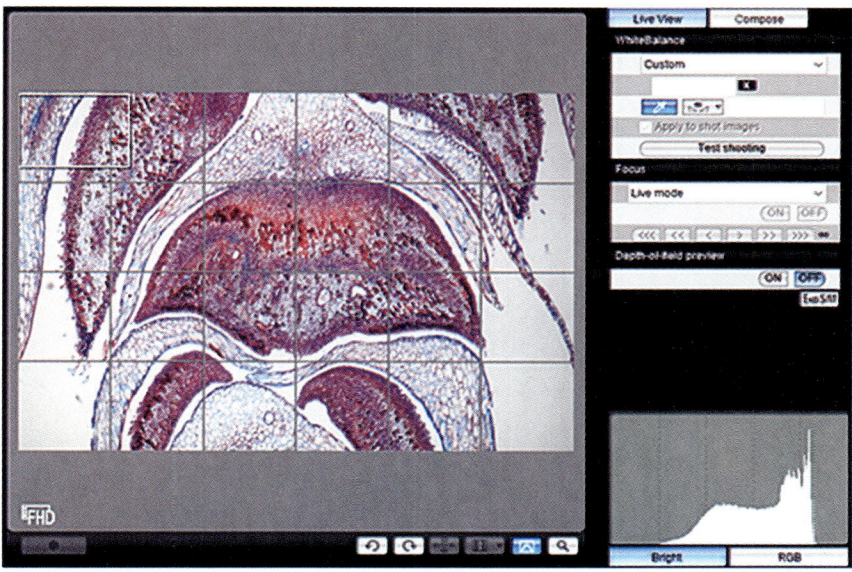

Fig. 24.7: Canon EOS Utility remote control software.

mechanical shutter to first close (or the mirror to flip down) to make an exposure. For the microscopist, silent Live View mode is desirable for higher magnification photomicrography. More information and results of trials on Silent Live View can be found on *Charles Krebs' website*.

- The ability to record video. Several DSLRs released in the last year now have the ability to digitally record high-quality video. This can be of great use to the microscopist working with live specimens or dynamic processes such as crystallisation.

A potentially interesting and very recent development is the availability of digital cameras with removable (interchangeable) lenses that have eliminated the DSLR mirror. Whilst the author has no practical experience of these systems, the absence of a mirror would appear to significantly reduce shutter vibration. Both Olympus and Panasonic have released a number of models; whilst there is as yet no accepted terminology for this design of camera, both manufacturers' cameras are marketed as 'Micro Four Thirds' system, which relates to a sensor size of 18.0 × 13.5 mm; field of view will therefore be smaller than with an APS-C DSLR of similar cost.

Dedicated digital cameras are made by the major microscope manufacturers as well as dedicated specialist imaging companies. These systems tend to be of lower resolution and higher price than consumer-level cameras but are of very high sensitivity to cope with the low levels of fluorescence and confocal microscopy. Such systems (by virtue of their price) are generally not used outside of specialist laboratories but are slowly becoming available on the second-hand market. Potential buyers should be very cautious of the risks involved with these electronic systems, sold outside of the manufacturer's support or warranty. Lower cost dedicated microscope cameras are available via Internet auction websites but reports of image quality from some systems have been disappointing. Table 24.2 summarizes the relative merits of each class of digital camera.

Table 24.2 Comparison of cameras for photomicrography

Camera	Sensor	Typical cost	Advantages	Disadvantages
Webcam	1.3 MP	£60–80	Low cost Ease of use	Low resolution Can only be used with computer attached Cannot be used for other photography
Compact (fixed lens) camera	6–12 MP	£100–400	Relatively low cost Useable for normal photography Some can be used 'tethered' to computer	Coupling fixed lens to camera eyepiece Potential for 'blooming' or hot spots from fixed lens
Digital single lens reflex(DSLR)	8–21 MP	£400–2,000+	Removable lens Large sensor size Flexible operation and exposure modes Remote control via computer	Cost Potential vibration from mirror
Dedicated digital microscope camera	2–8 MP	£2,000–6,000	High quality and light sensitivity	High cost Relatively low resolution

CONNECTING CAMERA TO MICROSCOPE

The simplest solution to bring camera and microscope together is the dedicated microscope webcam. These directly replace the microscope eyepiece and drop into the microscope tube; adapters are usually provided for wider diameter stereomicroscope tubes.

DSLR cameras have a removable lens and usually require a T2-mount adaptor which connects to the camera lens bayonet and provides a universal female M42 thread. This can then be coupled to a conventional microscope camera adapter with relay or projection optics (Fig. 24.8).

Many microscope adapters consist of a simple tube without relay optics; it is inappropriate to use an eyepiece designed for *visual* use with these adapters as these are designed to project an image at infinity (and brought to a focus by the lens of the eye). To project a *real* image onto the camera sensor, refocusing will be necessary which introduces spherical aberration, and degrades the image. The correct solution with these simple tube adapters is to use a specially designed *projection* eyepiece. In many ways, the ideal solution for coupling a DSLR to the microscope is to use a second-hand photomicrographic system; these can often be bought on Internet auction sites at very low cost. They were provided by the main microscope manufacturers and are often

specific to fit microscopes from the same manufacturer. They generally consist of a shutter body with prism for diverting light to a graticule eyepiece for focusing, or vertically to a 35 mm film camera body (Fig. 24.9).

Replacing the film body with the DSLR provides the benefit of much greater precision of focus (using the eyepiece rather than the camera or computer screen) and can effectively eliminate the vibration problem with DSLR cameras. This is achieved by first triggering the DSLR shutter, with a shutter speed of several seconds, waiting two or three seconds for vibration from the mirror to subside, then

Fig. 24.9: DSLR mounted in place of 35 mm film body on a photomicrographic camera system.

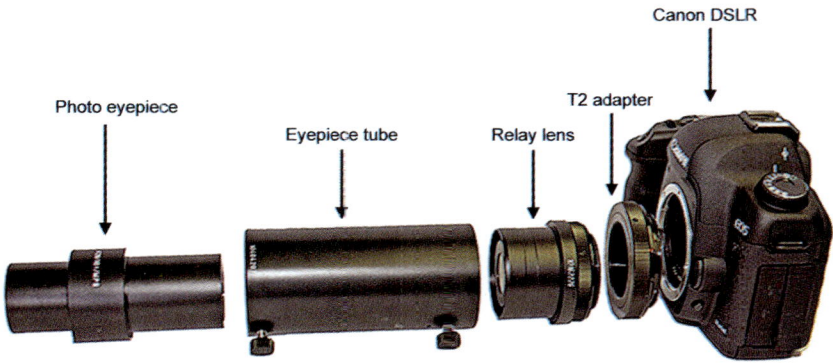

Fig. 24.8: Camera attachment system with relay optics.

triggering the photomicrographic system shutter, timed for correct exposure.

The lens on compact digital cameras however, cannot be removed and some ingenuity is required to connect these to a microscope. With fixed lens camera, it is appropriate to use a *visual* eyepiece on the microscope; the camera lens performs the same function as the eye's lens in focusing the image onto the photosensitive receptor (the eye's retina or the camera sensor). The eyepoint of the eyepiece should be at least 15 mm above the upper surface of the eyepiece in order to avoid vignetting in the image. Often these 'high eyepoint' eyepieces are designed for use by spectacle wearers and are marked with an icon of a pair of glasses. The most widely-adopted solution is to use a microscope eyepiece that has a screw thread, allowing eyepiece and camera to be directly connected (in most instances) via a screw thread adaptor. The Leitz Periplan ×10/20 GF eyepiece (marked with model number 519 815) serve this purpose well, having a plastic eye-cap which is screwed to the top of the eyepiece. Removal of this reveals a 28 mm male screw thread which can mate directly to a female lens filter mount on the camera of the same thread, or to an interconnecting 'stepping' adapter. Other models of Periplan eyepiece are also suitable, but may have a different screw thread and the upper surface of the lens projects above the screw thread, requiring a short spacing tube between the eyepiece and camera lens to prevent the two optical surfaces coming into contact. Periplan eyepieces are no longer available new but can be obtained from Internet auction sites (although prices are rising dramatically). Adapter rings can also be purchased from Internet auction sites or from specialist companies such as SRB-Griturn. Other suitable eyepiece systems are also available secondhand and it can often be a matter of trial and error to establish a workable system. Care should be taken with any system to evaluate images carefully for flare, ghosting or hotspots in the final picture.

SOFTWARE FOR DIGITAL PHOTOMICROGRAPHY

The digital photomicrographer need not have any software (or indeed a computer) to record digital images; files can be printed out directly from camera storage media on many domestic printers or at high street processors. However, to achieve the full benefits and potential that the digital image offers, a range of software tools is essential.

GETTING STARTED FOR DIGITAL PHOTOMICROGRAPHY

Two important considerations for photography through the microscope in achieving satisfactory results are: *Sample cleaning* and *vibration control.*

Sample Cleaning

A clean sample simply makes a much more striking image. If one is investing the time required to take a high-quality photomicrograph, the results are dramatically improved if some time is spent for properly cleaning a sample prior to setting up the image. Typically, this can be accomplished with a stone cloth and a damp sponge to remove as much dust and finger oil as possible. This will save significant time in post-capture image processing and dust spot removal.

Vibration Control

Vibrations can cause loss of detail by blurring an image. This happens when subject movement occurs over the exposure time. Two solutions for this problem are: (i) Eliminating causes of movement or (ii) decreasing exposure time. By making the microscope more rigid or isolated from the environment, the photomicrographer minimizes vibration-induced problems. A *floating optical table* is best, but this is generally too expensive. To help eliminate vibrational movement, one can add shock-absorbing anti-vibration mounts or rubber pads underneath the microscope to help prevent vibrations from reaching the subject.

Another solution to help eliminate vibration-induced blurring is to decrease the exposure time so there is less chance for the subject to move. With digital photomicrography, you can increase the light sensitivity of your camera by adjusting the ISO setting, the modern equivalent to the American Standards Association's (ASA) "film speed." If your camera is more sensitive to light, then the required exposure time decreases; however, if the ISO is adjusted too high, image quality may suffer from what is known as "noise" or a grainy salt-and-pepper type appearance.

For users of DSLR cameras in particular, the mirror and shutter movements may cause some additional vibrations. If selecting a DSLR digital camera for photomicrography, it may be worthwhile to choose one with a "silent shutter" mode, which will greatly reduce camera-induced vibrations.

The Leica Microscopes for Photomicrography

Leica's DM500 and DM750 microscopes were designed specifically for medical related photomicrography. Both the DM500 and the DM750 offer **bright field, dark field** and **phase contrast** capabilities. The DM750 features a quite robust **construction** (9 kg). Using Leica's **LM digital adapters**, all current DSLR and system cameras (Fig. 24.10) can be attached to the DM500 and DM750 models quickly and easily in just a few simple and quick steps.

With its pre-centered, pre-focused condenser, tube-integrated eyepieces and additional viewing tubes, the DM500 is particularly suitable for **beginners**. The DM750 can also be **upgraded extensively** (Köhler field diaphragm, phase turret condenser for bright field and phase contrast, etc.) The four-position **nosepiece** (optional five-position nosepiece on the DM750) can accommodate the **Leica objectives** which are renowned for their quality.

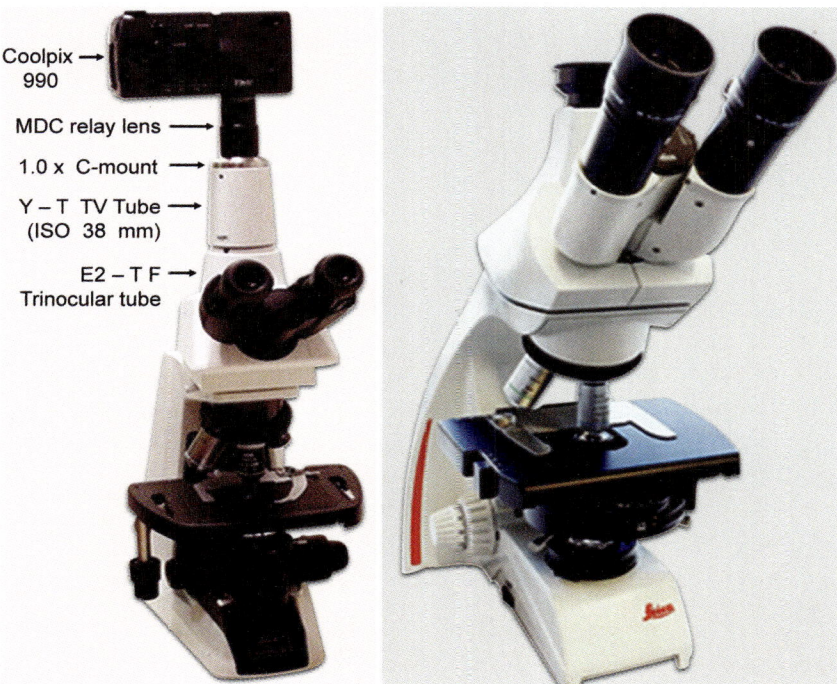

Fig. 24.10: Attachment of a digital camera with C-mount on a light microscope. On the right is shown a digital photomicrographic set-up of Leica camera.

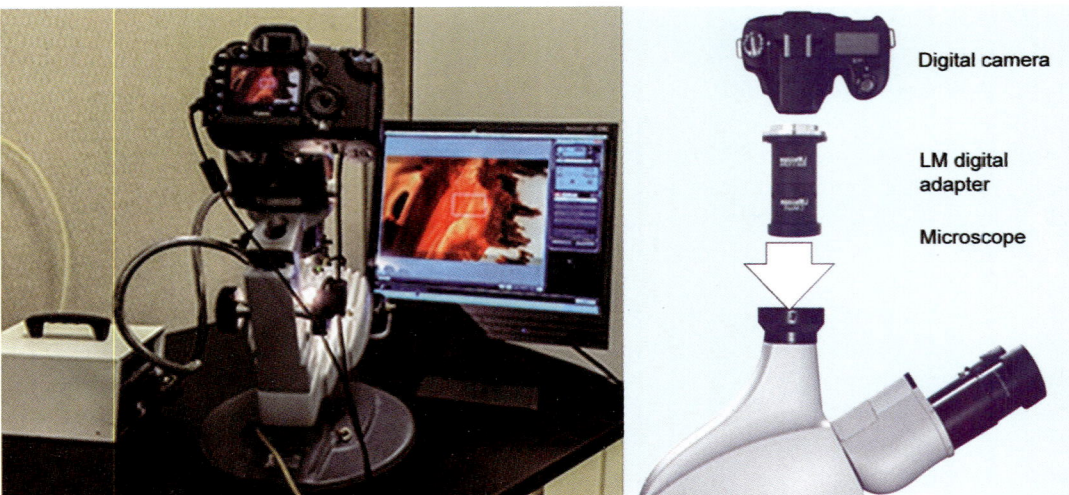

Fig. 24.11: A live view shown on a computer monitor is used for focusing and composition because the oculars are blocked by the camera. The camera is remotely controlled by the software, which helps prevent vibration-induced blurring that would occur by manually pressing the shutter button.

The microscopes of this series exclusively use **LED**s for illumination; no halogen alternative is available. Leica offers a large selection of rotatable **tubes** for its DM500 and DM750 microscopes [45° EZ tube, 30° EZ tube, 45° binocular, 30° binocular, 45° trinocular, and 30° trinocular].

Modern DSLR and system cameras (Fig. 24. 11) offer the latest technology and are generally very well suited for microscopy applications. Most of them can be controlled remotely via PC/Mac. Because of their high sales volumes, they offer an excellent price/performance ratio compared to special-purpose microscope cameras.

Features of top DSLR and system cameras are:

- Large, powerful full-frame sensors (36 × 24 mm)
- Sensor resolution of 50 megapixels
- High light sensitivity (*ISO* 100,000+)
- Extensive *dynamic range* (up to 14 aperture stops/f-stops)

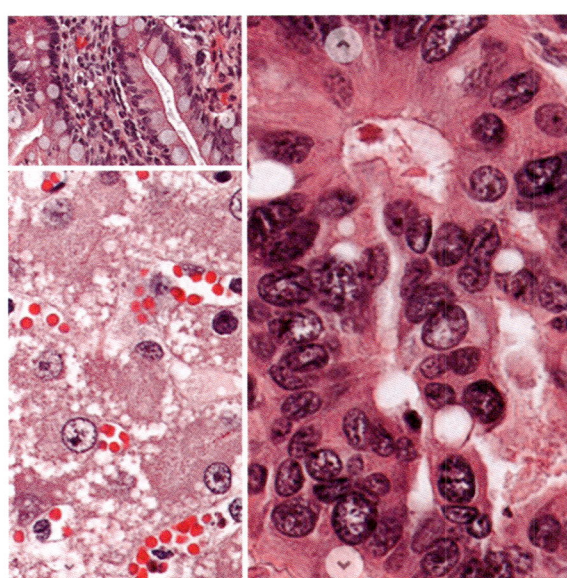

Fig. 24.12: Good quality photomicrographs are achieved with proper setting of a digital camera with C-mount on a light microscopes designed by standard manufacturers such as Leica, Zeiss, Nikon, Olympus, etc.

- *Short exposure times* (1/8000 second)
- *Full/Ultra HD* (4k) video function
- Live video capture on external monitors in ultra HD quality.

In most cases, these cameras are significantly more powerful than microscope cameras with smaller sensors (1/2" or 2/3"). Hopefully, the overall cost of digital camera equipment will be lower, allowing more users to acquire better equipment and capture better images (Fig. 24.12).

Basic Software

The most basic software requirement is the driver file for the camera. This enables the camera and computer to communicate when linked by cable (or wireless transmission) and enables basic functions such as image download. This software is supplied with the camera from new and is often available as a free download from the manufacturer's website. Although it can be possible to download images without the relevant camera driver, other camera functions will not be controllable from the computer without it. Those buying secondhand digital cameras should ensure that the driver software is included, or at least downloadable from the Internet.

The most beneficial software is the image editing program. It is not the intent of this article to review the many packages that are available, varying in cost from free shareware to advanced and complex systems costing several hundred pounds; general books and magazines on photography should be consulted to review the market. However, the author uses Adobe's Photoshop Elements, and like many other users, has found this package to be relatively simple to use, reliable and more than adequate in functionality. The image editing program provides a wide range of functions; those important to the photomicrographer include image cropping, exposure management, color management, cloning to remove dust marks and sharpening. Many

editing programs also include photo album organizers—electronic management of digital images into album-like folder structures. Unless many thousand photographs are to be stored, the author has found these to be unnecessary as the required level of management and backup can be achieved with a simple folder structure, rigorously maintained. A further advantage of image processing software is the 'Panorama-Merge' function. Designed to blend a number of adjoining pictures of (typically) a landscape into a single panorama, this can be used to great benefit in photomicrography where the field of view is insufficient to record the whole specimen, or where it is desirable to use a higher numerical aperture objective to achieve high resolution. Systematic imaging of the specimen according to a grid pattern (using the X and Y controls of a mechanical stage), ensuring overlaps of each image, allows large specimens to be recorded at high resolution.

Many digital cameras can be controlled directly from a computer via a USB link and this offers many benefits to the photomicrographer; preview of the image on the computer screen, control of basic camera functions such as ISO setting, color temperature setting, metering mode, exposure control, remote triggering of the camera shutter and near-instantaneous download of the image for immediate viewing. Some manufacturers (e.g. Canon) include this software 'bundled' with the camera; others provide this as an accessory, often at high additional cost. Third-party 'remote capture' software is also available for purchase via the Internet, of which the best-known system is Breeze Systems' DSLR Remote Pro software. In the author's experience third-party software can provide much better functionality and interface than the camera manufacturer's own software.

Advanced Software

The possibilities that current software provides to the amateur digital photomicrographer could not have been imagined 10 years ago. Advanced software is now available, either at low cost or sometimes for free download that can manipulate digital image files to achieve amazing results. This includes image stacking software that combines multiple exposures taken at different focal depths into a single combined, in focus image. Other software will combine single image files taken at timed intervals into a video sequence; time lapse sequences that were once the province of professional television can now be achieved by the amateur with software costing only a few pounds. Furthermore, the compromise between exposing for 'shadows' or 'highlights' can now be solved by different softwares that combine several exposures into one 'blend'.

For the photomicrographer, and especially for high magnification photomacrography, image stacking software has to be one of the most significant advances in recent years, enabling stunning images with high depth of field to be easily produced. Professional, and expensive, systems that process multiple images have been available for some years;

the functionality of these systems is now available in free or low cost software for the amateur. Systems currently available include Combine-Z, Helicon Focus and Zerene Stacker. The author uses Helicon Focus which provides a very easy to use interface, with effective results. Other Zerene Stacker as providing images greater flexibility in use.

CONCLUSIONS

Digital imaging has made the recording of high-quality photomicrographs accessible to all at low cost. Cameras have now evolved to a stage where virtually any digital camera will produce an image suitable for printing at A3 size or greater, and at high resolution. A range of software is now available to the amateur that enables images to be generated that previously were not possible, and all at relatively low cost. Whilst the field continues to develop rapidly, the 'pixel-war' of ever-increasing sensor resolution has probably reached its peak. However, new software will continue to be developed that further increases the possibilities for digital imaging with the microscope.

SUMMARY

- Universal availability of digital cameras has made the recording of high-quality images available to all microscopists irrespective of skill or experience of the operator. Instant viewing of results allows errors to be quickly corrected and a new image captured for all but the most active of subjects.

- The heart of digital imaging is the **image sensor**, which consists of an array of light sensitive **receptors**, embedded into a **microchip** containing the *wiring* and *circuitry* necessary to record light levels captured from each receptor.

- The receptors, termed *pixels* (an abbreviation of picture elements) generally consist of **photodiodes** embedded in a well. The photodiodes convert photons of light striking the sensor into electrons in a proportional relationship (*the more photons striking, the more electrons generated*). The charge generated is measured by the microchip circuitry, converted to a digital signal and processed by in-camera software.

- An **integrated circuit** (also referred to as an **IC**, a **chip**, or a **microchip**) is a set of electronic circuits on one small flat piece (or "chip") of semiconductor material, normally silicon. The integration of large numbers of tiny transistors into a small chip results in circuits that are orders of magnitude smaller, cheaper, and faster than those constructed of discrete electronic components. The IC's mass production capability, reliability and building-block approach to circuit design has ensured the rapid adoption of standardized ICs in place of designs using discrete transistors. ICs are now used in virtually all electronic equipment and have revolutionized the world of electronics (possible by the small size and low cost of ICs).

- ICs were made possible by experimental discoveries showing that semiconductor devices could perform the functions of vacuum tubes, and by mid-20th century technology advancements in semiconductor device fabrication. Since their origin in the 1960s, the size, speed, and capacity of chips have progressed enormously, driven by technical advances that fit more and more transistors on chips of the same size—a modern chip may have several billion transistors in an area the size of a human fingernail.

- ICs have two main advantages over discrete circuits:
  i. Cost is low because the chips, with all their components, are printed as a unit by photo-lithography rather than being constructed one transistor at a time. Furthermore, packaged ICs use much less material than discrete circuits.
  ii. Performance is high because the components of ICs switch quickly and consume comparatively little power because of their small size and close proximity. The main disadvantage of ICs is the high cost to design them.
  iii. Fabricate the required photomasks. This high initial cost means ICs are only practical when high production volumes are anticipated.

Choice of Digital Camera

- Any digital camera can be utilized to record images from the microscope.

- If the camera is to be used for general photography as well as photomicrography, a conventional consumer level compact camera or **digital single lens reflex** (DSLR) is necessary to provide a lens function. Both sensor *resolution* and sensor *size* have a major impact on cost of the camera.

- The cheapest starting point for digital photomicrography is a dedicated webcam.

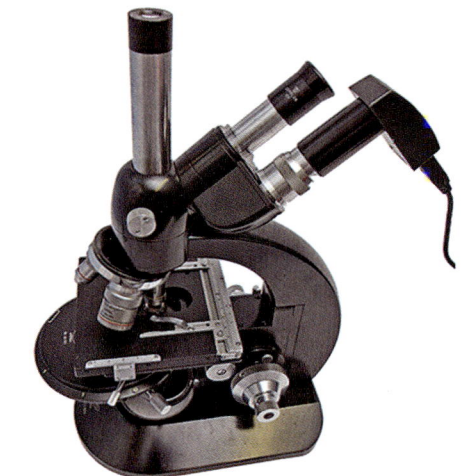

Annexure
Photomicrography

Questions	Answers
What is a photomicrograph mean?	A picture taken by a light microscope that is an enlarged reproduction of a minute object made either by drawing with free hand or through photographic process.
What is photomicrography?	It means the art of reproducing pictures of microscope images generally of a tissue section.
Do the two terms *photomicrograph* and *microphotograph* mean same?	NO! The term *microphotograph* is often used wrongly for photomicrograph; it actually means a minute photograph which must be looked with a microscope to reveal its details.
What is a photomacrograph?	Photomacrograph means a photographic image of a relatively large object magnified only a few times but not over.
What is an electron micrograph?	A micrograph prepared using an electron microscope.
What is a digital micrograph?	A digital picture obtained either directly with a microscope or by scanning of a photomicrograph. Digital micrographs are commonly obtained using a USB microscope attached directly to a home computer or laptop.
Define projected image diameter	The diameter of the circular image of the field as it is projected on the focusing screen by the camera used for photomicrography.
What is a micron bar?	The micro bar (also termed scale bar) is a bar of known length, used for measurements, displayed on a picture/photograph/photomicrograph.
What relation the field diameter has with photomicrography?	A term used for the actual diameter of the illuminated circle seen when looking into a microscope through the top lens of a microscope eyepiece. The field of view number is normally 18 for most of the eyepieces except in wide field eyepieces when this number may be up to 30.
What is a depth of focus?	The depth of focus is defined as the **sum** of distance on both sides of plane of sharp focused image. Larger the aperture size of the photographic lens, shorter is the depth of focus, and the vice versa.
What relation exists between the numerical aperture of an objective lens of a microscope and the resolution of the microscope?	Numerical aperture (NA) is regarded as the minimum distance between the two points, whose images can be seen as distinct from each other. The approximate resolving power of microscope can be expressed by formula: $$R = \frac{NA}{0.61 \times \lambda}$$ Where, R = Resolving power (in nm) λ = Wavelength (in nm) of light NA = Numerical aperture of objective

Index